THE CHINESE REAL ESTA

This is the first book to fully present, analyse and interpret the Chinese real estate market.

Junjian Albert Cao examines the Chinese real estate market's growth trajectory, unique governance and factors affecting values and investment in the context of reforms, rapid economic growth and urbanisation.

The Chinese Real Estate Market provides essential insights into the institutional change surrounding the development of the property market, government intervention at local and national levels, taxes and other regulatory changes, and factors such as market practices, economic changes, government policies and social changes that affect the value of real estate.

Furthermore, the book analyses academic and policy debates on issues such as:

- commercial property investment
- housing price inflation
- property rights protection
- affordable and social housing
- market practices and regulation
- environment and sustainability
- taxation
- property-led growth
- the reliance of local economic growth on the property sector.

The book offers a comprehensive, in-depth and up-to-date account of the Chinese property market and presents a full assessment of the investment potential of Chinese real estate. It is a must-read for students, academics and real estate professionals interested in this fascinating real estate market that has implications for China's and the world's economies.

Junjian Albert Cao is a Senior Lecturer in the Department of Real Estate and Construction at Oxford Brookes University, UK. He previously lectured at Sun Yat-Sen University in China before coming to the UK to study for his PhD, and he has vast research experience on the property market and urban development in China. In addition to supervising PhD students, Dr Cao has carried out several funded research projects on China and has acted as both an internal and external examiner for PhD students in the UK.

International Real Estate Markets

The Chinese Real Estate Market
Development, regulation and investment
Junjian Albert Cao

'This is a landmark text which provides a rare and detailed investigation into many facets of the Chinese real estate market. The range of topics is extensive and the reader benefits from a well-planned layout and professional writing style. Considering the relevance of the Chinese market to the global real estate sector, this text is very timely and is an essential addition to every property researcher's library.'

Richard Reed, *Chair in Property and Real Estate, Department of Finance, Deakin Business School, Australia*

THE CHINESE REAL ESTATE MARKET

Development, regulation and investment

Junjian Albert Cao

LONDON AND NEW YORK

First published 2015
by Routledge
2 Park Square, Milton Park, Abingdon, Oxon OX14 4RN

and by Routledge
711 Third Avenue, New York, NY 10017

Routledge is an imprint of the Taylor & Francis Group, an informa business

British Library Cataloguing in Publication Data
A catalogue record for this book is available from the British Library

Library of Congress Cataloging-in-Publication Data
Cao, Junjian.
The Chinese real estate market : development, regulation and investment / Junjian Cao.
pages cm. – (International real estate markets)
Includes bibliographical references and index.
1. Real property–China. 2. Real estate investment–China. 3. Commercial real estate–China. 4. Residential real estate–China. 5. Property development–China. 6. Real estate business–China. I. Title.
HD926.5.C36 2015
333.30951–dc23
2014038822

ISBN: 978-0-415-72312-1 (hbk)
ISBN: 978-0-415-72315-2 (pbk)
ISBN: 978-1-315-85785-5 (ebk)

Typeset in Bembo
by Cenveo Publisher Services

To Tess, Melody and Miranda

CONTENTS

FIGURES

TABLES

BOXES

PREFACE

In only two to three decades, the urban landscape of all Chinese cities has been transformed from a humdrum juxtaposition of dull, low-rise residential buildings and polluting factories to clusters of glittering skyscrapers, grandiose shopping malls and countless stylish, high-rise residential tower blocks. The reforms and the opening-up campaign, initiated 36 years ago, have led to the marketisation of production and consumption, and the commodification of urban space in China. Established in the mid-1980s, the real estate market in China has fundamentally altered the way urban land and buildings are developed and allocated. Property-led urban growth has provided a much-needed new driver of economic growth as Chinese cities build to alleviate the shortage of buildings and to accommodate a large number of migrants from the countryside. The real estate market has raised the housing standards of urban dwellers and created wealth for property owners and investors. With its impact on fixed asset formation and its demand for domestic as well as overseas raw materials and manufactured products, the Chinese real estate sector not only is important to the Chinese economy but also influences the world economy. Recently, the Chinese housing market has received worldwide attention because of the expectation that a slump in the market there may result in a sharp slowdown of the Chinese economy, the largest powerhouse of the world economy. A holistic review and in-depth analysis of the Chinese real estate market are thus useful to people who are interested in understanding China's real estate market, urban development and economic growth, and need to make decisions relevant to the Chinese property market.

Due to its size, uniqueness and speed of change in the past 100 years, China has presented an intellectual challenge to anyone who seeks to comprehend its past developments and future directions. Over the years, many books have been written in English by both Chinese and non-Chinese authors to provide explanations and interpretations of various aspects of this most populous country with an unbroken culture for 5,000 years. The Chinese real estate market, however, has not been covered by such efforts so far, even though the market has become increasingly and in some cases profoundly important to the political, economic, social, legal and environmental changes in China, and beyond.

This book is written to fill this void. It provides a review of the past and current developments of the Chinese real estate market, an integrated analysis of regulation and investment in that market and a discussion on many topical issues related to the market, including informed predictions of the future directions. It aims to enable readers to understand why the Chinese real estate market has developed in its current form and what its future development will be under institutional change brought by the reforms and opening-up campaign and globalisation. By reading this book, readers are in a better position to make informed decisions linked directly or indirectly to the Chinese real estate market.

A recurring theme in this book is that the Chinese property sector has been in gradual transition from administrative and non-market allocation of resources to market allocation. Gradualism, a key characteristic of the Chinese reforms, has allowed time for step-by-step introduction of market-supporting institutional arrangements from the 1980s, which were acceptable to the economy, the government and society at the time of their introduction. Institutional change in the property sector has thus been led by both top-down and bottom-up approaches, and has been 'learning by doing' in nature. Major changes in national rules governing the market have often had local trials. Competition among cities generates innovation and allows better practices to prevail. In this way, production has not been disrupted, but rather continually enhanced with piecemeal incorporation of new practices, innovation and governance. Such an approach is responsible for the continuous expansion of output and rising productivity in the economy as a whole and in the property sector in general. On the other hand, lack of knowledge and experience in governing and operating in a market economy reinforces the path-dependency nature of institutional change, with excessive administrative measures to rein in price changes by the central government and widespread short-termism by market actors. Likewise, the piecemeal nature of institutional change has slowed down the exit of administrative resources allocation, which was used by local governments to promote their versions of urban development and real estate administration, distorting market practices. The slow exit of administrative power was conducive to an exploitation of power and access by some to obtain gains at the expense of public interests. A new wave of major reforms, started in 2012, promises to promote further marketisation and minimise government command in the market.

Another recurring theme is the need for reforms to increase the supply of market-supporting institutional arrangements. At the start of the reforms, there was a dearth of institutions that protected private property rights and supported market exchange. The creation of a private property rights system in land in 1990 and the subsequent legal changes have enhanced the protection of property rights, facilitating the rapid growth of market exchange, productivity and economic growth. However, further reforms are needed to adjust the relationship between the state and the market and guide the development of the market on a sustainable path. In particular, the dependency of local governments on land sale revenues and taxes from real estate development and transactions has resulted in a series of undesirable market outcomes, for example rapid housing price inflation, potential oversupply of housing and commercial real estate, and urban sprawl. This dependency has led to local resistance of some institutional arrangements, such as housing size standards, which can alleviate market affordability of housing in key cities. Reform of the fiscal rules between central and local government is necessary to match the fiscal capacity and responsibilities of local governments to change this dependency. Furthermore, institutional change is required to improve market operations in property financing, housing transaction and property management in housing estates.

This book is based on my research on China's property market over the last 16 years, with the most recent fieldwork conducted in August and November 2013 and June 2014. A large number of graphs and pictures are provided in various chapters to illustrate the development of the real estate market in China. Wherever possible, data are drawn from national-level statistical agencies to ensure accuracy. Many case studies are used and efforts have been made to check all the details are correct. All but three of the pictures were taken by me, with the rest from reliable sources.

To supplement the author's coverage, 13 academics, researchers and practitioners, who are at the forefront of national and local policy-making and market operations, have shared their knowledge and insights on housing and commercial property markets, and on the provision of affordable and social housing. These independent observations appear in this volume as guest articles in boxes to further ensure a holistic picture of the huge and diverse Chinese real estate market.

ACKNOWLEDGEMENTS

The completion of this book would not have been possible without the support and help of a large number of people and organisations.

My thanks first of all go to my colleagues in the Department of Real Estate and Construction at Oxford Brookes University, Oxford. Among them, Professor Ramin Keivani, a long-time collaborator on my China research, has been very supportive and helpful throughout and has reviewed several chapters of the manuscript. Richard Grover, former Deputy Dean and currently Senior Lecturer, has carefully read through the whole manuscript, contributing significantly to the academic debate and the standard of English. Dr Mark Austin, Programme Lead for Post Graduate Programmes, has been very supportive and helpful throughout, and he provided comments on some chapters which contributed to the academic debate. Professor Joseph Tah, Head of the Department, has also been very helpful to this project.

I am particularly grateful to Tim Rowcliffe-Smith, guest lecturer in the Department of Real Estate and Construction, Fellow of the IRRV and Associate of the CIPD, who meticulously read through the whole manuscript, making significant contributions to the academic debate and the standard of English. He has been a source of energy and inspiration throughout the writing process.

Several institutions have played supporting roles in this project for which I am grateful. The Department of Real Estate and Construction, at Oxford Brookes University, provides a scholarly working environment for academic pursuit. The Royal Institution of Chartered Surveyors has been very supportive of my China research. The China Institute of Real Estate Appraisers and Agents has been very supportive of my research in China and has provided valuable assistance in several cities on several occasions. The Shanghai Academy of Social Science has provided valuable help to my fieldwork in Shanghai and organised academic exchanges that contribute to the academic debate.

I sincerely thank the editors, Ed Needle and Sade Lee, at Taylor & Francis. Ed inspired me to write this book and has been very encouraging and supportive throughout. Sade has been very supportive and helpful in the preparation of the manuscript.

A number of academics, professionals and practitioners both in China and in the UK have contributed short essays and pictures to this book, enriching and enhancing the content. These contributors are as follows (in alphabetical order):

- Qiang Chai, Executive Deputy President of China Institute of Real Estate Appraisers and Agents
- Stanley Chin, CEO, Treasure Capital Asia
- Xiaobo Dai, Professor and Head of Real Estate Research Unit, Shanghai Academy of Social Sciences
- William Dong, Deputy President, CURA Investment
- Yuanze Gao, PhD student, Department of Real Estate and Construction, Oxford Brookes University
- Junping Liao, Professor and Head of Centre for Real Estate Studies, Linnan College, Sun Yat-Sen University
- Min Liao, Chief Valuer and Deputy General Manager, Guangdong Seeing Appraisal Ltd
- Hongyu Liu, Professor and Deputy Dean, School of Civil Engineering, Tsinghua University
- Lin Liu, Head of Real Estate Research Unit, Investment Research Institute, State Development and Reform Commission
- Carlby Xie, Head of China Research, Colliers International
- Yuqing Xu, Lecturer, Department of Urban Planning and Design, Xi'an Jiaotong-Liverpool University
- Junshi Yuan, Analyst, Savills China
- Hongming Zhang, Professor and former Head of Real Estate Research Unit, Shanghai Academy of Social Sciences
- Richard Zhang, Head of China and Senior Director, CBRE UK.

ABBREVIATIONS

ALUP	Annual Land Use Plan
APL	Administrative Permission Law
ASH	Affordable and social housing
BSGS	Beijing, Shanghai, Guangzhou and Shenzhen
CAS	China Appraisal Society
CBD	Central Business District
CBRC	China Banking Regulatory Commission
CCP	Chinese Communist Party
CHFS	China Household Finance Survey
CIREA	China Institute of Real Estate Appraisers and Agents
CITIC	China International Trust and Investment Company
CN¥	Chinese yuan (or Renminbi [RMB], the currency in Mainland China)
CPA	Certified Public Appraiser
CPMI	China Property Management Institute
CREA	Certified Real Estate Appraiser
CREVA	China Real Estate Valuers Association
CSRC	China Securities Regulatory Committee
Decree 55	Interim Regulations Concerning the Assignment and Transfer of the Right to the Use of the State-owned Land in the Urban Areas
ECH	Economic and Comfortable Housing
F&B	Food and Beverage
FDI	foreign direct investment
GDP	Gross Domestic Product
GEA	gross external area
GFA	Gross Floor Area
GITIC	Guangdong ITIC

GP	general partner
GRES	Guangzhou Real Estate Society
GZITIC	Guangzhou ITIC
HKSE	Hong Kong Stock Exchange
HNWI	high-net-worth-individuals
HOPSCA	Hotels, Offices, Parks, Shops, Clubs and Apartments
HPF	Housing Provident Fund
HPIR	House prices to income ratio
IPO	Initial Public Offering
ITIC	International Trusts and Investment Company
LAL	Land Administration Law
LPH	Limited Price Housing
LRH	Low Rent Housing
LTACM	Long-Term Adjustment and Control Mechanism
LUR	Land Use Right
MLR	Ministry of Land and Resources
MLUP	Master Land Use Plan
MNC	multinational companies
MoC	Ministry of Construction
MOHURD	Ministry of Housing and Urban and Rural Development
NBSC	National Bureau of Statistics China
NDRC	National Development and Reform Commission
NPC	National People's Congress
NUFA	Net Usable Floor Area
PBoC	People's Bank of China
PC	People's Congress
PE	price to earnings ratio
PERE	private equity real estate
PPP	Purchasing Power Parity
PRH	Public Rental Housing
QFII	Qualified Foreign Institutional Investor
R&D	research and development
REA	real estate agent
REAL	Real Estate Administration Law
REIT	real estate investment trust
REA	Real estate agent
REP	real estate planner
REV	Real Estate Valuer
RIH	Redeveloped and Improved Housing
RMB	Renminbi (the currency in Mainland China)
SAFE	State Administration of Foreign Exchanges
SLAB	State Land Administration Bureau
SOE	state-owned enterprise
SSE	Shanghai Stock Exchange

SWV	South Wisdom Valley
SZSE	Shenzhen Stock Exchange
UFAR	usable floor area ratio
UREAL	Urban Real Estate Administration Law
WHPS	Wenzhou Housing Purchase Syndicate

PART I
Contextual factors

1

INTRODUCTION

1.1 Introduction

On a scale the world has never seen before, China's urban expansion and real estate development have recently been widely observed, reported and commented on. New buildings of all types have been emerging relentlessly in Chinese cities and towns. The urban landscape of Chinese cities features glittering skyscrapers, lofty residential tower blocks and grandiose shopping centres in the urban area (for example, see Figure 1.1) and countless commodity housing estates, retail outlets and industrial parks spreading out to the suburbs. The real estate market has been overheating and overshadowing the country's stock market and any other market, with frequent scenes of overnight queuing at the sales offices of commodity housing estates under construction. In 2013, over 10 million new commodity housing units were sold, consolidating the position of the Chinese housing market as the largest in the world. Acting as a growth engine, the real estate market has generated funds for Chinese cities to build state-of-the-art infrastructure: high-quality road networks, mass transit railways, ports and airports. Today Chinese cities are playing a crucial role in propelling the Chinese economy to become the largest in the world in the near future. The rest of the world stands to benefit from this urban miracle by supplying raw materials, machinery, luxury goods and money to the booming Chinese cities, and gaining from value-for-money manufactured goods, spendthrift tourists and, increasingly, money for investment. The Chinese real estate market is therefore important to the economic fortune of not only China but also of the rest of the world.

On the other hand, sceptics and pessimists have questioned the sustainability of China's urban growth, predicting 'crashes' or even a 'meltdown' of the real estate market and the subsequent 'collapse' of the Chinese economy. There is plenty of ammunition for such doomsday talk: mass protests against forced demolition and

FIGURE 1.1 Shennan Road East, Shenzhen
Source: photo by Albert Cao (2015).

relocation, congestion, pollution, unaffordable high prices, 'ghost towns', and high local government debts linked to massive urban expansion. To the cynics, the slowing down or even hard landing of the real estate market will weaken urban economies, which in turn will act as a drag on the country's economy. The collapse of the real estate market, according to them, will end the country's three-decade-strong economic growth. Indeed, Chinese GDP growth has slowed to 7.7 per cent in 2013 from 14.2 per cent in 2007 (see Chapter 2). Further slowdown, cynics say, will lead to mass unemployment and the eventual 'collapse' of the Chinese economy. What is not made explicit, however, is the impact that a Chinese slowdown or collapse would have on the world economy.

At the time of writing in the middle of 2014, it is clear that the Chinese real estate market has suffered a severe structural imbalance that needs to be addressed. The future of the Chinese real estate market is now a serious concern for businesses and governments in China and beyond. This book provides an in-depth understanding of how the Chinese real estate market has developed and gained such an important role in the Chinese economy and now is generating some serious structural problems. The reader will be able to understand how the regulators and market players interact to sustain the real estate market's contribution and why investment opportunities have abounded during the transformation. To set the scene, a brief history of the Chinese real estate sector is provided in Section 1.2.

1.2 Real estate in China: sixty years, three worlds

Real estate is important to the functioning of an economy and to the living standards in a society. It is a factor of production and in a mature market economy, a major category of investment assets held by individuals, families, companies and financial institutions. It is upon these assets that lending, pensions and social benefits depend.

For more than a century, the provision of real estate in China has fluctuated under drastically changing politico-social conditions that had strong implications for economic growth and living standards. During this period, three vastly different systems of real estate provision, referred to as three worlds in this book, were created under vastly different sets of political, economic and social conditions. These sets of conditions symbolise the three distinct periods in China's modern history: (1) the feudal and semi-colonial Qing Dynasty until 1911 and the turbulent Republic of China from 1911 to 1949; (2) the Communist People's Republic of China from 1949 to 1978; and (3) the reformist People's Republic of China from 1979.

The first world is characterised by real estate poverty under domestic unrest, foreign invasion and weak governance in the first half of the twentieth century. Housing shortages in major cities, particularly in the treaty ports,[1] was acute during late Qing Dynasty (from the late nineteenth to the early twentieth century) and was exacerbated by uncontrolled migration, uncoordinated construction, war destruction and slow rebuilding during and after the Japanese invasion (1931–45) and the civil war (1946–49). For instance, a net reduction of 10,177 residential buildings, mostly in multiple occupation, in Wuhan City, a major industrial and commercial city in central China, was revealed by building surveys conducted in 1936 and 1952. In the meantime, the urban population in Wuhan City increased from 0.9 to 1.3 million (Zhang, 2009). Due to lack of or delayed maintenance, most of the half-timbered buildings in the city were in a poor condition by 1950. A housing condition survey conducted in the city in August 1952, covering 633 buildings, concluded that 214 required minor repair, 235 needed major repair and 25 were beyond repair (ibid.). In Shanghai, the most developed and dynamic city in China, permanent housing was built to replace temporary housing in vast numbers in the first 30 years of the twentieth century (Figure 1.2). However, redevelopment of the very poorly constructed shanty houses and sheds resulted in riots by the residents in 1925. As a result, the spread of slum housing continued. By 1949, per capita living space was 3.9 m^2 (for the internal floor area, excluding the kitchen, the toilet and shower) in Shanghai, with 13.68 per cent of housing being shanty houses and sheds (Shanghai Chorogophy, 2014).

FIGURE 1.2 Yuyangli, Huaihai Road, Shanghai, an old housing estate built in 1917
Source: photo by Albert Cao (2013).

The establishment of the People's Republic of China in 1949 drastically changed the conditions for real estate provision and threw the country into a different world. Nevertheless, this different world still featured housing poverty and dilapidated commercial real estate but under completely different conditions: severely limited resources, inefficient public provision, and a controlled but still rapidly increasing urban population. Real estate in urban areas in Mainland China[2] was gradually nationalised as the country adopted a Soviet-style command economy and urban housing accommodation was considered the responsibility of the state. The real estate industry and the real estate market disappeared in Shanghai after 1956 (Zhang, 2012). Facing trade blockades and military tension,[3] first from the West, led by the United States, and later from the Soviet Union, the Chinese Government prioritised industrial projects in order to develop the small laggard industrial sector that it had inherited from its pre-1949 predecessors, at the expense of housing and non-industrial real estate (Wu *et al.*, 2007; see Chapter 2 for more details). During this period, China's urban real estate was provided mainly by the public sector through administrative rather than market allocation, with market trading of real estate being illegal. By charging very low rents, urban housing providers could not sustain housing provision. Urban housing was regarded as a drain on the resources that were desperately needed for more productive activities. The Cultural Revolution that disrupted economic growth seriously reduced housing construction from 1966 to 1976.

Because of limited resources being made available for housing and with rapid urban population growth (from 57.7 million in 1949 to 172.5 million in 1978) (NBSC, 2013), major improvements in housing standards were prevented, with urban housing conditions being poor throughout this period. By 1978, the national average per capita housing space in urban areas was 3.6 m^2 (Zhang, 1998), and most housing units were not self-contained. There was very limited availability of retail and leisure real estate. Because land was allocated without charge, wasteful use of land was widespread, with many industrial projects occupying far too much land in central urban locations.

Compared to the abrupt change from the first to the second world, the transition to yet another new world of real estate has been more gradual. In December 1978, the ruling Communist Party decided to carry out reforms to China's rigid and inefficient command economy to promote productivity, and to open up the country to international trade and other exchanges to take advantage of the gradual lifting of the blockade by the West after the visit of US President Richard Nixon to Beijing in 1972. The importance of real estate as a factor of production and as a company asset was soon recognised when China used land as its input into Sino-foreign joint ventures with overseas investors after 1979. The real estate market, once illegal and, in theory, non-existent under the command economy, re-emerged when real estate transactions were permitted between state-owned firms, foreign firms[4] and individuals in the early 1980s, initially in the Pearl River Delta, adjacent to Hong Kong, and in Shanghai. In 1987, open market land sales were conducted in Shenzhen, a 'Special Economic Zone' set up to experiment with economic reforms and opening-up measures.[5] This led to an amendment to the Constitution in 1988 to allow market transactions of state-owned urban land. It was expected that market allocation of land and buildings would improve the efficiency of land use and would generate more economic activities as firms and individuals sought to benefit from the provision and use of real estate.

The creation of private property rights on real estate started with the resumption of ownership rights over buildings in the 1980s. In 1990, a new land tenure system was established to allow private and foreign ownership of leasehold interests (Land Use Rights, or LURs) on state-owned land. Housing reforms, which changed urban housing from being a form of welfare from the state to a type of commodity, began in 1980 to relieve the state of the burden of housing provision and to promote the market as the main provider of housing. From the 1990s, the housing reforms became, in essence, a campaign to privatise the country's state-owned housing. Compared to the high-profile and dramatic changes in the former Eastern Bloc in the 1990s, the land use reform and housing reforms in China are 'quiet' property rights revolutions (see Section 2.1 for more details).

Since the 1990s, the role of the real estate market in economic growth and urbanisation has been widely recognised by both central and local governments in China. In 1998, the final phase of the housing reforms was rushed through, i.e. full privatisation to boost demand for commodity housing to help the economy fight spillovers from the 1997 Asian financial crisis. The reforms succeeded and the housing market took off. In 2003, the real estate sector was awarded the status of a pillar industry (key industry) to buttress rapid economic growth. With favourable policies and local government support, the real estate industry has become a key driver of economic growth, as indicated by its share of the national economy. China's nominal GDP in 2013 was 46.17 times its 1987 size while investment in market-oriented real estate development was 573.42 times greater. The share of GDP by real estate and construction industries rose from 2.19 per cent and 3.18 per cent respectively in 1978 to 5.62 per cent and 6.87 per cent in 2012 (NBSC, 2013).

Growing with the Chinese economy after the reforms and the opening-up campaign, this new world of real estate has finally ended the era of chronic real estate shortage due to abundant resource inputs, market provision and supportive policies. It has provided space for living, leisure and work to accommodate the largest urbanisation the world has ever seen. China's urbanisation ratio, expressed as the proportion of urban population to total population, grew from 10.64 per cent in 1949 to 17.92 per cent in 1978 and to 53.73 per cent in 2013 (Table 1.1), with the majority of the population living in urban areas since 2011. Housing production has expanded phenomenally, with total urban housing completion rising from 203.4 million m^2 (gross external area, GEA) in 1987 to 1,073.4 million m^2 in 2012. The housing market has dominated urban housing provision. In 2012, commodity housing as a proportion of total housing completions rose from 11.7 per cent in 1987 to 73.6 per cent. Per capita housing space in urban areas, a measure of housing standards, rose from 6.7 m^2 GEA in 1978[6] to 32.9 m^2 in 2012 (NBSC, 2013). Finally, this third world of real estate provision has created a dichotomy between homeowners and non-homeowners with implications for the country's politics, economy, society and environment. The former are increasingly wealthy as commodity housing prices have grown quickly, while the latter are increasingly falling behind in wealth.

In the past 60 years or so, China's real estate sector has changed from one form of poverty to another, and then to a world of relative plenty. This book is devoted to explaining how this world of plenty came about and what is likely to happen in the future.

1.3 China's real estate sector at the crossroads

During the period from 1979 to 2013, the real estate sector played a key role in China's economic growth. It has been a key growth engine for most, if not all, Chinese cities. As will be explained in later chapters, by using land to breed real estate development, a property-led development model has been adopted by the majority of Chinese local governments to generate economic and urban growth. Land sales revenues enable local governments to improve the local infrastructure so that more serviced land can be sold and industrial parks can attract inward investors. Real estate development generates jobs for the construction industry and stimulates demand for steel, cement, glass, electrical appliances, furniture and other sectors. It also creates demand for a variety of services, such as planning, development, agency work, consultancy, leasing, banking and investment. The space developed provides

TABLE 1.1 Urban population in China since 1949

Year	*Urban population (millions)*	*% of total population*	*Year*	*Urban population (millions)*	*% of total population*
1949	57.65	10.64	1993	331.73	27.99
1950	61.69	11.18	1994	341.69	28.51
1955	82.85	13.48	1995	351.74	29.04
1960	130.73	19.75	1996	373.04	30.48
1965	130.45	17.98	1997	394.49	31.91
1970	144.24	17.38	1998	416.08	33.35
1975	160.30	17.34	1999	437.48	34.78
1978	172.45	17.92	2000	459.06	36.22
1980	191.40	19.39	2001	480.64	37.66
1981	201.71	20.16	2002	502.12	39.09
1982	214.80	21.13	2003	523.76	40.53
1983	222.74	21.62	2004	542.83	41.76
1984	240.17	23.01	2005	562.12	42.99
1985	250.94	23.71	2006	582.88	44.34
1986	263.66	24.52	2007	606.33	45.89
1987	276.74	25.32	2008	624.03	46.99
1988	286.61	25.81	2009	645.12	48.34
1989	295.40	26.21	2010	669.78	49.95
1990	301.95	26.41	2011	690.79	51.27
1991	312.03	26.94	2012	711.82	52.57
1992	321.75	27.46	2013	731.11	53.73

Source: NBSC (2014b).

accommodation of acceptable quality and cost for businesses to occupy. Such real estate infrastructure has contributed to the rapid rise of Chinese cities to become regional, national, and even international business centres. In Beijing and Shanghai, the two largest cities in China, the first modern office tower blocks were built in the mid-1980s. In the first quarter of 2014, Beijing boasted 9.34 million m^2 Grade A office space, and Shanghai 5.3 million m^2 (Savills, 2014a; 2014b). Figure 1.3 contrasts the highest man-made structures in Beijing as the capital of the Ming Dynasty 500 years ago and the present day. The expanding modern office stock in the two cities lays the foundation for them to rival established international business centres such as Hong Kong, Singapore, Tokyo and Paris.

The rapidly improved infrastructure in Chinese cities is the most visible outcome of the benefits of the property-led development model. Take the development of the capital-intensive metro systems as an example. Significant amounts of current land sale revenues (see Chapter 3 for details) and borrowings secured on future land sales have been invested in the construction of metro systems, allowing many local governments to build a metro system on a scale that would otherwise have been impossible. At the time of writing (June 2014), 19 super-large Chinese cities[7] have 74 metro lines in operation, 7 operating lines under extension and 34 new lines in

FIGURE 1.3 Wangchunting (AD 1522) and Beijing International Trade Centre Phase III (AD 2007)

Source: photo by Albert Cao (2013).

construction. Before 1993, there were only three short lines operating in just two of these cities. Nineteen other cities are joining the metro club, with 13 cities building their first metro line and six cities having obtained building approval from the central government. A further six cities are applying for approvals. The most dramatic development in metro construction has taken place in Shanghai. In 1993, Shanghai Metro opened its first line for operation, but now has the world's longest underground railway mileage of any city.

The role of real estate in the economy and the system of governance in China combine to create a unique set of institutional arrangements in the real estate market. With the ultimate responsibility for generating local economic growth under fiscal constraints (see Chapter 2), many local governments depend on the budgetary and extra-budgetary revenues from the real estate sector to finance a significant number of initiatives and infrastructure projects, and make use of their administrative powers to shape the real estate market to maximise fiscal revenues. The Ministry of Housing and Urban and Rural Development (MOHURD), the national administrator of the real estate and construction sectors, is also dependent on the prosperity of both these sectors for prestige and income, and uses its administrative and regulative authority to maximise these benefits. Coupled with a lack of experience in real estate market development and regulation, real estate administration at local and national levels have exhibited insufficient capacity, or institutional inadequacy, to guide and regulate the real estate market to maximise long-term economic and social benefits. Later chapters will provide evidence and analysis to show that the real estate market governance system has been unable to adjust the imbalance in the real estate market. The following points set out the problems in China's real estate market and its governance system and illustrate imbalance and institutional inadequacy found therein.

First, local governments' dependence on land sale revenues and taxes from property development and transaction has generated a series of issues, including urban sprawl, housing prices inflation, loss of arable land, pollution and ecological degradation. Local governments use their monopoly position in the primary land market to maximise land sale revenues and expand the scale of the urban area so that more land is available for sale.

Second, the urban housing market has suffered from price inflation, empty homes and market disorder. Local governments are reluctant to dampen excessive demand for housing and tackle the structural imbalance in housing supply in case these result in a slowdown of the housing market. Speculative housing investment and multiple home-ownership are encouraged but not regulated. The commodity housing supply was biased towards the high-end market before 2006, which is where housing investment is focused, with the consequence of decreasing market affordability. Low holding costs, due to a lack of housing holding tax, act as a disincentive to the letting of housing by investors. Consequently, rapid housing price inflation coexisted with empty homes and a glut of high-end commodity housing. Unrealistic ambitions for urban expansion have caused an excessive oversupply of housing in some districts of certain cities, resulting in large numbers of empty homes. Such districts are referred to as 'ghost towns' by both the domestic and international media. In addition, there are problems such as the hoarding of land and housing stock by developers, frauds in sales and agency services, and problems with management and maintenance. Figure 1.4 shows a 43-storey high-end residential building at a prime location in Guangzhou, started in 1996 but left abandoned from 1998 to 2008. This building is reminiscent of the large-scale oversupply and project failures in the 1990s.

Third, the commercial real estate market, which includes offices, shops, factories and warehouses, has suffered from lack of coordination, poor investment decisions, and oversupply in many cities. Many prestige projects in cities, mostly involving commercial real estate development, perform poorly or even incur losses. In general, the commercial real estate market has not received sufficient policy attention. Currently only a small proportion of China's commercial real estate projects are of high investment value. The lack of financing for many commercial real estate projects has resulted in strata title sales, with implications for future management and values.

Fourth, the stability of the financial system has been threatened by irregularities in lending and fundraising, weak regulation, and through speculative investment activities targeting real estate development and investment. Lending by state-owned banks to the real estate sector has been subject to tight control. However, lending

FIGURE 1.4 Jingrongge (now Longzhuyuan), Yuexiu District, Guangzhou, derelict from 1998 to 2008
Source: photo by Albert Cao (2005).

by non-bank institutions to real estate development and investment by both authorised and unauthorised investment vehicles in real estate are abnormally large and carry excessive risks. Localised financial crises have occurred, leading to significant costs to investors, local and central governments.

Fifth, the administration and regulation of the real estate market, in particular, the housing market, have been excessive. Macro control measures to curb housing price inflation have been blunt and intrusive. Local governments tend to buttress the local market at the expense of the effectiveness of national policies. Infringement on private rights has occurred in macro control and forced demolition and relocation, sometimes causing small-scale unrest as the anger of the aggrieved owners flares up because of insufficient or delayed compensation. Irrational town planning, over-ambitious and unrealistic urban design, and unchecked greed by some developers aggravate traffic jams, reduce the amount of open space, damage urban cultural heritage and cause environmental and ecological degradation. Corruption and malfeasance by local government officials bring additional risks to the market because they encourage excessive risk-taking and disrupt good governance and market order.

Sixth, the reliance on the real estate market by many local economies has increased the downside risks of a fall in housing prices affecting the property market and the local economy. Local market collapse has been found to depress local growth for years in some cities (see examples in Chapter 4 and Chapter 10) in the 1990s and 2010s. Wenzhou City in Zhejiang Province, once an exemplar of economic dynamism, has been struggling to cope with declining housing prices (see Chapter 9 for details).

All of the above problems originated from a halfway reform in the transition from a command to a market economy and incomplete institutional arrangements established to enable the smooth running of a market economy. The government at the central and local levels has too much power to intervene in market operations, distorting resource allocations and fostering unrealistic market expectations. Lack of institutional arrangements in both the housing and commercial property markets has led to a structural imbalance in housing supply and a large extent of empty housing, and excessive oversupply of offices and industrial land in many parts of the country. An ineffective allocation of funding and responsibilities to local governments has caused local fiscal reliance on land sale receipts and taxes from property development and transactions, leading to weak policy implementation and market excesses.

China has entered a decade of far-reaching reforms marked by the ruling Party's reform blueprint released in November 2013 to bring further marketisation and better governance. Restructuring the economy will reduce the reliance on fixed asset investment (including real estate development) for economic growth and increase the contribution of services and consumption. To achieve this aim, fundamental changes to real estate governance are required to shape the future directions of the real estate market. However, a major market correction, started in 2014, tests the determination of the government to refrain from old-fashioned market intervention and allow the market to allocate more resources.

The Chinese real estate market is at a crossroads. To continue its role in powering economic growth and urbanisation, the real estate sector has to change to make housing more affordable and must reduce waste from inefficiency and embrace governance that is more effective. A more mature system of institutional arrangements is needed to support the healthy growth and sustainability of the real estate market. The author is of the opinion that a thorough review of the past and current operations of the real estate sector could provide food for thought for policy-makers and practitioners working to improve the efficiency of the real estate market. To achieve this end, a systematic analysis of the development and regulation of the real estate market as a whole and in sections is undertaken in the following chapters, supplemented by an analysis of current and future investment opportunities. This book has been written with the aim of influencing the future of the Chinese real estate market.

1.4 Aims of this book

It is clear that during the past 20 years, the gradual establishment of the real estate market in China has fundamentally altered the mode of economic growth and boosted the competitiveness of Chinese cities. The real estate market has created wealth and raised the living standards of the Chinese people. With its impact on fixed asset formation and its demand for overseas raw materials and manufactured products, the Chinese real estate sector is important not only to the Chinese economy but also has implications for the world economy. Yet there is a shortage of texts that provide an in-depth analysis and holistic review of the Chinese real estate market. This research-based book will fill the gap in that knowledge.

The principal objectives of the book are to inform readers of the past, current and possible future developments of the Chinese real estate market, and to provide an integrated analysis and synthesis of those developments to assist readers when making informed decisions linked directly or indirectly to the Chinese real estate market. In particular, the book does the following:

1. provides research-based knowledge on the Chinese real estate market, its regulatory regime, and the wider socio-political and economic environment;
2. develops perspectives to appreciate the status quo and anticipate future changes in the Chinese real estate market;
3. explains the institutional and non-institutional factors that have shaped and continue to influence the development of China's real estate market;
4. offers insights on assessments and decisions relevant to the Chinese real estate sector, including investment in Chinese real estate.

The terms 'real estate' and 'property' are used interchangeably in this book. China in this book refers to Mainland China only, excluding other autonomous Chinese territories, i.e. Hong Kong, Macao and Taiwan.

1.5 The analytical framework

The establishment and development of the Chinese real estate market have been a process of profound institutional changes that were shaped by the decisions of the market players and the administrative and regulatory agencies of the government. To make sense of those changes, this book adopts an institutional perspective, which regards institutions, or institutional arrangements, as rules of the game and their enforcement mechanisms (North, 1990). Institutional arrangements govern the choices that players (both organisations and individuals) make and can either support or hinder market or non-market transactions. Market-supporting institutional arrangements facilitate information transmission on goods, participants and market conditions. The arrangements also enforce contracts, protect property rights and maintain orderly competition in markets. They work collectively to lower the costs of market transactions. By contrast, market-weakening institutional arrangements

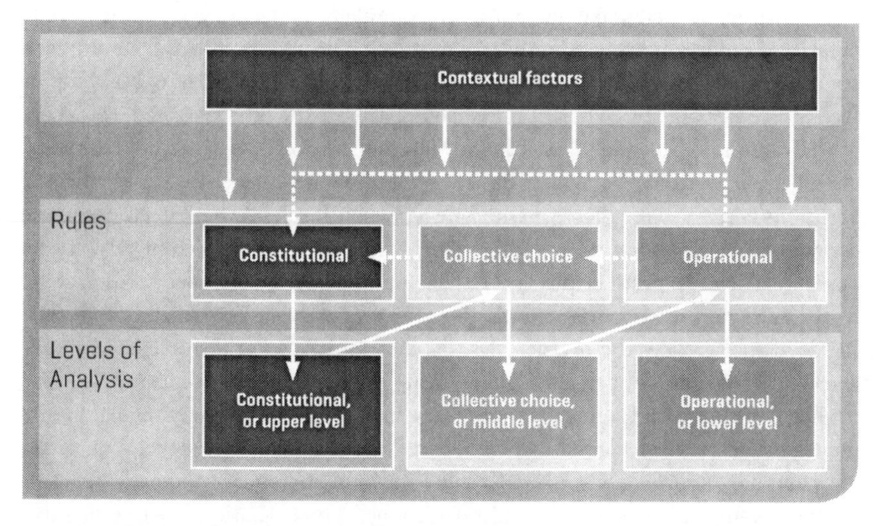

FIGURE 1.5 Three-level analytical framework and the contextual factors
Source: Cao and Keivani (2014b).

increase transaction costs, infringe on property rights and impede the development or even the establishment of markets. They are the major barriers to productivity increases and economic growth.

To address the myriad of institutional arrangements, an analytical framework based on new institutional economics (ibid.) and the meta-theoretical synthesis of institutional approaches (Kiser and Ostrom, 1982) is employed here to classify institutional arrangements surrounding the real estate market in China into three levels: operational, collective choice and constitutional (Figure 1.5). Organisations and individuals operating in the real estate market and its governance structure are agents who either persist or change existing institutional arrangements. They normally operate at different levels of institutional arrangements according to their objectives and capacity, as outlined in Table 1.2.

An outline description of the three levels is provided here. The operational level is the lowest level where individuals, firms and the operational arms of administrative

TABLE 1.2 Actors (organisations and individuals) at different levels of institutional analysis

Levels of analysis	*Actors (organisations and individuals) at different levels*
Constitutional	National People's Congress, State Council (Central Government); Ministries
Collective choice	Provincial, Municipal and County/District Governments and their relevant departments and bureaux
Operational	Government management offices; developers; real estate management firms; agents; banks and other financial institutions; investors; buyers; tenants; other stakeholders

and regulatory agencies take decisions, for example real estate development, investment, and enforcement of administrative orders. The actions at this level produce concrete results, or physical changes, such as the construction of new buildings or the enforcement of law. The collective choice level is the intermediate level where officials administering and regulating the real estate market take decisions, for example setting the criteria by which the administrative arms issue presale certificates. Decisions taken at the collective choice level establish rules to bind decisions at the operational level, which do not affect the physical changes *per se*, but rely on the operational level to produce physical results. For instance, a developer needs to abide by the rules on presale of commodity housing set by the collective choice level. If the developer violates the presale rules, the decision-maker at the collective choice level is able to enforce such rules by requiring the developer to conform.

The constitutional level is the top level, where decisions taken by regulators and politicians become rules, binding decisions at the collective choice level. For example, central government sets rules to require the local governments to provide housing for low-income households at affordable rents. These decisions are the rules that officials at the collective choice level need to obey when they, in turn, formulate rules or regulations for the operational level. For example, local governments set regulations for the provision of housing at affordable rents for both government operational units and developers to deliver.

Similarly, actors at the constitutional level are constrained by rules set by a country's political system, courts and cultural traditions. However, analysis on the rules governing decision-making by actors at the constitutional level is beyond the scope of this book.

The three-level framework allows analysis of institutional arrangements and their impact on transaction costs, property rights and incentives at each level and across levels. For example, insufficient protection of property rights of existing homeowners at the operational level incentivises developers to acquire land through forced demolition and relocation. Local governments' operational arms sometimes do not act to prevent developers from forced demolition and relocation because of the desired economic growth generated by the development and sometimes because of bribes received from developers. The problem of forced demolition and relocation feeds through to local governments at the collective choice level. However, increased economic activities and tax revenues incentivise local governments at the collective choice level to covertly or overtly tolerate this behaviour by developers and the operational arms of local governments. The local governments sometimes enact policies and regulations to institutionalise forced demolition and relocation and stop short of feeding through the problem to the constitutional level, which is kept in the dark with no reaction.

In the framework, actors at lower levels can report to their counterparts at higher levels to affect decisions taken at higher levels. For example, the problem of forced demolition and relocation eventually feed through to the constitutional level from the operational level by affected owners petitioning the central government or ministries. Concerned that the injustice in such forced demolition and relocation could result in social instability, the central government or its ministries at the constitutional level changes the rules for demolition and relocation, and requires local governments to enforce them. The local governments at the collective choice level, in turn, make their own rules according to the new responsibility imposed by the constitutional level, for the operational arms to enforce. In this way forced demolition and relocation are outlawed and enforcement of the new rules gradually reduces the occurrence of forced demolition and relocation.

The three-level analytical framework is an open system. Relevant factors outside the property market are taken into account by the framework (Figure 1.5). For example, the opening-up policy brings in foreign investors, who demand high-specification offices. Such demand creates the office leasing market. More foreign investors arrive and their leasing demand pushes up prime office rents, incentivising developers to take risks to develop schemes that seem out of touch with market conditions at the time. The resultant office development boosts economic growth and encourages local governments to provide collective choice rules to incentivise the development of high-specification office buildings by the market, to designate central business districts (CBDs) and to expand development zones (i.e. industrial parks and business parks).

Thus, this analytical tool facilitates a scrutiny of incentives, rules, actions and outcomes for a multi-scalar system. It is particularly useful in analysing complex systems like property development in China where different actors interact within and outside the rules set by national and municipal regulatory bodies.

1.6 The structure of the book

The Chinese real estate market consists of three main parts – the land market, the housing market and the commercial property market – each of which has its own characteristics, yet is closely connected to the other two. In particular, the urban land market is unique due to state ownership of all urban land and local governments' monopoly in land supply (see Chapter 3 for more details). Thus the land, housing and commercial real estate markets have their own characteristics and merit separate coverage.

Activities in the Chinese real estate market can be grouped into development, regulation and investment. Property development, in particular commodity housing development, has dominated activities in the real estate market so far, with the transaction volume of new real estate higher than that of existing stock. Consequently, the housing prices in most Chinese cities are set by new housing. Regulatory activities are intense in China's housing market due to the dynamism of the market, price inflation, social and environmental concerns and the legacy of an interventionist government. Real estate investment, covering both commodity housing and commercial real estate, has become increasingly active and prompted the growth in the real estate consultancy sector.

To account for the past and present and to explore the future of the Chinese real estate market, this book is divided into five parts and 15 chapters. Part I sets the scene for the book in two chapters. Chapter 1 provides an overview of the state of real estate in China and the current problems facing the market actors and regulators, and presents the objectives of this book and the analytical framework adopted to analyse issues and topics associated with the Chinese real estate market. Chapter 2 explores the contextual factors surrounding the development of the real estate market, breaking them down into political, economic, social, technological, legal and environmental factors for analysis, and concluding that good fundamentals and renewed reforms will sustain economic growth at the current growth rates.

Concentrating on the development of the Chinese real estate market, Part II describes and analyses the development of the land, housing and commercial real estate markets. Chapter 3 examines the institutional change to China's land ownership system that paved the way for the land use reform, the development of the land market and the evolution of the administrative system that gave rise to the unique primary land market that is monopolised by the local authorities. Chapter 4 starts with the housing reform, an institutional change comparable and parallel to the land use reform. It then surveys the development of the housing market, the attempts by the central government to rein in housing price inflation by macro control, and the creation of new institutional arrangements for commodity housing development, transaction and investment. Chapter 5 explores the rise of the commercial real estate market and the different development pathways of the office, retail, industrial and logistics real estate markets.

Part III concentrates on land administration, planning and development control, macro control of the housing market to rein in housing price inflation, regulation of the real estate finance market and taxation on real estate. Chapter 6 delves into the current land supply mechanism and market regulation, and scrutinises the property-led development model and its impact on urban planning and development control. Chapter 7 reviews the arduous process of building housing market institutions through administrative measures and policy shocks to reduce housing price inflation, and analyses the mixed outcomes on imposing rules on size standards and purchase restrictions. Chapter 8 discusses issues in real estate financing and its regulation, and probes the use of taxation as a macro control tool to curb housing price inflation and the market responses to such tools.

Focusing on market dynamics and changes, Part IV investigates the state of investment and opportunities in direct and indirect real estate investment in China and the investment activities by Chinese firms and individuals overseas. Chapter 9 studies investment in commodity housing, covering changes in residential land development and land banking, sophistication in housing development, the dynamics of keeping housing as an investment, and property management and agency. Chapter 10 covers commercial real estate investment by investigating the factors affecting the current and future performance of the office, retail and logistics markets, and analyses case studies. In contrast to the previous two chapters focusing on direct real estate investment, Chapter 11 explores the rise of indirect investment vehicles such as shares, trusts and funds, and evaluates their risks and potential. Chapter 12 traces foreign direct investment in Chinese real estate and probes the international expansion of the Chinese real estate sector into favoured destinations in Asia, Europe and North America.

After the market-based analysis, Part V provides a synthesis of current and forthcoming regulatory and market changes that shape the future direction of the Chinese property market. In Chapter 13, the rebalancing of the housing supply between the state and the market is discussed, and the efficiency of China's massive affordable and social housing programme is evaluated. Chapter 14 examines the start of a property downturn unfolding in 2014, and evaluates macro control on efficiency and equity grounds. It explores the long-term adjustment and control mechanism in conception and early development, which embraces more market allocation of resources and less direct intervention. The chapter also briefly explores sustainability in China's real estate sector, a future direction for the sector. Finally, in Chapter 15, the development of the Chinese property market is evaluated on economic principles of efficiency and equity with considerations for future policy formulation.

Notes

1 Treaty ports were created after China in the Qing Dynasty was defeated by the European powers. One of the concessions made by China was to lease land permanently to those powers for them to build ports for trade.
2 The civil war from 1946 to 1949 ended with the Kuomingtang (the Nationalist Party) fleeing to Taiwan and the Communist Party taking over Mainland China. As a result, China split into Mainland China under the rule of the Communist Party, Taiwan under the rule of the Kuomingtang, Hong Kong under the rule of the UK (until 1997) and Macau under the rule of Portugal (until 1999).
3 The blockade by the US-led West targeted only Mainland China.
4 Such firms and individuals were 'foreign' because they were not from Mainland China. In fact, in the 1980s most of these firms were Chinese firms from Hong Kong.
5 Such measures were part of the attempts to improve economic efficiency in the reforms and opening-up campaign that began in December 1978. See more in Chapter 2 on the campaign.
6 See MOHURD (2011).
7 Excluding Hong Kong, Macau and cities in Taiwan.

2

POST-REFORM POLITICAL AND SOCIO-ECONOMIC CHANGES

The property market in China has been the result of a series of institutional changes taking place in many closely related arenas of the country during the reforms and opening-up campaign from 1978. After scrutinising the reforms and the opening-up policy from 1978 to 2014, this chapter categorises these changes into six sections on political, economic, social, technological, legal and environmental arenas, which constitute the external, or contextual, factors of the real estate market. Important changes, in particular, changes that have implications for the development of the property market, in these arenas are the subject of analysis. Added to the analysis is an examination of a renewed 'reforms and opening-up' campaign promised by the Third Plenary Session of the Eighteenth Congress of the Chinese Communist Party (CCP), held in November 2013, discussing their impact on the property market.

2.1 The reforms and opening-up campaigns

On 1 October 1949, the People's Republic of China was established in Mainland China by the CCP after the Kuomingtang, which was the ruling party in the previous government, fled to Taiwan. The economy in Mainland China at that time was in a very bad shape after years of war against the Japanese and civil wars. For example, in 1949 the country had a population 3.6 times that of the United States but a steel output only 0.2 per cent and a grain output 71.9 per cent of that of the United States.[1] This dire economic situation underlines the first world of real estate provision in China outlined in Chapter 1.

The new China led by the CCP began with immense appeal to a people torn by over a century of foreign invasions, corrupt governments and domestic turmoil, and was heralded as a forceful authority able to rebuild the nation. The enthusiasm of the people allowed the new government to change the economic institutions of the country without much protest. Along with many other Third World countries, China wrongly believed that by focusing on heavy industry, the country could catch up faster with the developed countries. On the contrary, China should have made use of its abundant and cheap labour, its competitive advantage, to concentrate on the development of labour-intensive agriculture and light industry. By focusing on capital-intensive heavy industry, China adopted a competitive strategy that ignored its innate advantage for economic development because the country was suffering from an acute shortage of capital and technology (Lin, 2012). To pool the very limited resources for heavy industry, China quickly adopted the Soviet-style command economy and imposed nationalisation of urban land, industry and commerce and the collectivisation of rural land and agricultural production. As a result, China built a heavy industry pretty much from scratch, and by 1970 it had the sixth largest industrial output in the world (Mellows-Facer and Maer, 2012). It successfully tested an atomic bomb in 1964 and launched a man-made satellite in 1970.

However, the achievements in heavy industry and defence-related technology were gained at the expense of living standards and economic efficiency. Living standards stagnated as most of the surplus created went to heavy industry, which generated very limited jobs. Total real agricultural output grew 70.1 per cent from close to subsistence level in 1952 to 1978 while the population grew 77.71 per cent. Real per capita income of urban residents grew 17.65 per cent from 1957 to 1978, an average year-on-year growth of only 0.78 per cent on the very low base in 1957 (NBSC, 1999). Lin (2012) shows that real total consumption of urban residents grew 112 per cent from 1952 to 1978, but that of rural residents grew only 58 per cent. In 1978, per capita GDP was CN¥379 (NBSC, 1999), equivalent to US$241 according to the overpriced official exchange rate at that time, and merely 0.8 per cent of that of the United States.[2] In the same year, per capita GDP in Taiwan was US$1,604,[3] which was a sharp contrast to the low living standards in Mainland China. The picture for economic efficiency was no better. GDP growth from 1952 to 1978 was disrupted by three policy-induced recessions[4] (Figure 2.1). Lin (2012) indicates that total factor productivity growth in China from 1952 to 1981 was 0.5 per cent a year, lower than the average of 2 per cent in the developing world. With the trade blockade and low domestic innovation, most of the industrial sector in the 1970s still used the 1950s technology from the Soviet Union.

The poor economic performance during this period led to the low housing standards in the second world of real estate provision (see Chapter 1). As China started to open up to the outside world from 1972 following the lifting of the trade blockade by the West, Chinese people gradually learned they were living in poverty. Demand for change was rife.

2.1.1 Early reforms and the opening-up policy

As a response to the low standards of living in China, the economic reforms and opening-up campaign were initiated by the CCP in its Third Plenary Session of the Eleventh Congress in December 1978 to improve the performance of

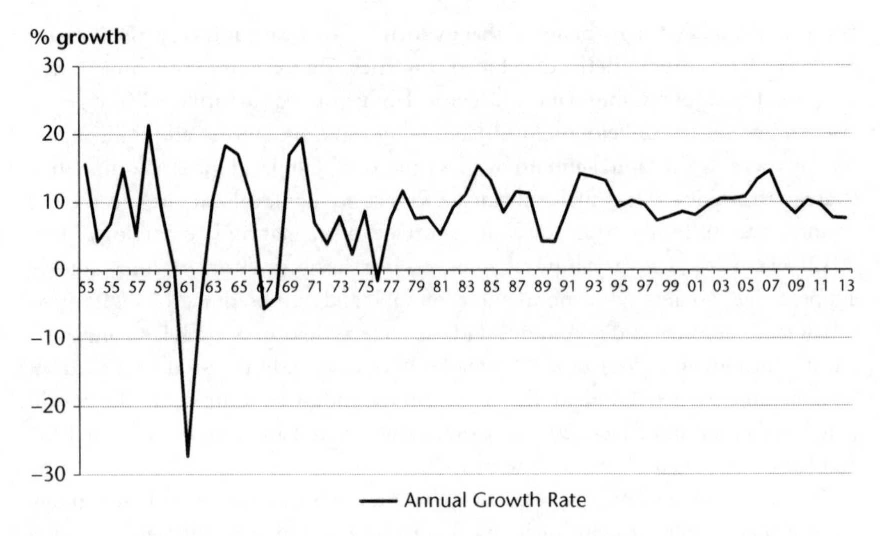

FIGURE 2.1 Real GDP growth of China from 1953
Source: NBSC (2103, 2014b).

the centrally planned economy in Mainland China (Nolan and Ash, 1996). Some scholars regarded the reforms, which were mainly economic, as necessary to justify the legitimacy of the CCP to continue its rule (for example, He, 1996). Without clear objectives, let alone a visionary blueprint, the reforms in the 1980s had taken a gradual, evolutionary approach, best illustrated by Deng Xiaoping, China's late leader, as 'crossing the river by feeling the stones underfoot'. In October 1992, the Fourteenth Congress of the CCP defined the goal of the reforms as establishing a 'socialist' market economy. In November 1993, the Third Plenary Session of the Fourteenth Congress of the CCP produced a blueprint for the reforms, the *Decisions on Issues of Establishment of a Socialist Market Economy System*. The document contained 50 articles of plans to reform the country's institutional arrangements to enable markets to take more important roles in resource allocation in Mainland China. Such a blueprint has been followed till today.

Compared to the 'big bang' approach adopted by the former Soviet Union states in the change from a command economy to a capitalist one that brought economic disaster (Walder, 1996), China's gradualist approach has led to success in economic development during the transition. Among the many reasons offered to explain China's success, two arguments are of particular relevance to the real estate market development. First, the reforms and opening up have enabled the country to develop the economy according to its comparative advantages (abundant labour and cheap resources), a strategy following a comparative advantage (as opposed to defying it) (Lin, 2012). In the early stages of the reforms and the opening-up policy, China combined its abundant low-cost labour and cheap land with foreign capital and technology (both scarce domestically) to become competitive in exports. Second, the gradualist reforms allowed time for old institutional arrangements to adapt and new ones to develop (Walder, 1996). Market economy institutions, for example the stock market and the property market, were established by trial and error and by overcoming the resistance from the old guard of socialism.

2.1.2 The property rights revolution

The reforms and opening-up campaign ushered in a new system of property rights best illustrated by the four amendments to the 1982 Constitution. In the early 1980s, Hong Kong investors bought housing and commercial property and established *de facto* private ownership of those properties. The first amendment to the Constitution in 1988 legitimised private ownership of land use rights (LURs) and properties, leading to the creation of a land tenure system that allowed long leaseholds on state land to be privately owned and transacted in the market (see Section 3.1 for more details). In 1993, rapid marketisation was endorsed by the second amendment to the Constitution, to legitimise the market economy and give autonomy to state-owned and collectively owned enterprises to make all their own business decisions. Further growth of private ownership and private economic activities prompted the third amendment, in 1999, to declare privately owned businesses an integral part of the socialist market economy and to honour private ownership by state protection. In 2004, the fourth amendment to the Constitution afforded the same status to private property rights and public property rights, and announced that the state encourages the development of the private economic sector. Private property rights have become legally inheritable since then.

2.1.3 State-owned enterprise reforms

Reforming the state-owned enterprises (SOEs) has been symbolic of the determination and direction of reforms. Under the command economy system, SOEs were told what to produce and how, were given all the inputs and were not responsible for distributing the finished products. The opening-up campaign has ushered in a series of reforms on SOEs. In 1979, SOEs were allowed to retain some profits and were incentivised to produce more than the plan dictated. Later, SOEs were given autonomy in some decision-making, such as the procurement and sale of products. Pay to SOE employees was linked to productivity and managers entered into contracts with the state on management responsibilities. From 1986 in some SOEs, shares were issued to staff in an attempt to reduce short-termism brought about by management contracts. Shareholding companies were created and became the most popular ownership structure for SOEs. From 1993, the establishment of modern management systems became the main direction of SOE reform. In 1995, the proposal to concentrate management and supervision in big SOEs and to let go small SOEs was adopted, which resulted in the bankruptcy or privatisation of most of the 70,000-strong SOEs in the country and millions of workers being laid off from 1996 to 2000.

The SOE reforms from 1996 to 2000 dramatically changed the roles of SOEs and threw Chinese society into a turmoil. In this campaign of dividing up the state's assets, the winners were the managers and others who had the cash, with management buyout being the most frequent fate of small SOEs. The losers in this campaign were those staff members who were old and poorly educated. Many workers had been overpaid under the old SOE regime. After leaving the SOEs they could not obtain similar pay and thus became impoverished. On the other hand, the efficiency of the remaining SOEs, normally the giants in key industries and sectors, was greatly improved. Since the 1990s, most of the remaining SOEs have been listed in stock exchanges in China (the Shanghai Stock Exchange and the Shenzhen Stock Exchange) and beyond (the Hong Kong Stock Exchange, the New York Stock Exchange and others), raising significant capital and acquiring much expertise in modern enterprise management. By 2013, 89 Chinese firms, most of which were SOEs, were in the Fortune 500, with three in the top 10. This is a big change compared to 2008 when only 34 were in the list of the 500 largest companies (Forbes, 2013). However, there is still some preferential treatment from the state for SOEs, especially for sectors in which China has no comparative advantages (Lin, 2012). For example, SOEs have easy access to cheap credit offered by state-owned banks. As a result, SOE reform has yet to be completed.

2.1.4 New initiatives in reforms and opening up: the Third Plenary Session

In the era of economic reforms and opening up, China has developed into a regime of a Stolypin-style combination of political control and economic freedom. The central theme of the reforms, and indeed governance, was commodification – labour, healthcare and space, to name a few. However, the reforms slowed down or stalled after 2000. As a result, marketisation has only been partially completed and the government at all levels retains a significant amount of power to intervene in the market.

In many sectors, such as electricity and finance, those in power, close to power and able to buy power gained competitive advantages as they wielded their power for private gains. To some extent, large government ministries and interest groups captured much of the gains released by the last round of reforms. When power could be exchanged for money, corruption was rife. Recent investigations uncovered dozens of senior local and national officials gaining a total of tens of billions of dollars from selling power. On the other hand, opportunities were not equal. Non-SOEs were not working with the same advantage regarding loans from state-owned banks. Entry barriers for some markets exist at provincial and even municipal boundaries. The economic, social and natural environments deteriorated due to neglect and malfeasance by the responsible industries and the government. Fraud and dishonest behaviours abounded. Income inequality, represented by the Gini Coefficient, rose to a very high level of 47.4 in 2012, close to the 47.7 of the United States in 2011. There are two social strata in the country, i.e. the urban registered households, having an annual average per capita income of CN¥26,955 (in 2013), and rural households whose annual average per capita income is CN¥8,896 (in 2013) (NBSC, 2014b). There is widespread discontent expressed on the internet, where quasi-freedom of expression is tolerated. Concerns about whether China would fall into the Middle Income Trap were real because power had a higher return than productivity and innovation. This situation would stifle innovation and productivity growth, leading to China's manufacturing and exports losing competitive advantages compared to both the high-technology West and the low-cost developing countries. The widening income inequality would stall domestic demand growth. If that all happened, China's economic growth would stagnate.

The response from the CCP is to bring forward a new round of reforms to tackle the above-mentioned problems accumulated over the years when economic growth was prioritised at the expense of many other issues. Xi Jin-ping, who assumed power as the General Secretary of the Party in November 2012 and became president of the country in March 2013, has become the architect of a new round of reforms and opening up. His anti-corruption campaign has resulted in the fall of dozens of provincial or ministerial-level officials and some national-level officials. Abundant anecdotal evidence indicates a sharp reduction in consumption in high-end restaurants and in luxury goods has occurred due to the dramatic fall of on-the-job consumption by government officials. The scale of the anti-corruption campaign could have some macro effects, estimated to be over US$100 billion of lost output a year (BBC, 2014).

The Third Plenary Session of the Eighteenth Congress of the CCP, held in November 2013, put forward a blueprint of further reforms and opening up in the communiqué, the *Decision of the Central Committee of the Communist Party of China on Some Major Issues Concerning Comprehensively Deepening the Reform* (CCP, 2013). The document contains 60 articles of plans for reforms in economic, political, cultural, social, ecological and civilisational fields and national defence and military arenas. The reforms are to be completed by 2020. The future directions of reforms in the economic arena, as outlined by the communiqué, can be summarised as follows (Caixin, 2013):

1. Markets should play a decisive role in resource allocation, with the state playing its part to ensure the smooth running of the market.
2. The ownership system should feature the coexistence of different forms: the state sector ownership to strengthen its influence, and the non-state sector ownership to increase its dynamism and creativity.
3. A modern market system should be established to allow firms autonomy in decision-making, fair competition, free choice by consumers, and free movement of factors of production.
4. Macro control must be more rational and governance more effective (see Chapter 7 for a discussion on macro control in the housing market).
5. A modern fiscal system should be established (see Chapters 6 and 14 for a discussion of local government fiscal systems).
6. A new urban-rural relationship should be established to enable the industrial sector to assist the agricultural sector; urban areas to assist the development of rural areas; and urban and rural areas to eliminate their division.
7. The country's economy should be made more open.

It is expected that the reforms outlined above will increase market competition, improve governance and rationalise intervention in all markets, including the property market (see Chapter 14 for more discussion).

Some progress has been made on the reforms. For example, the reform task specified in Article 9 of the communiqué is to establish fair and transparent rules for the market to operate. Work has started to remove local trade barriers and monopoly in certain industries. The reform tasks specified in Articles 17 and 18 are to improve budgetary management and taxation. The reform packages on fiscal and taxation reforms were promulgated in early July. Property tax has been on the agenda, which will change the costs of holding property and affect housing market investment (see Chapter 8). The reform task specified in Article 23 is to complete the mechanism for healthy urbanisation, which means reforming the two-tier household registration system that divides the country's population into urban and rural households.[5] On 30 July, the reform package for the household registration system was announced (Xinhua, 2014a), which paved the way for 100 million peasants to become part of urban households, a change that will increase demand for properties in the cities and towns affected. The reform task specified in Article 36 refers to strengthening of the anti-corruption system, which is relevant to the property sector where corruption is rampant. As mentioned above, the anti-corruption campaign started in late 2012.

Reform tasks that have a direct impact on the property market are yet to be initiated at the time of writing. For example, the reform task specified in Article 11 aims to open up the primary land market to include collectively owned rural land, thus breaking the state monopoly (see Chapter 3 for details of such a monopoly). In December 2014, a detailed proposal was approved by the central government to set up trials to explore ways for collectively owned land to gain the same status as state-owned land in LUR sale, leasing and conversion into company assets. The reform task specified in Article 12 covers the development of a more comprehensive financial market. The reform is to introduce more foreign as well as domestic private financial institutions, thereby increasing competition and expanding the range of financial products (see Chapters 11 and 12 for more details).

2.2 Changes in politics and governance

Since the reforms and the opening-up campaign, China's political system has been stable, with the rule by the CCP being maintained. Political stability in a country undergoing rapid transition is regarded as essential for economic growth, and the success of China's gradualist reforms is often attributed to the preservation of key existing institutions while introducing market institutions. Nevertheless, there have been significant changes in the political arena and the way the government is run, even though the political system has been maintained.

2.2.1 From ideology to pragmatism

The most fundamental political change is the transition from ideology to pragmatism. The Third Plenum of the Eleventh Congress of the CCP in December 1978 changed the core task of the CCP from 'class struggle' to economic construction. Since then, the communist ideology has been put aside and pragmatism embraced as the central doctrine. The colour of the cat no longer matters as long as it catches mice.[6] The long debate on socialism and capitalism ended in the discovery of the commonality of both: socialism has a market element and capitalism has a plan element. It does not matter to socialism if it uses the tools invented by capitalism as long as they promote productivity and efficiency for socialist production. Under such a doctrine, market institutions like the stock market, the labour market and the property market are tools that can be used to promote economic growth and raise living standards.

Pragmatism and a focus on economic growth as the Party's new ruling doctrine have underpinned the transformation of the Chinese government from an anti-market totalitarian state into a largely pro-business authoritarian one. It enables Beijing to tolerate, encourage and even plan innovation and experiments in economic development at the local levels, and to adopt those that are successful at the national level. For example, rural reform started in an impoverished small village of Anhui province in 1978, where peasants spontaneously divided up land in collective ownership for private cultivation. The peasants, becoming owners of *de facto* leasehold interest in farmland, were allowed to retain all the surplus output after honouring their responsibility, namely to provide a certain amount of grain to the state (hence the term 'family responsibility system'). It is *bona fide* privatisation of the means of production, the opposite of communist ideology, in a bottom-up institutional change. It met strong resistance but was later accepted by the Party because it greatly raised productivity. This rural property rights reform, after extending all over the country in 1983, virtually turned tens of millions of peasant families into private businesses. The reform raised grain output in China from 304 million tons in 1978 to 407 million tons in 1984, ending the long history of food shortages in the country.

Another example, an institutional change by a combination of bottom-up and top-down initiatives, was the creation of the Shenzhen Special Economic Zone as a test bed for innovation and experiments, which pioneered many of the market-oriented reforms in China. The creation of the land market in Shenzhen led to a land use reform, with leasehold interests of state-owned land sold for private ownership. The idea of 'selling land to breed development' was successfully tested in Shenzhen and provided the Chinese state with a mechanism to cash in its vast land holdings in the urban areas and thus solve the funding shortages of local governments, in particular for infrastructure (see more in Chapter 3).

2.2.2 Information costs and monitoring

A number of mechanisms have substantially lowered the costs of information for the Chinese government and society on alternative modes of development, technologies and other knowledge. The system of *liu xue*, or studying overseas, has proved to be extremely useful for China. Through the unprecedented scale of studying overseas, China has been able to learn and made use of experiences and expertise in developing as well as developed countries to build China's economic success. From 1978, the Party and the government have become eager learners from successful market economies with thousands of officials sent to North America, Europe, Japan and Hong Kong for all sorts of visits, exchanges, conferences and courses. There is a saying circulating in China that best describes the scale and intensity of such learning – the Party School of the Central Committee of the CCP has two campuses: one in Beijing and the other in Harvard University. Of course China learns not just from the USA. Over the years thousands of officials have been involved in such trips to the UK alone. Even more significant learning has been done by Chinese students studying overseas, who were first sent by the government, and then mainly by universities and firms, and later mainly by their parents. In the last decade the number of students going abroad grew rapidly due to the increasing wealth of Chinese families. From 1978 to 2012, 2.64 million have been sent abroad and 1.09 million have returned.[7]

The media and the internet have played a significant role in reducing information costs and monitoring the performance of the government and the behaviour of officials. The domestic media has become much more open in discussing sensitive issues and commenting on policy failures, and the government has shown much greater tolerance. On the other hand, the internet has been a driving force for change in governance, with a significant amount of corruption cases, policy mistakes, local government wrongdoings and injustice being disclosed by net users in blogs and micro-blogs. Many online public forums are nearly free on expression. As time goes by, internet users are becoming more mature, making the internet an effective tool of public monitoring of government. In response, the government has built channels to collect public opinion through the internet, becoming better informed of public opinions on various issues and policies.

2.2.3 The relationship between the central and the local and the rise of property-led development

Since the reforms started, significant changes have taken place in the relationship between central and local governments. Before the reforms, local governments were part of the central plan and had very limited autonomy to make decisions on local affairs. By dismantling the institutions of the command economy, the central government delegated power to local governments, which resulted in a significant decentralisation of economic decision-making in the 1980s and the first half of the 1990s. A tax-sharing arrangement was made for a fixed amount of taxes collected by local governments to be paid to the central government. This arrangement incentivised local initiatives in economic development but encouraged exploitation by local government who took advantage of information asymmetry. To retain more tax revenues, local governments played down local economic growth, leading to the shrinking fiscal capacity of the central government. The falling capacity of the central state had a negative impact on the infrastructure investment and other areas that required central funding. In 1994, the central government signed a tax-sharing deal with all local governments, with the central government taking a bigger share of the country's tax revenues (Herschler, 1995). The centre took control of the fiscal revenues again.

The recentralisation of tax revenues had major implications for local governance. As the reforms progress, the Chinese central state has increasingly assumed the role of regulator and retreated from direct involvement in many sectors. Local governments, on the other hand, have taken greater roles in local economic and social developments, with local top officials' career prospects and terms in office closely linked to local economic performance. There is a mismatch of a smaller stake in the nation's fiscal revenues and a bigger share of tasks by local governments, which led local governments to make use of the property market to obtain revenues to fund the completion of their tasks. This fiscal dependency has a profound impact on local governments' behaviour and their local policies for the property market.

To promote economic growth, the local state has become a 'developmental state'; a concept based on the behaviour of the bureaucratic Japanese state, which made a strategic alliance with and provided financial support to selected big firms to foster key industries to compete with their counterparts in the West (Castells, 2000). This model has also been adopted by other newly industrialised Asian economies such as South Korea. In China's case, the involvement of the local governments in economic growth has been more extensive. Many SOEs are headed by local officials and are given local financial and other support to carry out various economic functions, with or without competition from their private or foreign counterparts. In this sense local governance in China is 'local corporatism' (Oi, 1996). Recently, the Chinese local state has been moving away from direct involvement in market activities, but is still an entrepreneurial type of market actor in urban development and mega projects (Wu *et al.*, 2007). An example of a city-scale project is the development of an out-of-town central business district by Zhengzhou (Cao, 2009).

To attract footloose international and domestic capital, Chinese cities are competing intensely with each other. The key to capturing mobile capital is the strength of the local agglomeration in terms of population, infrastructure and the business community. Local governments make use of their city planning power to expand urban areas and carry out key projects, which require vast amounts of capital. With limited tax revenues at their disposal, local governments turn to land resources that are at their discretion, selling land in the largest quantities allowed by the central government and highest prices achievable in the market, to raise funds for local development. Projects like central business districts and metro lines not only provide modern city images but also are platforms to launch a series of real estate projects to bring in new investment opportunities. To maximise net land sale revenues and the speed of development, infringement on existing property rights has occurred frequently through low compensation for properties taken and forceful demolition to quicken delivery. Furthermore, local governments have been prioritising economic policy over social policy and economic development over the environment, with traditional communities often broken up during relocation and insufficient social housing provided.

A trend opposite to decentralisation, the reconstruction of state control, has to some extent restored the state's governing capacity and reduced the local state's autonomy to regulate economy. The Party and the central government have tightened up appointments of high office at both provincial and local levels. Certain delegated rights have been reclaimed by the centre. For example, land sale quotas are now set by the Ministry of Land and Resources, a central government agency for land administration (see Chapter 6 for more details).

2.2.4 Local government debts and deleveraging

The need for funds by local governments in their ambitious development targets and plans prompted the use of local financing platforms, i.e. a number of local state-owned companies, and other fundraising channels such as some projects to borrow large amounts of money. Land sale receipts and future tax incomes were often used as guarantees for the repayment of loans. In the past few years, debts incurred by local governments have become a concern to both domestic and international observers. Local governments raised debts to fund their rescue packages (see Section 2.3.6) and to invest in infrastructure. An audit by the China Academy of Social Science in 2011 revealed the scale of those debts in the China National Accounts 2013 (*South China Morning Post*, 2013). According to the Accounts, the total government debts in 2012 were CN¥27.7 trillion, equivalent to 53 per cent of the GDP in the same year. Local governments' debts were CN¥19.9 trillion, up from CN¥10.7 trillion in 2010. In particular, about CN¥13.5 trillion local government debts were raised through shadow banking, which is outside the country's banking regulation. This indicates that local government debts are mainly short term and risky in the face of the current economic slowdown. The rapid increase in local government debts also led to scepticism by international financial commentators. For example, Sharma (2014) hinted a credit crisis was likely to occur in 2014 due to this 'big cloud of debt hanging over China'.

The problem of local government debts was serious. As discussed earlier in this section, local governments have to perform a range of roles that local fiscal revenues are insufficient to support. Property-led development is a useful source of revenue. However, the scale of the local infrastructural development is so large that local governments had to borrow from banks and shadow banks. Attempts have been made by local governments to borrow directly from local communities, but there is a problem of mismatch of returns from infrastructure and interest rates paid on such borrowings. The average return from infrastructure is only 3 per cent (Xu, 2014), but local governments found it hard to borrow at a 6 per cent interest rate from local communities.

Nevertheless, pessimism about China's ability to solve its local governments' debts is unfounded. First, local governments' debts have created high-quality assets, i.e. the local infrastructure. It is true that the local infrastructure in many Chinese cities is ahead of their stage of economic growth. However, with great growth potential (see the rest of this chapter), readiness in the infrastructure will facilitate growth. Thus local infrastructure projects generate higher rates of return for the city as a whole than that at the project level. Second, change in financing mechanisms will allow a restructuring of the local government debts. For example, the central government has allowed local governments to issue bonds in 2014, providing long-term debts at lower interest rates to local governments than direct borrowing from the community. Third, the financial position of the central government is sound and can support the local governments. The ultimate source of funding is central fiscal revenues. In addition, the central government has a low debt to GDP ratio, and it can borrow more cheaply than local governments at the sovereign debts market. The central government in China has a track record of bailing out state-owned banks in 1998 when over 30 per cent of their assets were non-performing loans. The production of the China National Accounts indicates that local debts have received due attention from the central government.

2.3 Economic growth, imbalance and restructuring

In the past 34 years, China has created a miracle in its continued, high-speed economic growth, which is the ultimate source of demand for property in the country. The following survey of the growth records, the imbalance in such growth and the need for restructuring provides an insight into understanding the development of the property market in China.

2.3.1 GDP growth and per capita GDP

From being a very backward and war-torn economy in 1949, China has made huge strides in economic development, especially after the reforms and opening up of 1978. Average annual growth was 6.15 per cent from 1952 to 1978, but 9.76 per cent from 1979 to 2013. The Chinese economy has become the second largest in the world in nominal terms since 2010, overtaking Japan which held that position for over 40 years. In 2013, China's GDP was 55.3 per cent of that of the USA in nominal terms, a significant change from 8.30 per cent in 1983.[8] In purchasing power parity (PPP) terms, China's GDP was 86.88 per cent of that of the USA in 2011 and is expected to overtake the USA in 2014 according to research commissioned by the World Bank (International Comparison Program, 2014). This is a significantly higher estimate than that made in 2005 by the International Comparison Program. In cumulative terms, China's economic growth in the last 35 years is far higher than any other major economy in the world (Figure 2.2).

Being a lower-middle income country, China's per capita GDP is still low compared to developed countries. In 2013, nominal per capita GDP was merely US$6,959 (NBSC, 2014b). Figure 2.3 shows per capita GDP at PPP of a selection of countries from estimates on 2013 results by the International Comparison Program. China's annualised growth is 8.81 per cent, second only to Russia's 14.05 per cent. Taking the total amount of GDP at PPP, China's gain, the biggest in the world, was about 60 per cent more than the USA during this period.

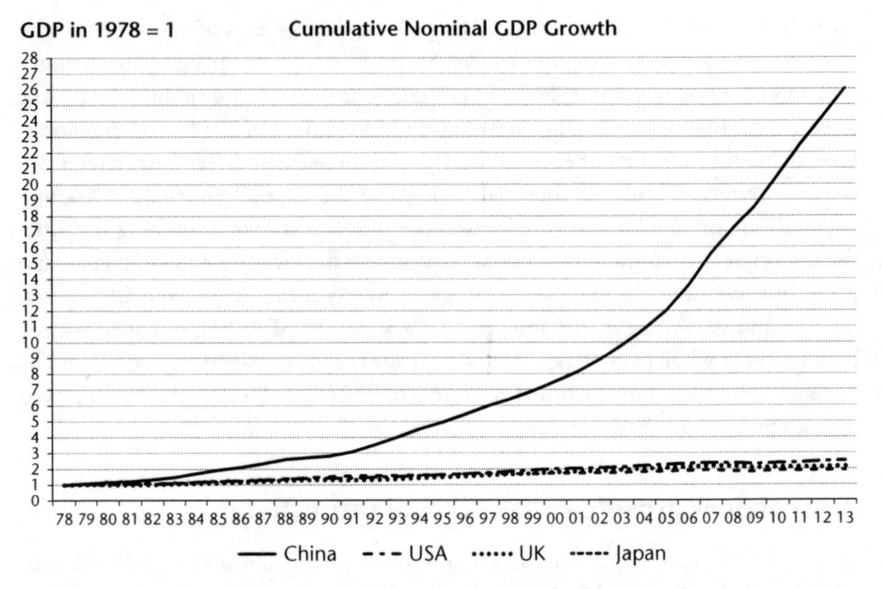

FIGURE 2.2 Cumulative growth in GDP of selected major economies, since 1978
Sources: NBSC (2013, 2014b).

2.3.2 Inflation

China's inflation was high in the 1980s and the first half of 1990s due to shortages in most sectors of the economy. It has been mild since the mid-1990s due to over-capacity in most sectors and an inexhaustible pool of cheap labour released by migration from the countryside and made redundant by the reforms of the SOEs.

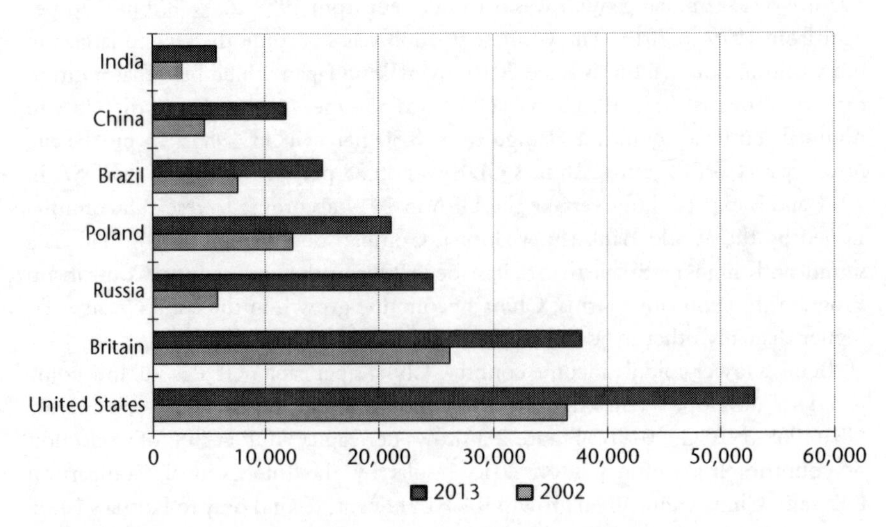

FIGURE 2.3 Comparative GDP per capita in purchasing power parity in 2002 and 2013
Source: World Bank.

Yet from 2007 inflation started to pick up because of resource bottlenecks, for example the rising international prices of commodities and a shortage of cheap migrant workers, and because of policy changes, for example fast rising wages due to minimum wage policy. Figure 2.4 shows China's inflation since 1978.

2.3.3 Manufacturing

The isolation of China by the West in the 1950s to 1970s, and by the Soviet Union in the 1960s and 1980s, forced China to prioritise manufacturing using out-of-date technologies. After the reforms and opening up, China's manufacturing sector acquired Western technologies and management knowhow through the purchase of machinery and technologies, and through foreign direct investment in manufacturing. At the same time, the Chinese government continued to support certain strategic manufacturing sectors in which China had no comparative advantages. As a result, China's manufacturing sector has enjoyed rapid expansion and become the most comprehensive in the world, capable of producing more product lines than any other country. In recent years China has been labelled 'factory of the world', which was vindicated in 2010 when China's manufacturing output overtook that of the USA for the first time, ending America's position as the largest manufacturer in the world in over a hundred years (1895–2009). China's manufacturing has a large share in the world, producing nearly half of steel and television sets, over 70 per cent of mobile phones and about 90 per cent of integrated circuits. Table 2.1 lists some of China's main industrial outputs.

It is worth noting that China is moving towards medium to high technology manufacturing as the country increasingly gains competitive advantage in capital-intensive industries and starts to lose competitive advantage in labour-intensive

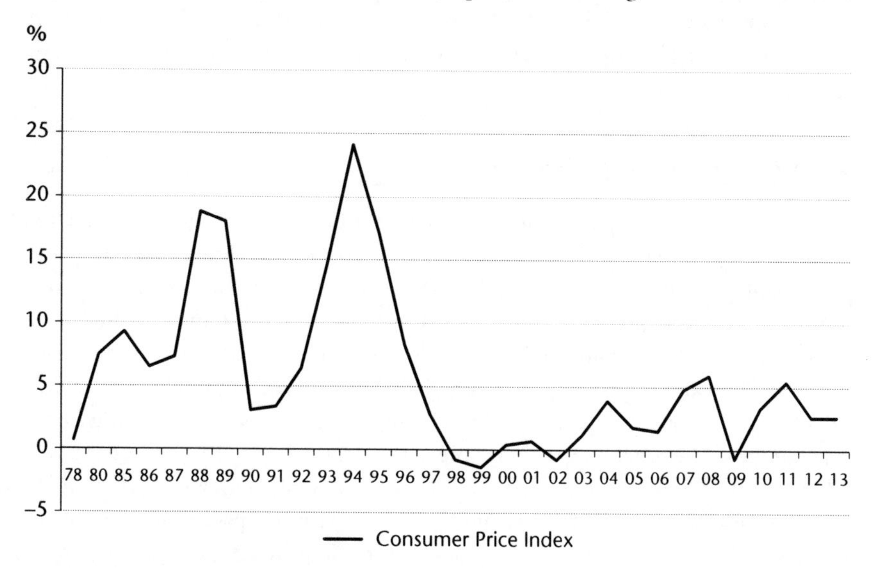

FIGURE 2.4 China's inflation since 1978

Sources: NBSC, various years.

TABLE 2.1 China's main industrial products

Products	*1980*	*1990*	*1995*	*2000*	*2005*	*2010*	*2012*	*2013*	*2013 World Ranking*
Steel (million tons)	31.78	66.35	95.40	128.50	397.00	637.23	723.88	779.04	1
Motor vehicles* (million)	0.01	0.51	1.50	2.10	5.70	18.27	19.28	22.12	1
Passenger cars (million)	0.00	0.00	0.10	0.60	2.80	9.58	10.77	12.10	1
Personal computers (million)	0.00	0.00	1.40	6.70	80.80	254.84	354.11	336.61	1
Television sets (million)	0.00	10.33	35.00	39.40	82.80	118.30	128.24	127.76	1
Mobile phones (million)	0.00	0.00	10.00	84.50	303.50	998.27	1,181.55	1,455.61	1

*four wheels or more

industries. In 2013, China's equipment manufacture output accounted for over a third of the world total. For example, China accounted for the manufacture of 60 per cent of electricity generators, 41 per cent of ships, 38 per cent of machine tools and 25 per cent of vehicles in the world in 2013 (Chinanews, 2014a). Such a trend can alleviate the decline of total labour availability as population growth slows down and the average age rises.

2.3.4 Imports, exports and FDI

Since the reforms and opening up, China has become part of the global production chain through foreign direct investment (FDI) and joint ventures, and an active participant in international trade. In the 1980s and early 1990s the country exported food and low-tech products to earn hard currency for the import of high-tech machinery, precision equipment and vehicles. After 1993, it started to enjoy a trade surplus and exported more products of high technology and high added value while importing raw materials, parts for assembly, high tech equipment and luxury consumer goods. After obtaining membership of the World Trade Organisation in 2001, China's exports soared (Figure 2.5) and, since overtaking Germany in 2009, it has been the largest exporter of merchandise. In 2013 China replaced the USA as the world's largest trading nation in goods.[9] With improving technology and an expanding economy of scale, China's exports are now competitive. Its demand for raw materials and commodities like crude oil and iron ore is strong due to an

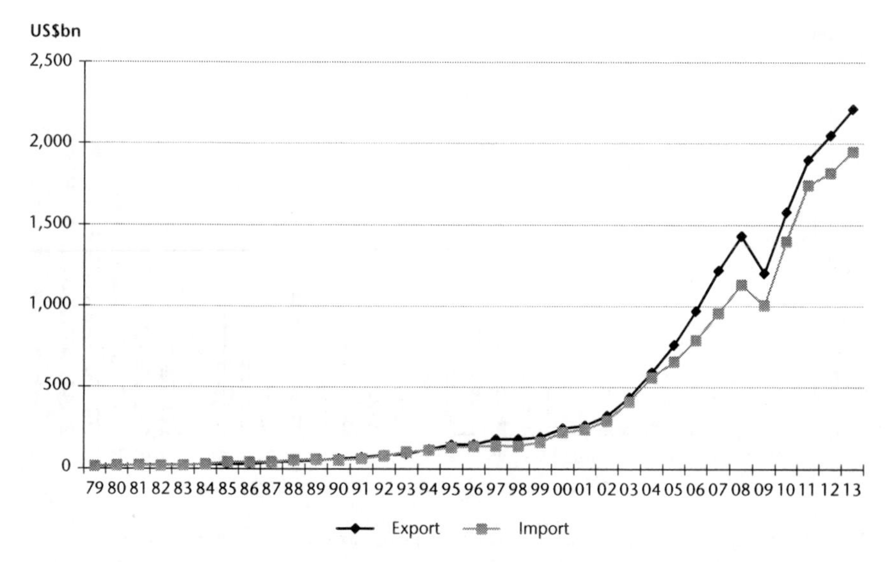

FIGURE 2.5 China's merchandise trade since 1979
Sources: NBSC (2013; 2014b).

insufficient domestic supply. Meanwhile, China is a major trader in services, with a total volume of trade at US$539.6 billion in 2013 (NBSC, 2014). Unlike merchandise trade that generated an annual surplus over US$150 billion in the past eight years, China's service trade has been in deficit, which is US$118.5 billion in 2013.

China's exports have been linked to the country's success in attracting inward FDI (Figure 2.6), which has greatly increased the competitiveness of products made in China and upgraded technology and management knowhow. From 1979 to 2013, China received a total of US$1.4 trillion in FDI. The country has been one of the largest destinations of FDI in recent years, with the lion's share going to manufacturing and creating a large foreign manufacturing sector inside the country. About half of the exports of the country were by foreign firms producing in China, with 49.9 per cent in 2000 and 49.0 per cent in 2012 (NBSC, 2013). Since 2003, Chinese firms, especially those in the mining and manufacturing sectors, have started to invest abroad to secure raw materials and to expand their market share. With the appreciation of the Chinese currency (CN¥), Chinese outward investment has expanded into other sectors such as real estate (see more details in Chapter 12).

2.3.5 Exchange rate control

To ensure the stability and competitiveness of the export sector, China has been adopting a policy of exchange rate control, pegging the CN¥ to the US dollar since a major devaluation of the CN¥ in the early 1990s. The exchange control had been subject to a long-standing criticism by the USA that China had gained trading advantages by

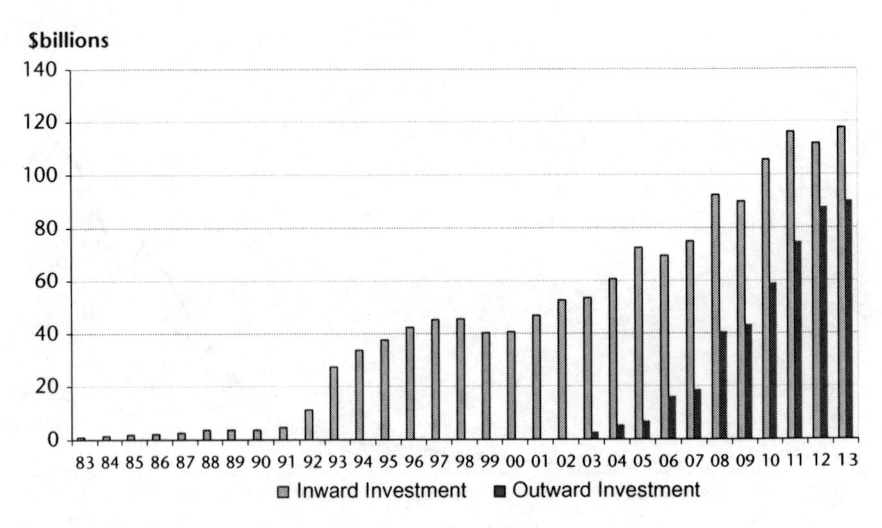

FIGURE 2.6 Chinese non-financial inward and outward investment after 1982
Sources: NBSC (2013; 2014b).

fixing the exchange rate at a low level. Since July 2005, China has allowed a controlled appreciation of the CN¥, which had risen in value by 35.7 per cent against the US dollar by early 2014. To implement its exchange policy, the People's Bank of China, the central bank, has to purchase all the foreign currency that enters China through trade surplus, net FDI and other means. As a result, the Bank had accumulated a foreign currency reserve worth US$3,821.3 billion at the end of 2013 (NBSC, 2014). The side effect of this exchange policy is the difficulty for the Bank in controlling the money supply in China, which has produced a long period of abundant liquidity and fuelled asset price inflation. To reduce the pressure of mounting foreign currency reserves, China has been relaxing control on overseas investment by Chinese families and firms, resulting in the rapid expansion of outward direct investment from China.

2.3.6 Fixed asset investment, rescue packages and economic restructuring

Since 1978, China's economy has had three main drivers: fixed asset investments; exports; and consumption. Fixed asset investment has been the main driver of economic growth in China recently, with gross fixed capital formation reaching 76.7 per cent of the GDP in 2013 (Figure 2.7). Such a high ratio enables China to build very expensive infrastructure like the high-speed railway, motorways, ports, airports, metros, and a large number of new buildings. The rapidly improving infrastructure paves the way for future economic growth. However, the return to investment for major infrastructure projects has been declining due to the gradual exhaustion of easy and profitable opportunities. Like exports, the room for further growth of fixed asset investment is limited, raising worries on whether China can sustain its

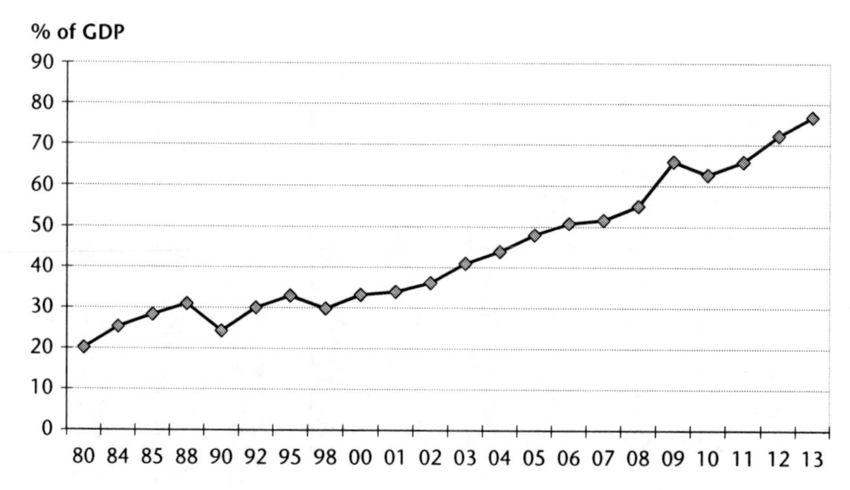

FIGURE 2.7 Gross capital formations as percentage of GDP
Sources: NBSC (2013; 2014b).

high-speed economic growth and whether infrastructure investors, mainly local governments, can repay the large amount of debts incurred in funding the infrastructure investment. These fears are largely unfounded because local government debts are manageable (see Section 2.2).

The rescue package at the end of 2008 for China to avoid a major economic slowdown comprised a total of CN¥4 trillion from the central government to invest in infrastructure and other projects. However, local governments added their own rescue packages with a total of CN¥18 trillion on various projects, mostly on fixed assets, pushing the Chinese economy back on a high-speed path. The reliance on fixed asset investment for economic growth, however, creates an imbalance in China's economic structure with investment taking too large a share. Recently such an imbalance has become the main target of economic restructuring, to boost consumption and thus make the economic growth more sustainable. In 2013, the total output of the service sector for the first time exceeded that of manufacturing and construction, indicating that a major shift in China's economic structure has occurred (Figure 2.8). Unlike exports and fixed asset investment, the level of services in China is rather low and there is much room for growth. Continued growth of the service sector as per capita income rises will rebalance China's economic structure and power the economy for further growth.

2.3.7 Regional disparity in development

Another area of imbalance is the expanding disparity of regional economic growth (Figure 2.9). Before 1978, this was much less marked than today due to China's regional policy that drew resources from rich regions to poor regions. After 1979,

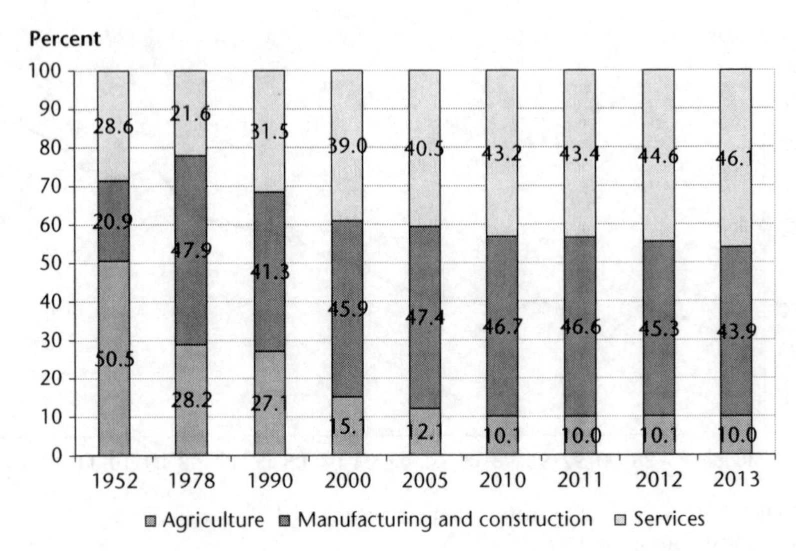

FIGURE 2.8 China's GDP sector ratio
Sources: NBSC (2013; 2014c).

special economic policies were set to favour regions with potential and resource allocation has become increasingly decided by the market. As a result, coastal areas benefiting from China's open door policy have received more foreign and domestic investments and have grown faster. To balance the regional inequality, China

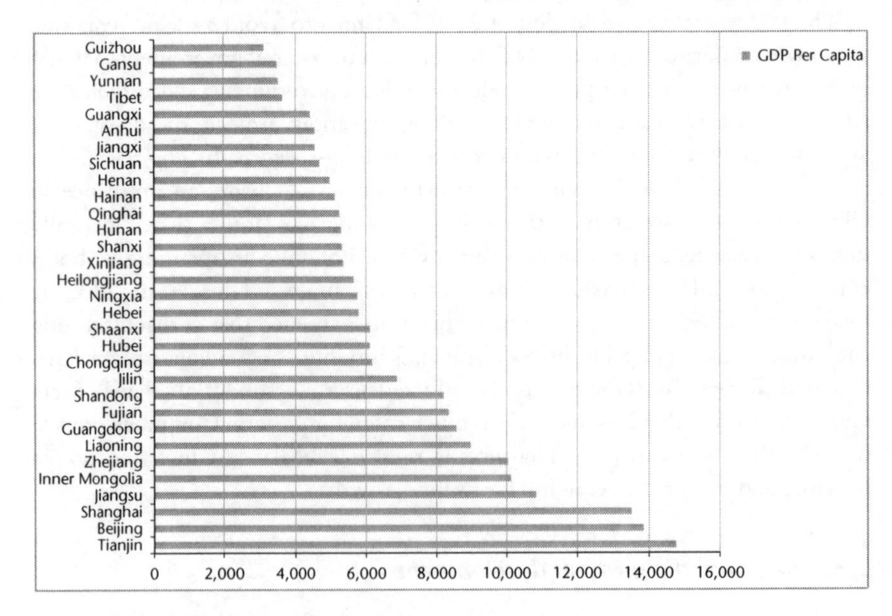

FIGURE 2.9 GDP per capita (CN¥) in all provinces in 2012
Note: The average US$ to CN¥ exchange rate was 6.315 in 2012.
Source: NBSC (2013).

initiated a series of new regional policies like 'Go West' in 1998 to boost economic growth in the western part of the country, 'Boost Central' in 2004 to accelerate economic growth in the central regions, and 'Rejuvenate Northeast' in 2006 to rejuvenate the northeast provinces that had been the industrial heart of the country but had fallen behind recently. Since then, the vast non-coastal areas have had faster economic growth, but regional disparity remains.

2.3.8 Ownership structure

Since 1979, China has been transformed from a rigidly planned economy into a mixed market economy, with the state sector, the private sector, the mixed sector (mainly state and private) and the foreign sector all contributing to the economy. In agriculture, the private sector has overwhelming dominance. In manufacturing and construction, state-owned firms accounted for 19.74 per cent of total assets and 8.94 per cent of total profits, privately owned firms 19.85 per cent and 32.61 per cent, foreign-owned firms 22.43 per cent and 22.56 per cent, and mixed-owned firms 37.98 per cent and 35.89 per cent respectively in 2012 (NBSC, 2013). The state sector has a dominant position in a few sectors like steel making, automobiles and the petrochemical industries. In domestic trade (wholesale, retail, hotel and catering), the private sector has a dominant position. Only in a few sectors in the service industry such as banking and financial services does the state sector have a dominant position.

2.4 Urbanisation and social changes

Before 1979, urbanisation was strictly controlled by the household registration system that limited population movement. At that time, China's policy regarding urbanisation was state-led, extensive industrialisation, devoting resources to industrialisation with urbanisation as a by-product. Under such a policy the pace of urbanisation was very slow. Since 1979, the country's policy on urbanisation has changed to urban-based intensive urbanisation. The new policy sees cities, rather than industrial projects, as growth poles and prioritises cities as the nodes of economic growth. Cities were delegated more power to orchestrate economic development and rural counties were subject to the leadership of the central cities to expand the resource pool for cities to lead economic growth. Rural to urban migration, which was virtually banned before 1979, was allowed through expanding the coverage of household registration and tolerance of a floating population, i.e. allowing those who have no local household registration to work and stay in cities and towns. With their welfare benefits still provided by the local governments of their origin, those migrants were attracted due to job opportunities and higher pay, despite difficulty in getting healthcare cover and education for them and their children. Massive migration from rural to urban areas eventually led to the historical shift when the urban population surpassed the rural population in 2011. In 2014, the government published the *National Plan on New Urbanisation* (State Council, 2014), which sets new targets for urbanisation up to 2020. For example, the urbanisation level is to be raised to 60 per cent in 2020, which requires transferring about 100 million rural population to cities and towns between 2014 and 2020.

Under the new urbanisation regime, cities and towns with faster economic development were able to expand rapidly. In the 1980s and 1990s, coastal areas, in particular the Pearl River Delta and Yangtze River Delta where foreign direct investment in manufacturing is concentrated, attracted large numbers of migrant workers, mostly from rural areas of the inland provinces. By 2012, the floating population, i.e. those with their household registration separated from their place of abode (not in the same city), totalled 236 million (NBSC, 2013). Such separation creates demand for housing in cities favoured by migrants, who buy housing for themselves and sometimes for their parents. Rapid housing price inflation ensues, resulting in modest-income households in those cities being priced out of the housing market.

Rapid urbanisation has caused a number of social problems in urban China. First, urban poverty looms large. The availability of vast numbers of migrant workers has slowed down wage growth in Chinese cities in low-skilled jobs, widening income inequality. Second, social segregation emerged, with migrants in low-skilled jobs concentrating in suburban, poor-quality residential clusters, according to their places of origin. Such clusters sometimes harbour illegal trades and are difficult to manage. Third, many new urban communities have been created by urban redevelopment breaking down existing urban communities. Such new communities have lower levels of belonging, cohesion and cooperation, leading to challenges for local governments in urban governance.

The nature of Chinese society has changed after 1979 from a rigid and administratively 'mechanical' society into a fast-changing, informally liberalised and organic society. In general, the Chinese population has become more mobile and better connected. In 2013, the number of times people travelled inside China was 3.26 billion and travelled outside China was 98.2 million (NBSC, 2014), far more than the 0.69 billion and 8.4 million respectively in 1998 (NBSC, 2003). The number of mobile phone users in 2013 was 1.23 billion and internet users was 618 million (NBSC, 2014). The number of private cars rose to 64.1 million in 2013 from 6 million in 2003 (ibid.).

With the success of the one-child policy in urban areas, population growth in China slowed down from 1.04 per cent in 1978 to 0.48 per cent in 2010 and 2011 (NBSC, 2013). The resultant population structure (Figure 2.10) triggers alarm about the aging population and the pension crisis. Cautious adjustments are being made to the one-child policy. A two-children policy was enacted for families with both husband and wife being the only child throughout the country by 2011. In 2014, the same policy was extended to families with either husband or wife being the only child. It is worth noting that the young age population has started expanding as age groups between 0 and 4 and 5 and 9 are larger than the 10 to 14 age group. Population growth rate bottomed in 2011 at 0.479 per cent and rose to 0.492 per cent in 2013.

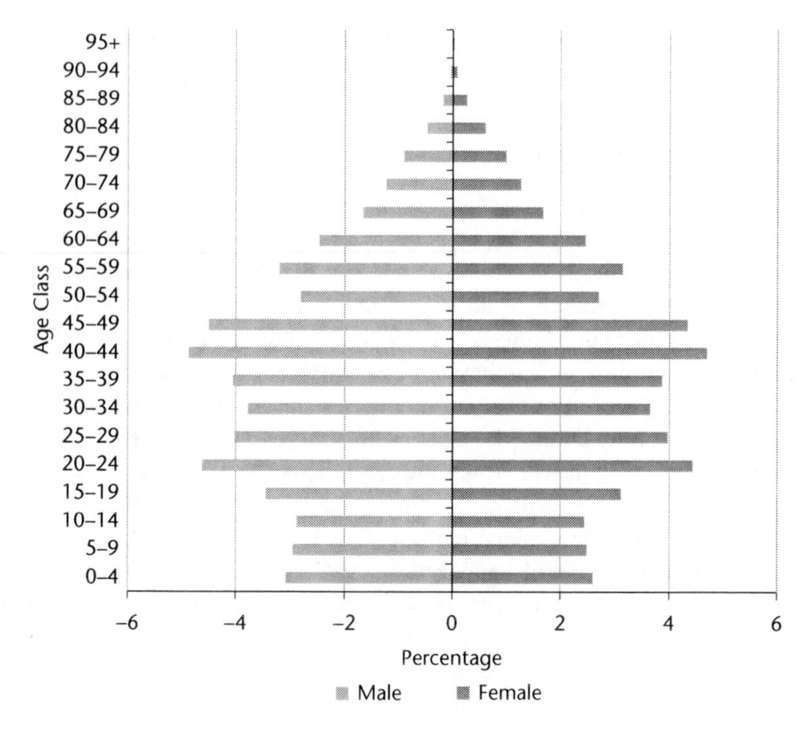

FIGURE 2.10 Population pyramid of China in 2012
Source: NBSC (2013).

Another demographic trend caused by the one-child policy, is the significant increase in one-person households and two-person households. In 1982, average family size was 4.56 persons nationally and 3.77 in Shanghai. In 2012, average family size was 3.02 persons nationally and 2.35 in Shanghai (NBSC, 2013). Shanghai has consistently recorded the smallest average family size in China. Such a trend has a significant implication to housing as the number of housing units has to rise much faster than population growth.

Last but not least, there has been a reduction in the overall number of people in the age range 16–59. In 2011, the total number in this age group was 940.72 million but in 2013 it was 919.54, a reduction of 21.28 million or 2.26 per cent (NBSC, 2013, 2014). Such a rapid reduction has caused concern regarding the impact of reduced labour on economic growth and the future old age dependency rate. Nevertheless, the supply of skilled labour has been on the rise (see Section 2.5) so the reduction only occurred in unskilled labour. The resultant change in the labour structure corresponds to the change in the industrial structure as mentioned in Section 2.3. On the other hand, the reduction in total manpower will result in wage rises, which has been the case in the last ten years. Rising wages will facilitate expansion in the service industry and help rebalance China's economy, leading to continued economic growth centred on the service industry. The change in family planning policy is expected to increase new births, which will improve the old age dependency ratio in the future.

2.5 Education and technological changes

China was a latecomer to modern education, science and technology. In 1949, the country's illiteracy rate was above 80 per cent and graduates from higher education institutions were about 21,000, with a very low percentage in engineering. As a result, China could produce hardly any of the products an industrialised country could produce at that time. From 1949 to 1978, China achieved technological breakthroughs in certain areas, such as the atomic bomb (1964), man-made satellites (1970) and nuclear power generation (1970). Nevertheless, the general technological level in China was rather low. During the Cultural Revolution (1966–76) education at all levels was badly hit and many research and development projects were suspended. In particular, candidates for higher education were selected by political inclination rather than academic achievements in that period. In 1978, higher education graduates were only 165,000 out of a population of 963 million (NBSC, 2013).

After 1977, China's higher education sector resumed recruitment by national entrance exams and in 1978 recruited 402,000 students. Due to limited funding, the country's higher education grew slowly, with annual recruitment reaching 1 million in 1997. In 1998, the central government decided to expand the higher education sector to increase the coverage of higher education, reduce youth unemployment and promote economic growth. By 2008 annual recruitment exceeded 6 million (Figure 2.11). Although there are concerns about the quality of recruits, pressure on graduate employment and falling initial salaries for graduates, the expansion in higher education in China has substantially increased the pool of talent and supplied the country with educated manpower necessary for rapid economic expansion.

With increased funding and educated manpower, China has been expanding research and development (R&D) activities and has achieved significant results recently. The country's R&D intensity, or R&D expenditure as a percentage of GDP, has grown from 0.65 per cent in 1997 to 2.09 per cent in 2013 (Figure 2.12). In 2012, Chinese R&D intensity (1.98 per cent) surpassed that of the European Union (1.96 per cent) for the first time, but was still much lower than the major countries in EU such as Germany (2.92 per cent) and other major technology powers in the world, for example Japan (2.7 per cent) and the USA (3.2 per cent). With the huge demand for new technology and the blockade of high-technology exports from the West,[10] China's R&D intensity continues to rise. Another reason for a rising R&D budget is that more and more Chinese firms are at the technological frontier and require more resources and higher risk to increase productivity compared to the past when they could rely on technology and best practice of foreign enterprises (Conference Board, 2014).

Intensified research and development activities have resulted in surging patent applications in China. China's domestic patent applications, an indicator of innovation, were 526,412 and became the highest in the world in 2011, exceeding the

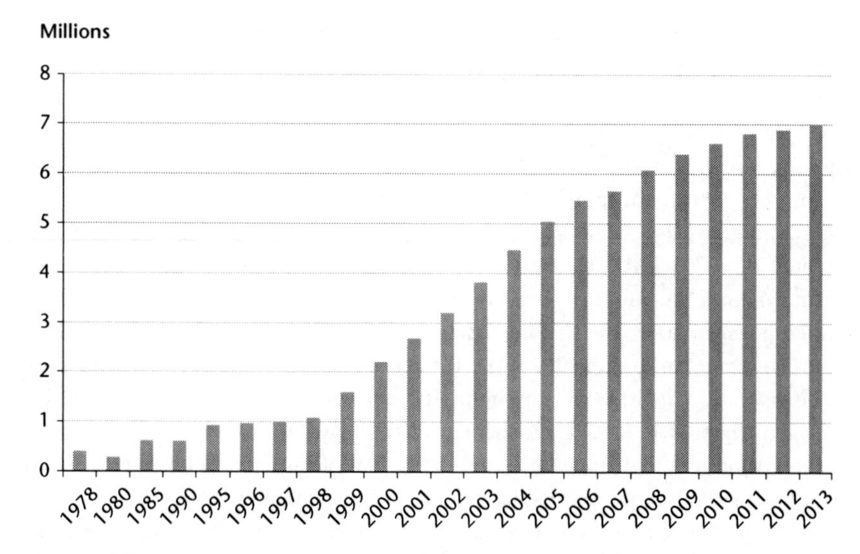

FIGURE 2.11 Annual recruits by higher education institutions in China
Sources: NBSC (2003; 2013; 2014b).

503,582 applications made in the USA in the same year (WIPO, 2012). In 2013, domestic patent applications rose to 825,000 in China. China's international patent applications, on the other hand, grew to 21,516 in 2013, rising to third place behind the USA (57,239) and Japan (43,918) (WIPO, 2014).

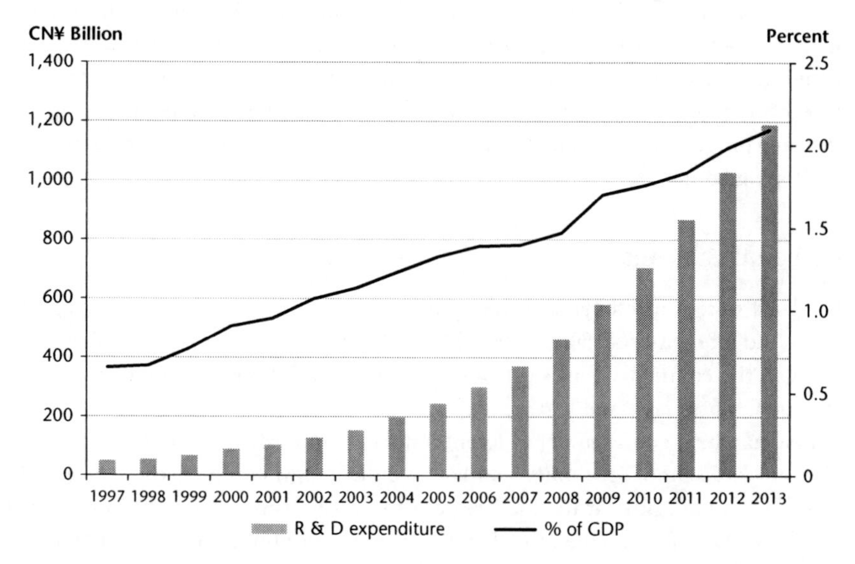

FIGURE 2.12 China's research and development spending since 1997
Sources: NBSC (2003; 2013; 2014b).

The rapid expansion and technological advances in high-speed railway in China are the first example of a major sector in China being ahead of the rest of world in technology and productive capacity. The first proposal for a high-speed railway was tabled in 1990 and by 2002 the first high-speed passenger train line (405 km in length) was completed, operating at a speed range of 200–250 km/hour. In 2004, the central government approved the first Medium to Long-term Railway Network Plan, aiming to have 12,000 km of passenger line operating at speeds above 200 km/hr. In 2007, China achieved operational speeds of 200–250 km/hour on six existing main lines, moving ahead of the rest of world at the time in improving existing lines. In 2008, the first line (Beijing to Tianjin) operating at 350 km/hour became operational. In June 2011, the 1,318 km Beijing to Shanghai passenger line started operation at 300–350 km/hour. Its passenger numbers had exceeded 100 million by February 2013 (Huochepiao, 2013). By December 2013, China has 13,000 km of new high-speed lines and 3,000 km of improved high-speed lines in operation, with all trains and railways made and built domestically.

In recent years China has entered a period of accelerated technological development. In consumer electronics, Chinese brands have occupied six places in the world top ten mobile phone brands, including the third and fourth places in the first quarter of 2014 (ZOL, 2014). In space technology, China succeeded in a manned space mission in 2003. In satellite navigation, China has been building its Beidou Navigation Satellite System, which became operational from 2012 and will provide global coverage by 2020. In the aviation industry, China is developing a regional airliner that has up to 100 seats and a medium-range airliner that has around 150 seats. China built two of world's fastest supercomputers, the Tianhe-I in 2010 and Tianhe II in 2013, respectively.

China is still behind developed countries in technology. But with rapid advances in technologies, it is expected in the near future China will change its status in manufacturing from 'Made in China' to 'Made by China'. Advances in technology have been supporting the restructuring of manufacturing towards more capital-intensive processes, with some labour-intensive processes being transferred to other developing countries.

2.6 Legal changes

The legal system in China is currently under gradual reform, with the Party, the public and international stakeholders all stressing the importance of establishing rule of law in the country. China's expanding foreign trade also requires transformation of some Chinese laws and legal practices.

Law in China comes under different influences in its development. Historically there are two schools of thought on law. Confucianism considers the state should lead the people with virtue and create a sense of shame that will prevent bad behaviour. Legalism, on the other hand, regards law as publicly promulgated standards of conduct supported by state coercion. Both schools sponsor a paternalistic idea of the state that knows more than its citizens and enacts laws to protect them.

This concept of the state persisted throughout China's thousands of years of feudalism and is still operating today.

After the fall of the Qing Dynasty in 1911, China was under the influence of Western laws, in particular German laws, and after 1949 was strongly affected by Soviet socialist laws. During the Cultural Revolution, the legal system completely collapsed. With the advent of the reforms, reconstruction of the legal system started with the 1982 Constitution, emphasising the rule of law under which even party leaders are theoretically held accountable. However, the legislative, executive, judicial, and pro-curatorial powers provided by the Constitution are all subject to the leadership of the CCP. The 1982 Constitution delineates extensive rights, including equality before the law, political rights, religious freedom, personal freedom, social and economic rights, cultural and educational rights, and familial rights.

The present legal system is provided for by the 1982 Constitution, which stipulates the National Peoples' Congress (NPC) as the supreme organ of state power and a structure of People's Congresses (PCs) at provincial, prefecture and county levels as the supreme organs of local power. The NPC and its Standing Committee have power to promulgate legislation, elect highest-level officials and set the budget, and local PCs have corresponding rights for localities but need to observe higher-level legislation and national administrative regulations.

Although far from perfect, the Chinese legal system has made tremendous progress in its reconstruction after 1976 (Peerenboom, 2007). First, more than 300 national laws and ordinances have been promulgated since 1979. These laws and regulations exhibit the influence of traditional Chinese views of the role of law, the country's socialist background, the laws in Taiwan that have a strong German foundation, and English-based common law used in Hong Kong. There has been consistent improvement to the laws and regulations. For example, the trademark law, first enacted in 1982, was amended in 1993, 2001 and 2013 to provide better protection to both domestic and foreign trademark owners.

On the other hand, legislation in a number of areas has been very slow. For example, it took over a decade for the Property Law to be enacted. A new amendment to the Land Administration Law has been in consideration for over a decade. At the time of writing, the national ordinance on affordable and social housing is still be drafted after years of consideration. This is not helpful because the government has carried out the world's largest building programme of affordable and social housing in 2011 without a proper national ordinance or law (see Chapter 13 for more details).

Second, there have been major attempts to develop and upgrade the professionalism of the legislature, the judiciary and the legal profession. The professionalisation of lawyers provides the basis for a legal services market, which was established in the 1980s. In 1992, the Chinese government opened the legal services market to foreign law firms, allowing them to establish offices in China. Figure 2.13 reports the number of law firms and lawyers in China in recent years and Figure 2.14 provides statistics of the civil lawsuit cases and administrative lawsuit cases handled by lawyers.

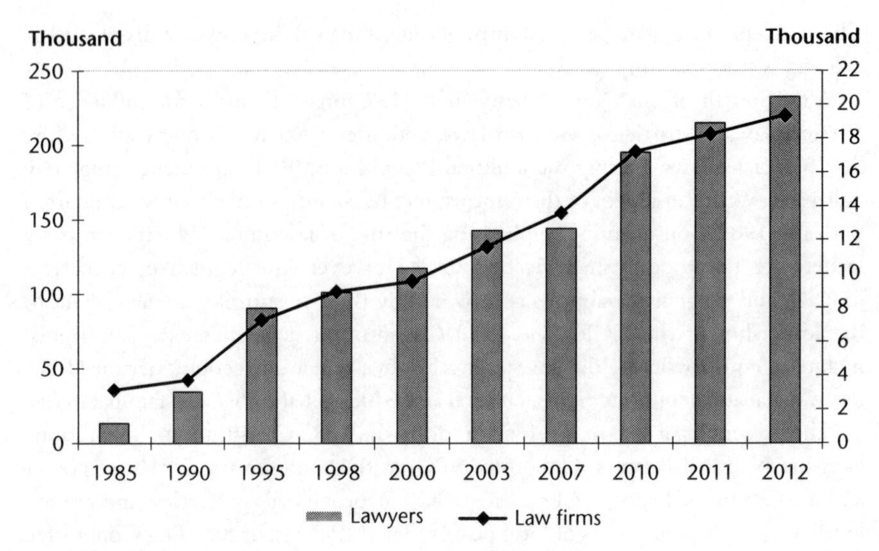

FIGURE 2.13 Number of lawyers and law firms in China
Source: NBSC (2013).

Third, laws to protect civilians against government abuse have been up and running. The 1994 Administrative Procedural Law was the first law to allow citizens to sue officials for abuse of authority or malfeasance. The 2003 Administrative Permission Law (APL) for the first time requires all laws and regulations that subject any civil act to approval requirements by the government to be published, and allows only laws enacted by the NCP and provincial PCs and regulations

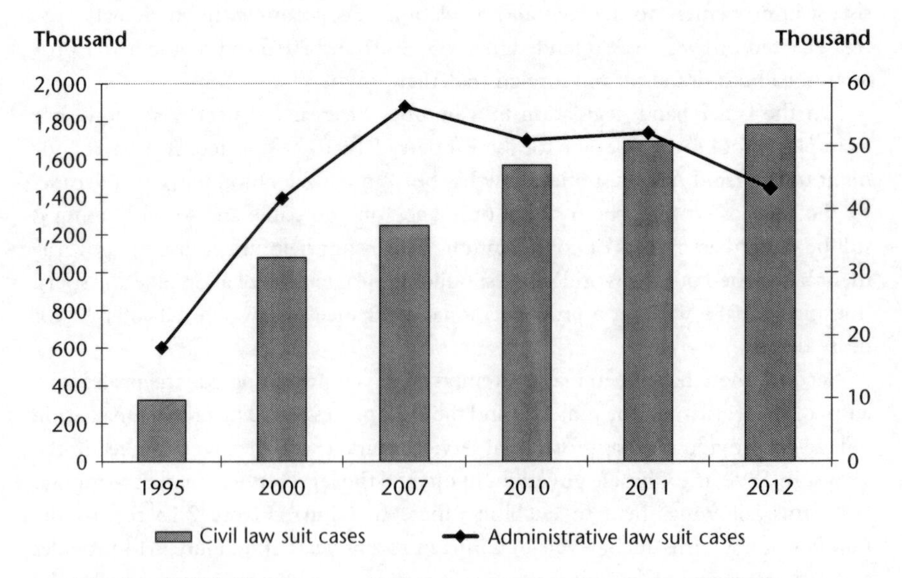

FIGURE 2.14 Civil lawsuit cases and administrative lawsuit cases handled by lawyers
Source: NBSC (2013).

promulgated by the State Council to impose administrative approvals. The APL represents an encouraging step forward to rationalising government power.

Fourth, there have been major efforts to increase transparency in the judiciary process to curb injustice and malfeasance in the police and court system. The Court Trial Live Broadcast Website provided live broadcasts of 45,000 court trials in 2013, and the Judicial Opinions of China website published 3,858 judicial opinions documented by the Supreme Court and 1,646,000 judicial opinions by various levels of local courts in 2013 (Zhou, 2014).

Fifth, the widely criticised system of re-education through labour has been abolished in 2013. Re-education through labour started in 1957 and was a form of administrative punishment by police, rather than through the judicial system, for minor offences. It has been accused of depriving people's liberty by arrest or detention without due proceedings before a court and thus violating some provisions in the Constitution. Its abolition is thus a positive step forward for human rights.

Despite the significant development of the legal system and the newly elevated role of courts, there remains some concern about the defects in China's legal system in regards to the rule of law. The judiciary is dependent on political pressure. There is corruption among public officials in the judiciary. However, direct intervention by the CCP has reduced in recent years, as has the direct influence of the CCP on the legislative process. The judiciary's attempts to increaese transparency indicates the direction of reforms in the country's legal system.

As part of the reform drive, the Fourth Plenary Session of the Eighteenth Congress was held in October 2014 to lay down tasks to comprehensively advance the rule of law. The main tasks include: (1) improving the socialist system of laws with Chinese characteristics, with the Constitution taken as the core to strengthen the implementation of the Constitution; (2) promoting administration by law to speed up the building of a law-abiding government; (3) ensuring judicial justice to improve the credibility of the judicial system; (4) promoting public awareness of law; (5) enhancing the building of a law-based society; and (6) sharpening the CPC's leadership in pushing forward rule of law. Such a blueprint may not satisfy some commentators who want to see the Party to be bound by law. However, these tasks are practical and much needed to usher China into a rule of law. Under China's current circumstances, a strong ruling party is necessary to lead those tasks to be completed.

2.7 Environmental changes

China is a vast country in terms of land area. A breakdown of the land area according to terrain reveals that 33 per cent of the country's land is mountains, 25 per cent plateaus, 19 per cent basins (geological depressions), 12 per cent plains and 10 per cent low hills. Such a terrain shapes the uses that can be sustained on land, with 32.6 per cent as grassland, 20.5 per cent as forest land, 13 per cent as cultivated land, 21.6 per cent as unusable land and the remaining 15.7 per cent as roads, mining, water and urban areas.[11] With a population of 1.36 billion, China is a country with severe shortages of usable land. In particular, per capita cultivated land in China is only a third of the world average. As a result, cultivated land protection dictates China's land use policy with implications for land use for property development.

Two chronic problems threaten to reduce the usable land area in China. The first problem is poor vegetation cover and the resultant soil erosion on the country's vast mountains and hills. In 1948, forest cover was only 8.6 per cent of the total land area, and by 1980 it was 12.7 per cent. Due to consistent effort in conservation and planting, forest cover has increased to 21.63 per cent in 2013 (Sciencenet, 2014). In particular, between 1999 and 2002, China converted 7.7 million hectares of farmland into forest in ecologically vulnerable areas. As a result, China now has the world's largest area of planted forest. The second problem is desertification, which is particularly serious in the west of the country. The advance of deserts, which covers approximately 27 per cent of China's total land area, has been aggravated by intensified human activities such as overgrazing and expansion of agricultural activities. At the time the reforms started, China initiated a 'Green Wall of China' project in 13 northern and western provinces covering about 4 million km^2 to plant a 'green belt' to hold back the encroaching desert. The project is possibly the largest ecological project in history and has now produced forests covering 264,700 km^2 (Xinhuanet, 2013). Nevertheless, desertification has not been totally halted and remains a serious threat.

Various forms of pollution have increased since 1978 as China industrialises, causing widespread environmental and health problems. For example, from 1995 to 2013, total waste water discharged (including sewage and industrial waste water) increased from 37.3 to 68.5 billion tons, sulphur dioxide discharged grew from 18.91 to 21.2 million tons, and industrial waste residue rose from 644.74 to 3,290.44 million tons (NBSC, 1996; 2013). China has responded with increasing environmental regulations and a build-up of pollutant treatment infrastructure that have produced improvements on some variables. Environmental protection has become a fundamental national policy since 1983. The provisional Environmental Protection Law, promulgated in 1979, became law in 1988. In 1988, the State Environmental Protection Bureau became independent from the Ministry of Urban and Rural Construction and Environmental Protection. In 2008, the bureau was upgraded to the Ministry of Environmental Protection. There have been areas of improvement in pollution control. For instance, soot discharged decreased from 14.8 to 12.4 million tons from 1995 to 2013 (NBSC, 1996; 2013).

Despite the above-mentioned efforts, pollution in general has become more serious as China becomes the largest industrial country and Chinese urban residents embrace high-energy consumption lifestyles associated with high-rise buildings, air conditioning and car ownership. For example, concentration of fine airborne particulates with diameter less than 2.5 micron rose as high as 993 micrograms per cubic metre in Beijing on 12 January 2013 (Bloomberg, 2013), while the World Health Organisation recommends no more than 25. It is worth noting that Beijing, lying in a topographic bowl and using coal as fuel, is subject to air inversions resulting in extremely high levels of pollution in the winter months. In February 2014, smog struck north China and at its maximum covered 1.43 million km^2, with fine airborne particulates as the main pollutants (Chinanews, 2014b). This incident prompted the government and society to act decisively to mitigate pollution in China. For example, Beijing Municipal Government promised to invest CN¥760 billion to tackle fine airborne particulates and other pollutions. Hebei Province, where much of the pollution in Beijing originates, has started to phase out the most polluting old factories and manufacturing processes.

The central government has also conducted a major energy policy review. More actions are expected to tackle the country's aggravating pollution problem. It is worth noting that pollution in China is partly caused by the country's development stage of rapid urbanisation, the high volume of fixed asset investment and large-scale manufacturing. The country's large number of exports of goods made by high energy consumption means it bears the pollution of other countries' consumption. Pollution will decline when urbanisation slows down and the service industry takes a bigger share of the economy, as indicated by the experience of the USA and other developed countries.

However, pollution in China is likely to deteriorate in the short term as continued urbanisation and large-scale manufacturing and construction expand the country's energy consumption. It is time for China to act swiftly to change its energy policy and embrace more sustainable development modes and lifestyles. The new amendment of the Environmental Protection Law that increases the responsibility of polluting firms on 24 April 2014 is an indication of the changing attitude towards economic growth and environment by the government and society.

2.8 Discussion: rise or collapse?

Having examined the political, economic, social, technological, legal and environmental arenas after the reforms and opening-up campaign, we can be sure that China has made great progress and become much stronger on all fronts. Although many problems exist, China has the political will, manpower, capacity and resources to solve those problems. In particular, economic growth is very likely to be sustainable, albeit not at a neck-breaking pace.

It should be borne in mind that lower GDP growth rate is inevitable as the size of the Chinese economy becomes bigger and bigger. For example, GDP growth in 1992 and 2007 was 14.2 per cent, the highest in the 1990s and 2000s, respectively. However, increments in GDP were CN¥511.11 billion in 1992 and CN¥2,221.46 billion (at 1992 prices) in 2007. It takes an increment in 2007 more than four times bigger than that in 1992 to have the same growth rate. In fact, the 2007 increment was close to the total GDP in 1992, which was CN¥2,693.73. By the same token, the GDP growth rate in 2011 was 9.3 per cent, much smaller than the 14.2 per cent in 2007, although GDP increments in 2011 of CN¥5,814.01 billion (at 2007 prices) were bigger than those in 2007 of CN¥4,949.59 billion (NBSC, 2013).

Another reason for the forthcoming slower growth in the Chinese economy is the transition toward a more consumer- and services-driven economy, which typically produces slowing productivity growth (Conference Board, 2014). In 2013, output from services for the first time exceeded that from manufacturing and construction (see Section 2.3).

Judgements that China's economy is at risk of 'collapse' are more or less based on an incomplete picture of the country. On the face of it, China is facing all sorts of problems. Rapid development and transition make those problems more acute. For example, it is not easy to gain permission from the police to hold demonstrations in China, but demonstration is a useful way to exert pressure on local governments on certain issues, such as more compensation on demolition and relocation. Unauthorised demonstration is usually termed 'disturbance to public order', which occurred 476,836 times in 2012 (NBSC, 2013). Even if 1 per cent of such 'disturbance to public order' gets reported in the media, there would be more than 13 reports every day. However, it would be misleading to conclude there is acute social instability in China from the seemingly countless unauthorised demonstrations. These issues are viewed very differently domestically, because most of them are over one-off disputes of benefits. China has 3,185 cities and counties, and 40,446 townships (NBSC, 2013), with most of those places experiencing rapid growth and inevitable disputes.

Another example is China's unique deflationary expansion during 1998 to 2002, when China had both modest deflation and above 7 per cent GDP growth rate. Equating deflation with falling demand, some scholars in the West considered the above 7 per cent growth rate to have been made up by the Chinese government. Yet they ignored that another cause of deflation can be supply increase, which is exactly what happened. Lin (2012) explains that due to rapid growth in fixed asset formation in the period from 1991 to 1995, China finally said goodbye to the shortage economy. Overcapacity kept prices low. Gang (2012) argues deflation is caused by an abundant labour supply from the countryside and SOEs laying-off workers, which kept wages unchanged or falling. Chapter 3 will add another factor, i.e. flat or even falling land prices. To the claim by Rawski (2001) that China made up growth figures because energy consumption fell in those years, Lin (2012) points out that increased competition due to overcapacity drove to bankruptcy many town and village enterprises which had backward technology, high energy consumption and poor product quality. The author observed that at least some of those bankruptcies were policy induced: to reduce pollution.

Jonathan Fenby, a former journalist with extensive experience of China, made the following comments on many such judgements:

> Sweeping judgements on China, positive or negative, make for snappy headlines and striking book titles, but they rarely accord with reality. The more I delve into China, the more convinced I am that the essential element in seeking to understand the country is to try to grasp the complexity…
>
> *(Fenby, 2012: xii)*

2.9 Conclusion

The reforms and opening-up campaign since 1979 were intended to improve efficiency in China's economic system, but they have generated profound and sweeping changes in all areas. This chapter has outlined the relevant developments in the political, economic, social, technological, legal and environmental arenas, which are relevant to the development and operation of the property market. The centre of change that has brought the transformation is the reforms and opening up, which started in 1978 without a clear goal but has aimed to build a market economy in China since 1993 by a top-down design. The 1993 reform blueprint has been responsible for the existing achievements in economic growth, but the reforms so far are still halfway in terms of marketisation (Wu, 2011), with administrative power still playing a key role in allocation of many resources. Widespread corruption threatened to cause the Chinese economy to slow down or stagnate and there is a risk that China would be stuck in the Middle Income Trap. Nevertheless, a new reform blueprint was enacted in 2013 to continue the market-oriented reforms to give the market the key role in resource allocation. A new wave of reforms has started to reduce government intervention in market operation. Hurdles to further urbanisation, for example, the household registration system, are on the reform agenda.

Taking a holistic view, we can be certain that the changes in China since the reforms and opening-up campaign started are positive and the Chinese economy has immense potential. As top-down reforms from 2012 gather momentum in the country, more profound changes are expected to take place in the right directions, which will transform China into a more productive and dynamic state on a sustainable path of development. It can be concluded that political and socio-economic developments in China are conducive to the sustainable development of the property market and the contextual factors impacting the property market are not likely to cause it to collapse. The downside risks of the property market are more from internal operations rather than external conditions. The next chapter embarks on the journey to scrutinise the operations of the property market.

Notes

1 Data from Chinese internet: http://xxw3441.blog.163.com/blog/static/753836242010112315631838/. In 1949, total grain output was 157.40 million tons in the USA and 113.18 million tons in China. In the same year, steel output was 70.74 million tons in the USA and 0.16 million tons in China (accessed 13 June 2014). In 1949, the US population was 149.19 million and the population in Mainland China was 541.67 million (NBSC, 2013).

2 Per capita GDP of the USA in 1978 was US$28,891 (http://useconomy.about.com/od/GDP-by-Year/a/US-GDP-History.htm, accessed 13 June 2014).

3 Data from the Chinese internet: http://xxw3441.blog.163.com/blog/static/753836242009275259520/ (accessed 30 March 2014).

4 The first recession was caused by the failed industrialisation policy and natural disasters in the Great Leap Forward from 1958 to 1960. The second recession was due to the disruption of the first two years, i.e. 1966 to 1967, of the government-induced political

infighting in the Cultural Revolution. The third recession was the result of political campaigns and natural disaster in 1976.

5 Household registration, or *hukou*, is required by law in Mainland China. A household registration record, issued per family, officially identifies persons in a family as residents of an administrative area and includes information such as names, parents, spouses, and dates of birth, deaths, marriages, divorces and moves of all members in the family. The system was developed almost 4,000 years ago, but was used in Mainland China from 1958 to prevent mass population movement, in particular from rural to urban areas. Now population movement has been free since the 1980s, and household registration has evolved into a sort of local citizenship, which confers political, economic and social rights in an administrative area upon the holders. If someone moves without a change in registration, there is no entitlement of the family to things like education and healthcare in their new residence.

6 This is a famous saying by Deng Xiaoping, the late leader who is accredited as the chief architect of the reforms and opening-up campaign from 1979.

7 Data from the Chinese internet: www.chisa.edu.cn/news/syyw/201309/t20130924_456191.html (accessed 13 June, 2014).

8 China's GDP in 2013 was CN¥56.89 trillion (NBSC, 2014), or US$9.25 trillion. The US GDP was US$16.72 trillion. In 1983 China's GDP was US$0.30 trillion and that of the USA was US$3.64 trillion (Wikipedia, 2014a).

9 China's total foreign trade in 2013 was US$4.16 trillion (NBSC, 2014), and the USA's was US$3.91 trillion.

10 The USA and Japan have never allowed export of high technology to China. The EU, on the other hand, was less restrictive since establishing formal ties with China in 1975. Nevertheless, an export embargo of weapons and high technologies was imposed in 1989 following the crackdown of students in Tiananmen Square. Nevertheless, this embargo has become less and less meaningful due to the great technological advances in China.

11 The total is 103.4 per cent, indicating some double or multiple counting.

PART II

Market development

3

THE LAND MARKET

The land market in China is unique due to the special land tenure system and the land administration system. This chapter first examines the system of property rights in land dictated by the Constitution and laws in China. It then explores the origin of privatisation of land use rights (LURs), or leasehold interests on state-owned land, and examines the current practices of government land sales that are crucial to land supply. The process of land market development and local government monopoly is examined and explained by the three-level analytical framework that analyses incentives and property rights at the different levels (Table 3.1). This chapter details the use of land by local governments to attract inward investment and raise funds for infrastructure investment; a process known as property-led growth by some academics. Such a growth model partially explains China's spectacular growth record. Price formation mechanism and risks in the land market are also analysed.

3.1 Property rights in land

China's property rights system in land consists of state ownership of all urban land and some non-urban land, collective ownership of most rural land, and private ownership of leaseholds interests on state-owned and collectively owned land. Only leasehold interests on state-owned land are tradable according to law and underpin private real estate ownership in the country.

China's public land ownership is legitimised by law. Article 10 of the Constitution[1] and Article 9 of the Land Administration Law (LAL)[2] state that land in urban areas of cities is owned by the state and land in rural and suburban areas is owned by peasant collectives, except for those tracts of land which belong to the state as provided for by law; house sites and private plots of cropland and hilly land

TABLE 3.1 Tiers of government and government branches in China

Tiers	*Governments*	*Government agencies*
National level	State Council	Ministries, Commissions, Bureaus
Provincial level	Government of Provinces, Autonomous Regions, Municipalities	Departments, Bureaus, Commissions
Prefecture level	Government of Municipalities, Prefecture	Bureaus, Commissions
County level	Government of Counties, Municipalities	Bureaus
Township level	Government of Towns, Townships, Streets	Administrative offices

are owned by peasant collectives. Article 2 of the LAL stipulates that China should practise socialist public ownership of land, namely, state ownership by the whole people and collective ownership by the working people. Furthermore, Article 51 of Property Law 2007 reaffirms state ownership of urban land. Thus there is no private ownership of land in the sense of freehold or fee simple absolute as in the UK and the USA and similar titles in other countries.[3]

The LAL details how the state executes its ownership rights in land. Article 2 explains that ownership by the whole people means that the right of ownership in state-owned land is exercised by the State Council on behalf of the state. Article 5 stipulates that the Land Administration Authority under the State Council is in charge of unified administration of and supervision over the land throughout the country. Here the Land Administration Authority refers to the State Land Administration Bureau set up in 1986, which was upgraded in 1998 to become the Ministry of Land and Resources. At the local level, land administration is executed by the land administration bureaux or departments of local governments at or above the county level. Provinces or equivalent autonomous regions and municipalities have the discretion to decide the rights and responsibilities of these bureaux or departments (Article 5). For land administration of collectively owned land, Article 5 requires that collectively owned land be used and managed by collective economic organisations of the village, or by villagers' committees, or by the rural collective economic organisation of townships (towns).

Transfer of ownership rights in land can only occur when the state acquires collectively owned land. Article 2 of LAL bans organisations and individuals from illegally transferring land ownership titles through buying, selling or other means. Owners of collectively owned rural land cannot acquire state-owned land or transfer their land to other private or collective owners. On the other hand, the state is allowed to lawfully expropriate or requisition land and provide compensation accordingly in the public interest, and can convert collectively owned land into state-owned land. Consequently state land ownership has been expanding because the state has been actively acquiring collectively owned land for urbanisation, industrial development and the provision of infrastructure.

The right to use state-owned land, or Land Use Right (LUR) on state-owned land, is the only legally tradable property right in land. Article 2 of the LAL stipulates that LUR may be transferred in accordance with law. However, the right to build is mostly restricted on state-owned land. Article 43 requires that all organisations and individuals needing land for construction purposes should, in accordance with law, apply for the use of state-owned land. Owners of collectively owned land can only build township or town enterprises or build houses for villagers and public utilities or public welfare undertakings of a township (town) or village after receiving lawfully obtained approval from the government. Such buildings cannot be transferred before land ownership is changed to the state. Article 2 of the Urban Real Estate Administration Law[4] (UREAL) requires collectively owned land within a newly designated urban area, to be requisitioned with due compensation and converted into state-owned land in accordance with law before leasing out to land users. On the other hand, those who acquire the right to use the state-owned land within the designated urban area, as stipulated by Article 2 of the UREAL, can engage in real estate development or transactions of real estate and exercise real estate administration. Article 8 of the UREAL allows the state to lease land to land users, by way of selling LURs for fixed terms to them; land users must pay the state for using the land.

LURs on state-owned land underpin the private ownership of land and buildings in China. Article 3 of the *Provisional Ordinance for Granting and Transferring Land Use Rights over State-owned Land in Cities and Towns* (hereafter referred to as Decree 55), enacted in 1990, stipulates that any company, enterprise, other organisation and individual within or outside the People's Republic of China may, unless otherwise provided by law, obtain the right to the use of the land and engage in land development, utilisation and management in accordance with the provisions of law. After paying the appropriate premiums, which is the market price of land, owners of LURs can occupy, use and derive income from using the LURs, and dispose of the LURs independently and without consultation with other parties.

Decree 55 is the first legal document to legalise private as well as foreign ownership of the right to use land in China. Property Law 2007 reaffirms such rights, with Article 4 declaring that the real rights of the state, collectives, individuals or any other right holder shall be protected by law and shall not be infringed by any entities or individuals.

LURs on state-owned land can be obtained by paying the premiums, which are mainly the present value of all rents for a fixed term in a lump sum, or by applying to the government for an indefinite term of use. The former is referred to as *granted LUR* and the latter an *allocated LUR*. Article 24 of the UREAL lists the land uses for which the allocated LUR can be obtained:

- land used by state organs for military purpose;
- land used for construction of urban infrastructures and public utility;

- land used for construction of the state-supported key energy, transport and water conservancy projects; and
- land used for other purpose stipulated by the laws and administrative rules.

Granted LURs can be legally transferred, leased, mortgaged and inherited. They are in fact building leases on state-owned land, with maximum terms specified in Article 12 of Decree 55 as follows:

- 70 years for residential uses;
- 50 years for industrial uses;
- 50 years for the uses of education, science, culture, public health and physical education;
- 40 years for commercial, tourist and recreational uses;
- 50 years for mixed or other uses.

Because LUR has only a limited term, there were doubts and anxieties over the renewal of the terms of LUR when existing rights are close to expiry. The Property Law provides a legal solution on the issue, as Article 149 states that the term of the right to use land for construction for dwelling houses shall be automatically renewed upon expiration, although there is no indication of how much the premiums will be. For other land uses, the leases shall be renewed according to legal provisions. Such legal provisions are provided in Article 22 of UREAL, which states that the leaseholder has to apply for an extension at least one year before the expiration of the term and the application shall be approved unless the land plot is needed in public interest. The lease is renewed on the condition that an appropriate premium is paid. At the time of writing, no residential LURs have reached the expiry date; the majority have over 50 years to run. No commercial and mixed use LURs have reached the expiry date; the majority have over 20 years and 30 years to run, respectively. However, a renewal exercise in Shenzhen to renew allocated LURs on special terms, which ran smoothly, provides insight on future granted LUR renewal (see Section 10.1).

The establishment of the LUR system effectively assigns allocated LURs to all existing land uses on state-owned land. The law allows the conversion of allocated LURs into granted LURs that are tradable. Article 39 of UREAL stipulates that if allocated LUR owners want to convert their allocated LURs to granted LURs, an application should be submitted to an empowered local government for approval and the applicant should pay the premiums. The Provisional Rules on Administration of Allocated Land Use Right (SLAB, 1992) provided for discounts for the premiums in converting allocated LURs to granted LURs to expand land market coverage. The discounted premium, however, should not be less than 40 per cent of the marked prices[5] of comparable land compiled by local land administration agencies.

3.2 1979–90: adoption of the pay-for-use principle

Before the reform started in 1979, China followed the Soviet doctrines of a command economy and banned market exchange of land in any form. With state ownership of urban land, the government made decisions on land uses and issued administrative orders to allocate land to land users. The allocation was free of charge according to the needs of the urban land users as recognised by the state plans. This allocation mechanism has three characteristics: (1) no payments by the land users for the use; (2) no definite terms of use; and (3) virtually no repossession attempted by the state. As a result, perverse incentives were created to occupy land for self-interest, either authorised by the plans or in anticipation of such an authorisation. When state land was undersupplied, collectively owned land was compulsorily purchased by the government and converted into state-owned land for administrative allocation to meet the insatiable demand generated by this land use system.

Such a land use system presented three problems to the central government at the constitutional level. First, the state derived no direct economic gains from its vast land holdings but the occupiers of land were able to run their possession to economic benefit. Consequently, the government only favoured land uses that contributed to new productive processes and discouraged other uses, such as housing and amenities, leading to stagnation and in some cases deterioration of living standards. This is in stark contrast to Hong Kong whose government generated huge incomes from its land holdings and used land sale revenues to build infrastructure and public housing to promote economic growth and living standards. Second, land was used in very inefficient ways, which was a major cause of the decreasing return to investment. Yet higher productivity and growth were essential to justify the Communist regime. Third, this system of land use resulted in an alarming loss of arable land; that in turn led to land-saving measures. These affected the property rights of and proved unpopular with existing and new state land users who were represented by different segments of the government. Using rural collectively owned land incurred much lower transaction costs for new land-related developments as peasants were poorly represented in the government. Furthermore, peasants generally welcomed the change of status from rural households to urban households to benefit from the state welfare system that treated registered urban households much better. Huge areas of quality arable land were lost in this way.[6]

The system of land provision, one feature of the second world of real estate (see Chapter 1), was perceived as very inefficient. A land reform was initiated to improve the inefficiency when the country started its reforms and opening-up campaign.

The urban land use reform initiated by the central government at the constitutional level since 1979 had three aims. First, the reform should bring in income to the state to alleviate fiscal pressure and provide funds for the much-needed infrastructure improvement. Second, land use efficiency should improve as land users are required to pay for the land they use. Third, loss of arable land should be mitigated through the establishment of a new land administrative mechanism to safeguard the food security of the country.

The development of the land market in China can be described in three stages (Table 3.2). This section focuses on the first stage of the land market, which featured the application of the pay-for-use principle to land, establishment of the land administration bureaucracy and legalisation of market land sales.

TABLE 3.2 The three stages of land market development in China

Stages	*Main developments*	*Time scale*
Stage 1: inception	Embryonic land market emerging in a few coastal cities; modest payment imposed for using land	1979–1990
Stage 2: expansion	Market coverage expanding to most new and existing uses; LUR sales method in the primary land market transferring from private treaty to open market disposal; active transactions in secondary land market	1990–2004
Stage 3: maturity	Open market disposal prevailing in primary land market; new LUR sales increasingly directed by national plan; accelerating land price inflation	2004 onwards

Several changes to traditional land uses occurred at this stage. First, the government used land as input into joint ventures between Chinese firms and foreign investors from 1979, signalling the re-discovery of land value by the Chinese state. Second, domestic firms were required to pay for the use of land in the form of the Land Use Fee, later called the Land Use Tax. Although the fee and later the tax were very modest, such a move changed the tradition of free land use and installed the institution of pay-for-use on land. This lowered the resistance to the later introduction of transaction costs that required all land users to purchase the LUR of the plots on which they operated. Third, a unified land administration system, headed by the State Land Administration Bureau, was set up in 1986 to end the fragmented land administration in the country. Fourth, LURs were sold on the open market to some developers and land users, both domestic and foreign, in a few coastal cities from 1987. The amendment in the Constitution in 1988 to legalise the market transaction of LURs laid the foundation for the development of the land market.

In the first stage the step towards pay-for-use was so small that local governments at the collective choice level and land users at the operational level were little affected. The battle ground was at the constitutional level to justify the change, a paradigm shift, against the orthodox socialist ideology. In the end pragmatism prevailed: land sales were conducted; the Constitution was amended; and Decree 55 was enacted.

3.3 1990–2004: transition to open market sales

The second stage is the expansion of the land market, featuring the establishment of tradable property rights in land, increasing the marketisation of land use and the formation of a three-tier land market to cover most of the new land uses. During this stage, the land market grew to become the main means of allocating land resources and a significant source of revenues for local governments (Figure 3.1).

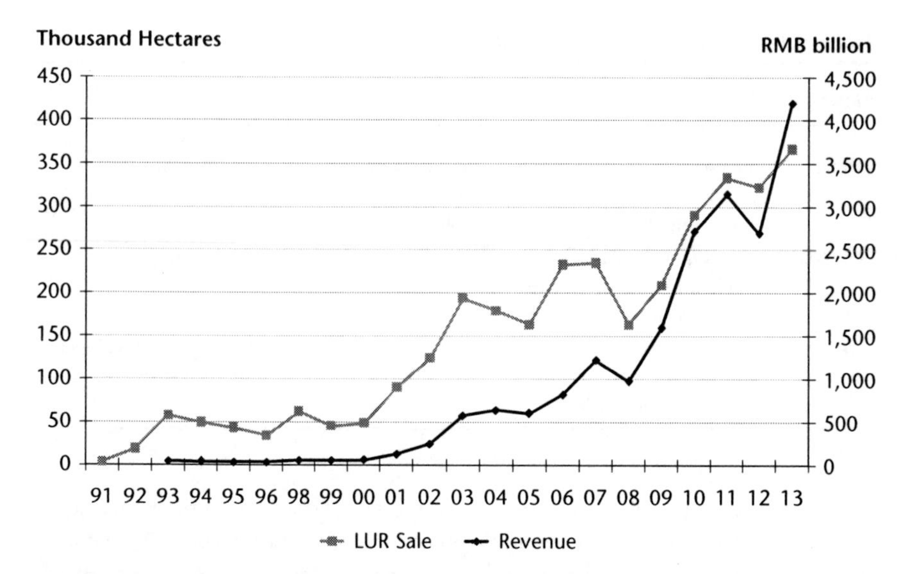

FIGURE 3.1 Scale of land sales and revenues generated, 1991–2013
Sources: MLR (2010; 2014b).

The main features of the land market at this stage are market expansion, transitional open market disposal and market building. Market expansion occurred after the promulgation of LURs, as the tradable property rights in land enabled from 1990 by Decree 55, provided incentives for land users and investors to buy land. Investment from Hong Kong, Taiwan and South East Asia, a contextual factor, played an important role in kicking off the land markets in many localities. In particular, land users and investors from Hong Kong, whose leasehold land tenure system was the inspiration for China's LUR system, demonstrated the merits of owning LURs in terms of gains from land price appreciation in a rapidly developing economy. Their behaviour contributed to the acceptance by the domestic land users to pay for the land in the land market. As a result, land sales rose dramatically from 1991 to 1993 (Figure 3.1).

The land market has a unique three-tier structure, i.e. primary, secondary and tertiary land markets (Table 3.3). At the core of the land market is the primary market, where sale of LURs by the state takes place. The secondary market was where developers serviced the land they bought from the local governments and sold to final land users and investors. The market among final users of land is the tertiary market where land owners/users engage in transactions involving purchase, sale, letting and mortgage.

3.3.1 Land sales, land prices and land use efficiency

During the second stage, there was a long period of transition from behind-closed-door deals determined at the officials' discretion, to open market sales determined by

TABLE 3.3 The three-tier land market structure in China

Markets	*Descriptions*	*Sellers*
Primary	The state sells LURs with fixed terms and planning conditions. The state has a monopoly position and can control the level of new land supply in the market.	County and city governments
Secondary	Developers buy LURs and develop land according to planning conditions for sale to residential and commercial land developers and users.	Developers
Tertiary	Land users and investors sell their residential and commercial land to other land users. Nobody has control of the market.	Land users, developers and investors

competition. At the collective choice level, local governments, in their capacity as representatives of the state, had multiple objectives for land sales: land sale revenues, relocating SOEs, attracting inward investments and urban regeneration. Maximisation of land sale revenues, best achieved through open market sales, was often not the key objective pursued by local governments at this stage. Instead, local governments often chose not to promote open market disposal so as to reduce land prices to promote their versions of development and attract inward investment. Consequently, they were not enthusiastic about enacting rules requiring land to be sold on the open market.

There are good reasons for local governments not to favour open market disposal of land. Most local land users, including developers, were short of money and found it difficult to pay full market prices for land. Footloose national and international investors, on the other hand, insisted on preferential treatments from local governments to reduce costs. Furthermore, local governments wanted the real estate market to drive local economic growth so offered large discounts on LUR prices to developers for market-oriented development at the time when real estate development was rather risky. This model of using land to breed development, or property-led development, aimed for subsequent local economic growth and future bigger tax revenues, which were believed to more than compensate for the initial loss of the land sale incomes. On the other hand, local officials at the operational level often used their position and authority to influence sales negotiations on particular projects for personal gain. Local governments were reluctant to change this informal rule of the game in land sales by promoting open market disposal of land.

As a result, land prices achieved in the primary market varied substantially and in most cases were lower than values inferred by developed property prices. Land sales were mostly conducted by private treaty. Corruption was rampant, with land buyers bribing officials to give them low prices. Land sale revenues remained low for a decade, curtailing the local government's ability to conduct infrastructural investment. Such a chaotic state in the primary land market led to claims that land use reform to build a land market was a failure (Xu and Yeh, 2009). Low land prices, however, were instrumental in the deflationary[7] expansion at the end of the 1990s (see Chapter 2, Section 2.8).

The local governments' property-led development model generally lowered land prices in exchange for more development projects and rapidly expanding GDP. It is not clear how much saving in land costs was offered to businesses conducting developments favoured by local governments. Anecdotal evidence on differential treatments, however, abounds. For example, in the early 1990s, developers in Guangzhou were required to pay only 20 per cent of the premiums for the sites they bought to obtain permission to start presale. A survey covering 80 housing projects by 31 development companies, conducted by the Guangzhou Municipal Land and Building Administration Bureau in 2004, found that 46 projects, started from 1993 to 1998, had outstanding LUR payments (Xinhua, 2004a). Buyers of these projects had difficulty in obtaining title deeds for resale or mortgage purposes because these projects had not cleared all their debts, including the premiums needed to be paid for the land.

Low land prices as a result of property-led development have a negative impact on efficiency in land use and loss of arable land, thus partly failing the aims of the land use reform to promote efficiency in land uses. The low initial land costs of property development encouraged many underfunded development firms to engage in market-oriented property development. Success was possible when the market demand was strong and buyers prepaid for the property (see Chapter 4 for more details). However, market demand subsided in 1993 in some cities due to the central government's macro control and in 1995 when the markets in Beijing, Shanghai and Guangzhou became oversupplied. Many projects were abandoned, leaving a large number of derelict sites and semi-finished structures in good locations and a significant number of non-performing loans in all state banks (see Chapter 5, Section 5.4.4 for details of failed projects and land dereliction).

3.3.2 Market building by the central government

During the transition, the national land market administrator, the State Land Administration Bureau (SLAB) and its successor from 1998, the Ministry of Land and Resources (MLR), had been actively involved in the creation of the land market. As a decision-maker at the constitutional level, SLAB/MLR was able to set the rules to require local governments at the collective choice level to sell LURs. To ensure fair prices were paid and thus safeguard the state's interests, SLAB developed and maintained a system of benchmark land prices to guide the sale of LURs. More importantly, SLAB/MLR had been actively promoting the means of open market disposal of LURs, which comprised public auction, open tender and public listing, rather than the secret disposal represented by private contracts. Open market disposal was preferred because it not only safeguards the interests of the state to maximise land revenues, but also minimises unfair competition and corruption.

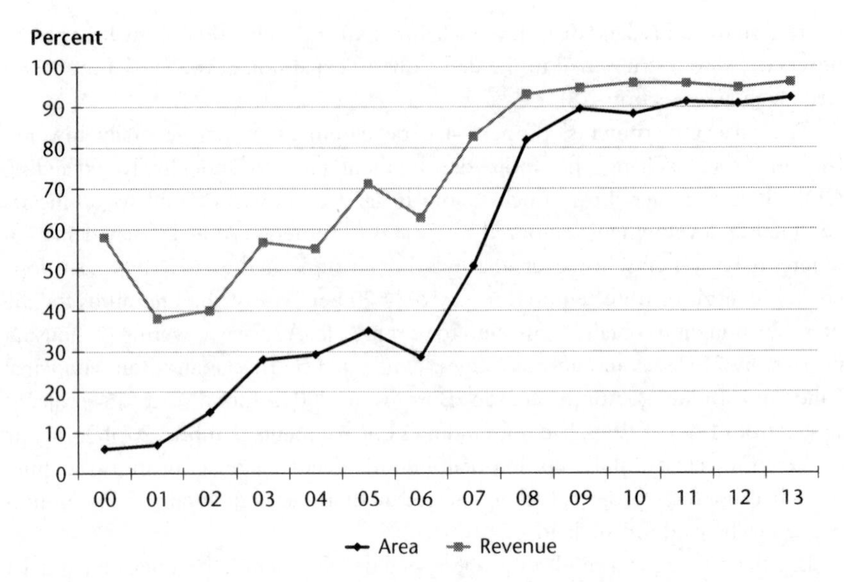

FIGURE 3.2 Share of open market sales in area and revenues, 2000–13
Sources: MLR (various years).

There had been strong local resistance to open market disposal. Such resistance was from local governments at the collective choice level and from local officials at the operational level. To impose open market dealing on the primary land market, the MLR enacted the Regulations on Tender, Auction and Listing of Land Use Rights in State Land in May 2002, banning all forms of LUR sales for community housing, offices, retail, tourist and leisure uses through private contract and requiring public tender, public auction and public listing instead (MLR, 2002). To curb violation and inaction by local governments, the MLR and the Ministry of Supervision jointly promulgated the *Circular on Continuing Enforcement and Supervision on the Tender, Auction and Listing of Land Use Rights for Market-Oriented Uses* in March 2004 (MLR, 2004). These two regulations were widely regarded as successful market building by the MLR to set up an open market disposal system in the country's primary land market. Institutional change brought about by the two regulations and their enforcement put an end to private contract as the dominant means of land sale (see Figure 3.2) and increased land revenues significantly (see Figure 3.1).

3.3.3 Industrial zones

Similarly, problems occurred with industrial land use. To attract inward investment, local governments set up thousands of development zones (mainly for industrial use, but also with other uses). Out-of-town development zones grew from 117 in 1991 to 1,951 in 1992 (SLAB, 1995). From 1991 to 1996, the number of development zones set up in the country was 4,210, with only 130 authorised by the central government. For the remaining zones, 989 were set up by provincial governments and 3,081 by local governments (Wu and Sun, 2000). These zones had a total planned area of 12,357 km², which is close to the total built-up area of 13,148 km² of the 467 cities in 1990 (Urban and Social Survey, 1991). With such ample supply, local governments had to compete with each other, with heavy discounts of land prices against the benchmark prices. As a result, the market for industrial land was distorted and prices and rents were basically controlled by local governments.

To tackle the expanding supply of industrial land, the MLR issued a directive: The Emergency Notice on Cleaning Up Land in All Parks and Strengthening Land Supply Control on 18 February 2003. A major clean-up ensued and ended in June 2004, with a striking discovery about the scale of industrial land supply before the clean-up. In the clean-up exercise, MLR found that in 30 provinces, there were 6,741 industrial parks of various kinds with a total planned area of 37,500 square kilometres, which was bigger than the total built-up area of 31,500 km² in the country in 2003. Also in the exercise, the Ministry revoked 4,735 parks and imposed stricter approval and check-up procedures (Xinhua, 2004b).

3.3.4 Incentives and the transition to market allocation

During this stage, a large amount of urban land for market-oriented use with allocated LUR titles was converted into granted LUR titles, which greatly expanded the coverage of the land market. Resistance was high because many existing land users could no longer enjoy very high-profit rents by operating on land with allocated LUR titles. Nevertheless, relocation of such land users due to urban redevelopment went ahead smoothly due to the windfall gains from converting allocated LURs to granted LURs at low cost, and selling the granted LURs at market prices. On the other hand, developers holding allocated LURs were forced to convert them into granted LURs by the MLR by August of 2004, to press ahead with open market disposal in the primary land market (MLR, 2004).

It is worth noting that some local governments pioneered the transition to open market pricing in the primary land market. For example, in 1997, Guangzhou Municipal Government decided to sell all LURs on the open market, which was done in 1999, and larger revenues generated from LUR sales helped the city to invest heavily in infrastructure and city image-building projects. The huge benefits of the open land market to local governments, as demonstrated in Guangzhou, encouraged other local governments to follow suit. Such incentives at the collective choice level lowered the transaction costs of enforcing MLR's constitutional decision on open market disposal of land. The establishment of the market disposal of land uses was sometimes referred to as the 'second land revolution' in contrast to the 'first land revolution' in the 1950s when land nationalisation and administrative allocation replaced the market allocation of land.

3.4 2004 Onwards: land price inflation and government-controlled land supply

The third stage of the land market development began in the second half of 2004 and features a system of open market land sales under government-controlled land supply, built by land sale regulations in 2002 and 2004. This system of land supply by open market disposal aims to achieve the following objectives: (1) protection of arable land by outlawing private negotiation between developers and land sellers and establishing government land sales as the only legal means to buy land; (2) an increase in the efficiency of land use by market allocation; and (3) the adjustment of housing prices through the manipulation of the land supply. Figure 3.1 shows the substantial increase in land sales and land sale revenues from 2002, and Figure 3.2 indicates the significant increase in open market disposal as means of land sales, which has stayed at over 90 per cent since 2008. As a new institution for the primary land market, open market land sales have resulted in marked changes in the behaviour of developers, real estate buyers and local government in the land market.

3.4.1 Inflation of land prices

Changes in the ways land was sold have affected the behaviour of developers. When land was mainly sold through private contracts, local developers were advantaged in obtaining land cheaply due to their local knowledge and their relationship to local government officials. Non-local developers, on the other hand, incurred much higher transaction costs in obtaining information and cultivating personal connections. With more land being disposed of on the open market, non-local developers have lower transaction costs in land acquisition and have seized the opportunities to operate regionally and nationally. More intense competition followed suit, resulting in a significant increase in land purchase by developers since 2002 (Figure 3.3). Developers who have stronger financial clout to bid for land and provide better products and services have been able to expand rapidly. In 2013, there were already seven developers with total annual sales over RMB100 billion (more details in Chapter 9). The national operations of developers have facilitated the transfer of best practice from cities where the real estate market previously operated and there were more foreign developers.

Since 2004, the strong demand for land, generated by the rapid economic growth and the propensity to invest in real estate, has led to escalating land prices. Figure 3.4 shows the national averaged land prices calculated from the total area of land sales and total revenues from land sales from 1993 to 2013, which provides a general picture of land price inflation over all Chinese cities and towns. Average land prices rose 52.7 per cent in 2002, 47.4 per cent in 2007 and 37.4 per cent in 2013, which corresponds to the property market sentiment in those three years when developers were confident of future property prices. Figure 3.5 presents land price indexes for commercial, residential and industrial uses in 105 cities monitored

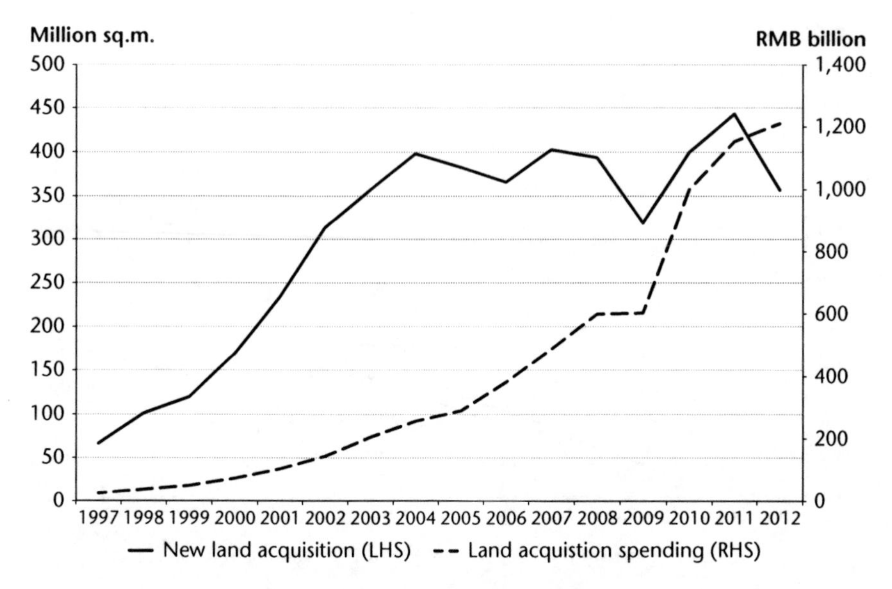

FIGURE 3.3 Land acquisition and spending on land acquisition by developers, 1997–2012

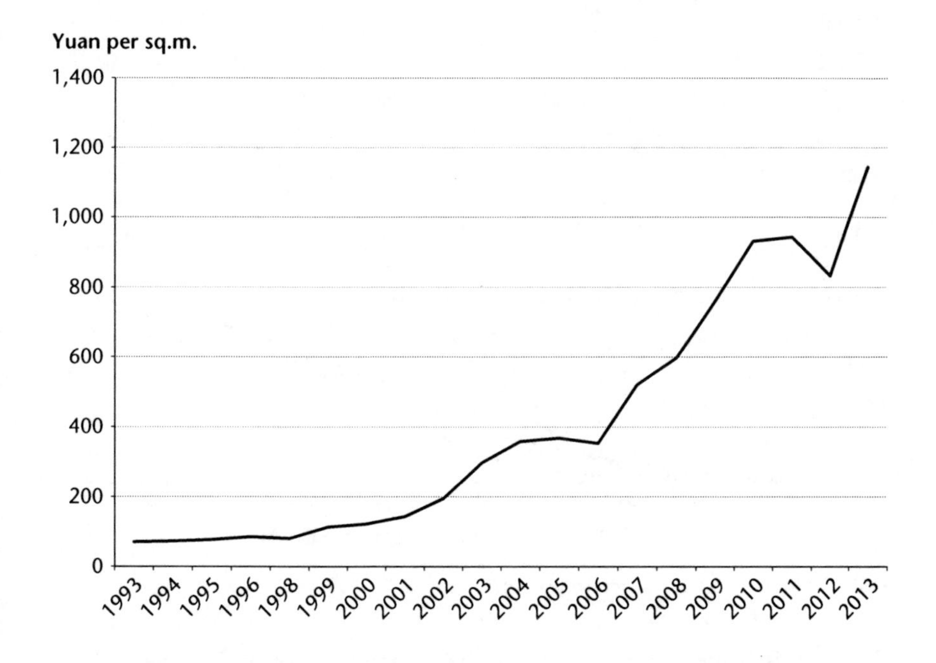

FIGURE 3.4 National average land prices, 1993–2013

Sources: MLR (various years).

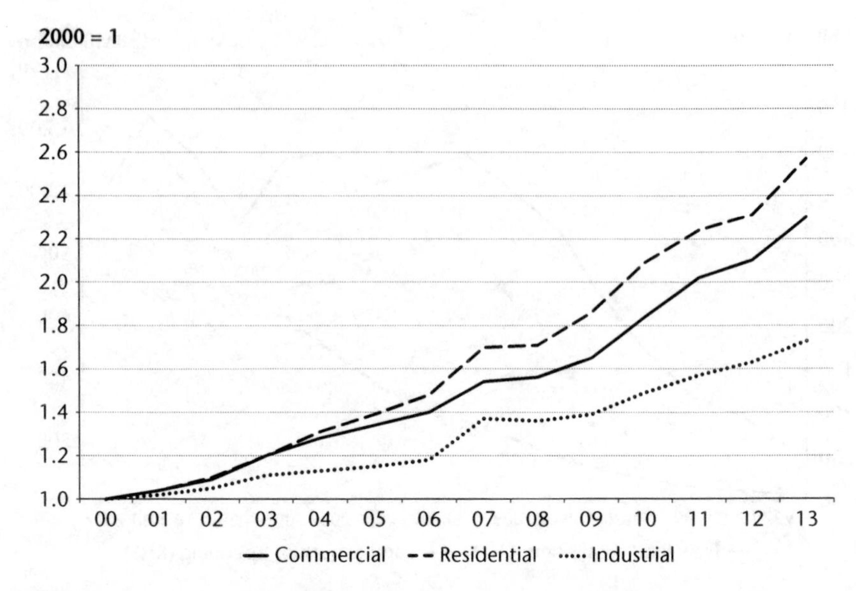

FIGURE 3.5 China Land Price Index for 105 cities in China, 2000–13
Source: Landvalue (2014).

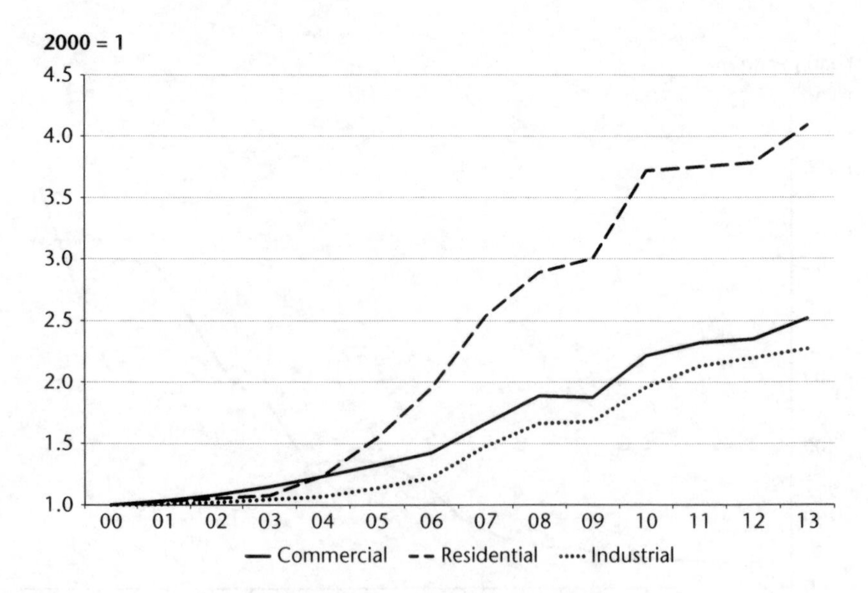

FIGURE 3.6 China Land Price Index for Beijing, 2000–13

by the MLR, with average annual growth rates at 6.62 per cent, 7.53 per cent and 4.31 per cent, respectively. Figure 3.6 illustrates the land price index for commercial, residential and industrial uses in Beijing, which shows higher price inflation than the 105 cities as a whole.

3.4.2 Land (price) kings

Sales of large plots of land became highly advertised in China and resulted in intense competition by developers fighting for sites and company images, which often led to those sites being sold at record land prices. Sites sold at record prices in such high-profile auctions were then referred to as 'land (price) king' sites, and used to indicate higher future property prices. Interviews in Beijing and several other cities revealed that land prices for land kings were often used as market evidence in valuations, resulting in higher prices in subsequent deals.

The first land king emerged in Beijing. Land plot 1, North of Huangcun Satellite Town, Daxing District in Beijing (308,100 m^2 with allowable housing space of 413,600 m^2) was sold on 18 December 2003 at RMB905 million (equivalent to RMB2,188 per m^2 development space) from a reserve price of RMB430 million. Such a price was a surprise to the development industry at the time (Sina, 2005). The fierce competition during the auction of land plot 1 prompted the auctioneer to warn the bidders that the high price was risky. Yet the deal brought instant fame to the successful bidder, Shunchi Real Estate of Tianjin City, and set a new benchmark for land prices in that part of Beijing; and as can be seen from Figure 3.6, even steeper growth in land values was yet to come.

Expectations of higher future prices prompted more and more land kings to emerge. The highest value land king in total price, at the time of writing, is Xujiahui Centre Project in Shanghai (99,188.8 m^2 with allowable development space of 584,203.6 m^2), which was sold at RMB21.77 billion (US$3.51billion) (equivalent to RMB37,624 per m^2 development space) on 5 September 2013 from a reserve price of RMB17.53 billion (Xinhua, 2013). The highest value land king in unit price, at the time of writing, is the National Agricultural Exhibition Centre site in Beijing (28,168 m^2 with allowable development space of 59,192 m^2) sold on 4 September 2013 for RMB4.32 billion (equivalent to RMB72,982 per m^2 development space) (Ifeng, 2013).

3.4.3 Regional variations in land prices

China's land prices show significant regional differences corresponding to the regional levels of economic development. Figure 3.7 contrasts land prices in the east, central and west of the country (see Chapter 2, Section 2.2 for regional development policies and economic growth in China). The highest land prices are found in the main cities in the east of the country, in particular, in the coastal cities. The land prices in the main cities in the west of the country are higher than the main cities in the central part of the country. Figure 3.8 depicts average land prices in the three main economic regions in China, i.e. the Pearl River Delta in Guangdong Province neighbouring Hong Kong and Macau, the Yangtze River Delta in the east and the Pan-Bohai Region in the north. Land prices in the Pearl River Delta, with Guangzhou and Shenzhen as the core cities, are the highest due to the very high level of economic development and the increasingly limited

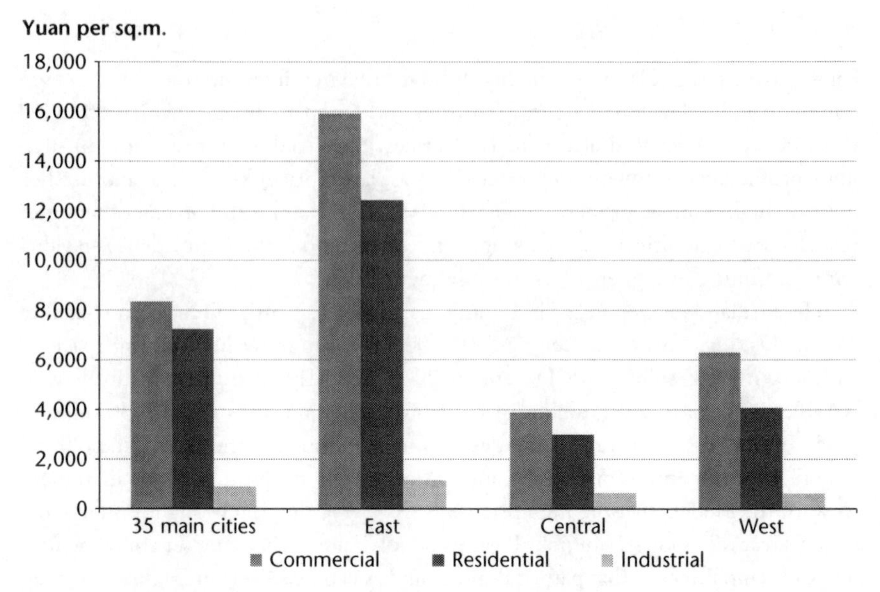

FIGURE 3.7 Land prices of 35 main cities in the east, central and west of the country, quarter 2, 2014

Source: CLSPI (2014).

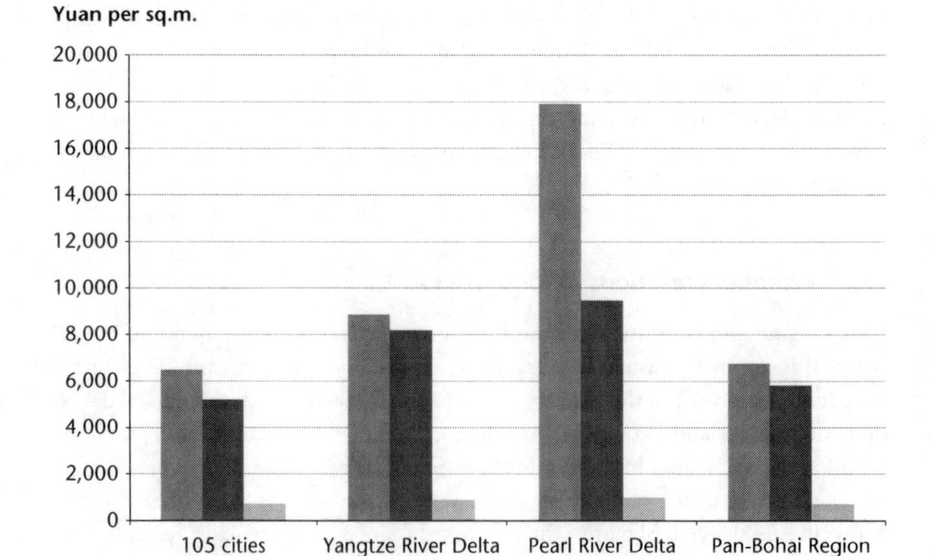

FIGURE 3.8 Land prices in 105 main cities and three main economic regions, quarter 2, 2014

Source: CLSPI (2014).

supply of land. In fact, new greenfield land supply has been exhausted in Shenzhen. Land prices in the Yangtze River Delta with Shanghai, Hangzhou and Nanjing as the core cities, are higher than those in Pan-Bohai Region with Beijing and Tianjin as the core cities. There is a problem of economic coordination in Pan-Bohai Region, with a belt of low economic development surrounding Beijing. Although land prices in Beijing are the highest in the country, average land prices for the region remain relatively low. The new plan to coordinate the development of Beijing, Tianjin and Hebei Province is causing land prices there to rise in the short and medium terms.

3.4.4 Land supply by the government

Since 2004 the Chinese government has controlled the supply of urban land to the real estate market by local governments monopolising the land supply at the primary land market. Existing land users holding allocative LURs were no longer able to sell their allocated LURs to buyers. Local government repossessed the allocated LURs for existing land users and then sold LURs. Land users could obtain land from the secondary and tertiary land markets. Figure 3.9 shows the receipts (values) of land transfer in the secondary and tertiary land markets. Total land transfer was CN¥40.1 billion in 2004 and CN¥81.9 billion in 2012. In contrast, land sales in the primary land market was CN¥640 billion in 2004 and CN¥2,690 billion in 2012 (Figure 3.1).

The role of local governments as the sole seller of land has raised concerns in the property industry. The first concern is about the multiple roles the government

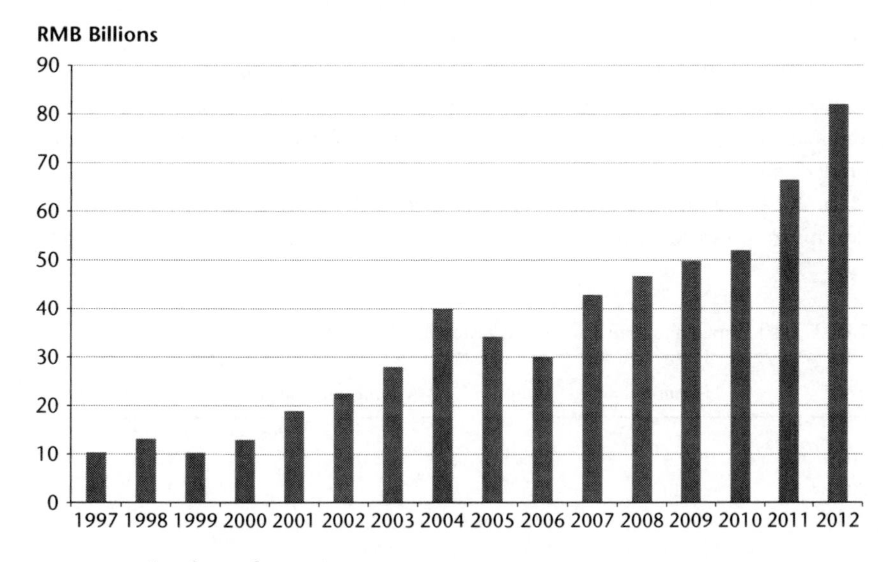

FIGURE 3.9 Land transfer receipts
Sources: MLR (various years).

plays. Apart from being the seller, local governments are also policy-makers and regulators of the land market. As a result, land sale decisions by local governments are not made according to monetary considerations alone. For example, in 1997, local governments were banned for almost a month from permitting the conversion of rural land to urban land. To dampen the demand for land and prevent the loss of arable land, the central government in 2004 strengthened the land administration and restricted local governments' power to approve new land uses and sell land at discounts (State Council, 2004). In 2005, local governments were required by the central government to reduce land sales to cool down the housing market, which was considered to be overheating and posing a threat to the country's financial system (see Figure 3.1 for lower land sales evidence). Developers blamed the reduction in land sales from 2004–05 as the main cause for housing price inflation in the period of 2004–06.

The second concern about the state monopoly on land supply rests on the suspicion that land supply by the local governments may not be responsive to land market price signals. Land supply is plan-led. To protect arable land and facilitate macro control, the MLR formatted a land supply plan every year according to past trends and the forecast of land use requirements throughout the country. For example, rapid housing price inflation has become a serious problem and one countermeasure is to increase residential land supply. However, the residential land supply plan had not been completed in the period between 2010 and 2013 when housing price inflation was rapid (Table 4.3). For 2010–11, the housing market was depressed due to macro control to mitigate housing price inflation, which affected demand for housing land. However, in 2011, the amount of housing land planned for sale was the highest ever. Because of lack of demand, housing land supply in the plans for 2012 and 2013 was reduced, but the housing market recovered from late 2012 and became very dynamic in 2013. Conversely, planned housing land supply was the lowest in the four years, recording a 91.64 per cent completion of housing land sales plan and record high land prices (see Figures 3.1 and 3.4). This mismatch of plan and market indicates the importance of making the plan more responsive to market price changes.

In 2014, the housing market slowed down with falling sales of new homes and downward pressure on prices, which is likely to filter through to the land market.

TABLE 3.4 Planned and actual supply of residential land (thousand hectares)

Years	*Planned*	*Actual*	*Unrealised supply*	*% of plan completed*
2010	187.4	125.4	62.0	66.92
2011	218.0	135.9	82.1	62.34
2012	172.6	135.9	36.7	78.74
2013	150.8	138.2	12.6	91.64

Source: Soufang (2014).

Developers who bought land at the peak of the land market of 2013 face the prospect of losing money if housing prices fall significantly. With reduced demand for land, land prices are likely to fall and local governments' fiscal balance will be under stress.

3.5 Discussion: are high land prices the cause of high housing prices?

Since 2004, rapidly rising housing prices have become an important issue in China. Rising land prices have been cited by the housing development sector as the key reason for high housing prices. Because the price of bread is based on the price of flour, people holding this view claim that housing prices are determined by land prices. The MLR, however, rejects this view as misleading and claims such a view affects efforts to tackle the real causes leading to rapid housing price inflation. In June 2009, a study conducted by MLR surveyed 620 real estate development projects in 105 cities and found that land costs accounted for 15–30 per cent of housing prices, with the average at 23.2 per cent.

Although land prices contribute to housing prices, economic theory asserts that housing prices are ultimately determined by demand for housing. This is manifested in land prices change in 2008–09 and 2011–12. During the period between 2008 to 2009 demand for housing slipped nationally due to the impact of the global financial crisis on China's economic growth, and during the period between 2011 to 2012 demand for housing fell in the major cities due to the impact of housing purchase restrictions on those cities (see Chapter 4 for more details). As a result, demand for housing land fell and prices of housing land fell (see Figures 3.3 and 3.4). In other years, strong demand for housing has resulted in rising prices of housing land.

There are many reasons for rapid housing price inflation. 'Shortage of land' for housing development that developers and others claimed was the key reason for housing price inflation is actually a reflection of the low efficiency of land use that reduces the amount of land available for housing. The total amount of land acquired by developers from 1997 to 2012 was 48.215 billion m^2 (see Figure 3.3) with normally over two-thirds for housing development. Assuming only 60 per cent was used for housing development and a low floor area ratio of 1.2, then 34.715 billion m^2 of housing could have been developed and could have provided a per capita housing area of 47.5 m^2 for the country's 731.11 million urban population. Yet total housing sales from 1986 to 2013 was 8.854 billion m^2 (see Figure 4.1) and in 2012 per capita housing area in urban areas was 32.9 m^2 (NBSC, 2013). Such a situation could be the combination of the following causes: (1) housing is developed in much lower density as a whole; (2) much of the housing land bought for housing development has not been developed due to poor location and developers' financial failure; and (3) a significant amount of land is hoarded by developers as land reserve. For major cities where housing prices are high and unused housing land is normally recycled quickly, raising development density and reducing hoarding are the ways to increase housing land supply.

To raise land use efficiency, in May 2014, MLR tabled the *Regulation on Saving and Intensifying Land Use*, to control the supply of new construction land in major cities. In cities with potential for more housing, land supply is to be increased (MLR, 2014a). According to MLR, 40 per cent of urban land and up to 5,000 km^2 of industrial and mining land (11 per cent of the country's built-up area) are in low efficiency usage. Floor area ratio (plot ratio) for industrial land is only in the range of 0.3–0.6. In particular, per capita construction land is 145 m^2 in urban areas and 240 m^2 in rural areas; well above the country's standards (Cheng, 2014). It is worth noting that per capita urban construction land was 82.4 m^2 for developed countries and 83.3 m^2 for developing countries (Lu, 2009) and China has an acute shortage of usable land (see Section 2.7). MLR is to set standards on industrial land use, real estate development density standards and floor area ratio standards.

The market's reaction to tightening supply by MLR in major cities is that such measures will raise land prices in those cities, which MLR has denied (Cheng, 2014). It remains to be seen whether land use efficiency in major cities can be further enhanced by MLR measures on density standards and floor area ratio standards. One issue is the disruption of policies on the provision of small size housing units by local governments (Lu, 2009), which is important to land use efficiency (see Chapter 4 for more details).

3.6 Conclusion

Since the 1980s, China's land use reform and land market development have fundamentally changed the way land is allocated and used in the country. The land use reform has created a functioning system of private property rights in state-owned land, enabling LURs of state-owned land to be traded on the market. To mitigate the huge demand and encroachment on land brought by economic growth and urbanisation, a system of land administration was established to regulate land conversion from agricultural to urban uses and protect arable land. An urban land market was ushered in, to allow the state to exploit the economic value of state ownership of urban land, leading to the formation of a three-tier system of land markets. To fully control land conversion and maximise land sale revenues, open market land sales were required in the primary land market.

The new land administration system, however, has to deal with the following problems. First, the efficiency of urban land use remains low after the establishment of the land market. Second, the system of land sales has to take into account the contribution of land prices to rising housing prices. In particular, tightening the land supply in major cities in China can fuel land price inflation and result in further housing price inflation if relevant institutional arrangements, such as floor area ratio and housing size standards, are not changed.

Notes

1 As amended in 1988, 1993, 1999 and 2004.
2 As amended in 1988, 1998 and 2004.

3 Land ownership by the Chinese state first originated from the confiscation of enemy estate during and immediately after the Civil War that brought victory to the CCP in 1949. All urban land was nationalised through compulsory purchase during the socialist transformation in the 1950s when the CCP promoted public ownership of land following Marxist principles and Soviet practices. In the rural areas, private ownership of land was ended by collectivisation in the second half of 1950s.
4 As amended in 2007.
5 The market price of land is normally calculated by reference to the market prices of comparable land. This is the basis for calculating the price of LURs, for taxation and for assessing the value of state-owned assets.
6 In 1957, the country had 111.8 million hectares of arable land (Peng, 1999). From 1958 to 1986, 40.73 million hectares of arable land was converted to non-agricultural uses such as residential and industrial uses (Bu and Li, 1998). During the period, there was a net decrease of 19.3 million hectares of arable land, almost the size of the UK, against a doubling of population.
7 From 1998 to 2000, there was deflation in China. During this period the Chinese economy grew over 7 per cent in real terms per year, which is referred to as deflationary expansion.

4

THE HOUSING MARKET

This chapter investigates the establishment of the housing market and its spectacular growth amid rapid urbanisation, fast economic growth and strong investment demand. It affords a brief account on housing reform that ended the welfare housing allocation system by state employers, and released demand for commodity housing. The demand side incentives and their impacts are explored and commented upon. The chapter then goes on to cover supply side incentives and constraints that work together with demand incentives to create a housing market that has moved away from meeting the needs of the mass of urban people, to satisfying the needs of the well-off and the investors. Market data on demand, supply and prices are provided on national as well as local levels, on a number of case study cities.

4.1 Property rights in housing

The property rights system in housing in China consists of two types of housing according to land ownership, i.e. housing on collectively owned land and on state-owned land. The former includes housing on house sites designated by county-level government and allocated to members by the collective. The latter comprises housing on sites with granted and allocated LURs and designated for residential use by urban plans. Such housing can be owned by the state, organisations or individuals.

Housing on collectively owned land is owned by members of the collective with the right to use house sites. Article 152 of the Property Law stipulates that the holder should be entitled to possess and use collectively owned land, and to construct residential houses and their affiliated facilities. The right to use house sites is free to members of the collective and can be removed if the collective ownership ceases to exist. The most likely case for removal of collective land ownership is by acquisition by the state. This housing right is not transferable, so housing on collectively owned land cannot enter the housing market, i.e. be transferred or mortgaged. Nevertheless, housing on house sites in the suburb of cities and towns has been widely let to migrant workers and sold in significant numbers to home buyers in some large cities, such as Guangzhou and Beijing. Yet the MLR has made it clear that the transfer of such housing to non-members of the collective is illegal. Much of such housing, though sold to buyers who are not members of the collective owning the land, remains outside the housing market if the collectively owned land on which such housing is built is not converted to state ownership. As a result, a buyer of housing on a house site only acquires the right to occupy such housing. Therefore, such housing is popularly referred to as housing with 'small property right'.

On the other hand, housing on allocated LURs on state-owned land owned by the state and organisations, which was the dominant form of housing prior to the housing reform in 1990s, can be converted to housing on granted LURs. Such conversion took place in mass quantity during the housing reform individually and collectively (see Section 4.2). Housing on granted LURs of state-owned land is commodity housing, the only housing tenure that is legally alienable in open market and can be acquired according to law by purchase, construction, gift, mortgage and inheritance. The Inheritance Law provides that commodity housing on granted LURs can be inherited by will or by the lawful successor.

Ownership of commodity housing on granted LUR consists of building ownership and LUR, with the latter subject to the remaining term of the LUR. Building ownership is a form of private property right protected by the state as stipulated in the Constitution and the Property Law. Article 64 of the Property Law stipulates that an individual is entitled to the ownership of his or her premises. Building ownership is permanent, subject to the length of the LUR. In fact, the recognition of private ownership of buildings by the state came earlier than the creation of the LUR tenure system. In January 1987, a standard building title deed was issued by the Ministry of Urban and Countryside Construction and Environmental Protection (later the Ministry of Construction and now the Ministry of Housing and Urban and Rural Development, or MOHURD) and formal building registration work started. Most cities set up separately a Building Registration Agency to deal with the registration applications, but some cities combine building registration to land registration and issued a joint title deed certifying building ownership and LUR ownership. The conflict arising from infinite ownership of building and finite ownership of land has been solved in principle by the Property Law conferring automatic renewal to residential LUR if the land is not required for public use. However, no stipulation is made to clarify if a premium needs to be paid and if compensation will be paid for the building if the LUR is not renewed (see Section 3.1).

The separation in registration of building ownership and LUR ownership brings some risks to uninformed home buyers, as there are cases in which some housing developers did not pay the full amount of land premium or used the LUR as collateral to raise loans but did not pay back those loans when they sold the commodity housing. Home buyers could be misled by the availability of building registration and building title deeds and suffer a loss due to the lack of LUR titles. They normally could not obtain the LUR title unless the developer pays the premium or clears all its debts.

Residential real estate in urban China is developed in most cases in the form of large buildings for multiple occupiers in housing estates. Some housing estates contain a mixture of multi-storey blocks and individual houses, or purely individual houses. The flats or individual houses are sold individually. Article 70 of Property Law stipulates that an owner of units in buildings of multiple ownership have ownership over the exclusive parts within the buildings, and have common ownership and the right of common management over the common parts. The management of the common parts is normally conducted by an owners' association, elected by all the owners, by appointing a management company. Housing ownership titles in China are often referred to as strata titles by practitioners in the English press, though there is no connection between the laws of China and Australia.

4.2 The housing reform

The system of public housing for urban residents on the eve of the reforms and opening up was established in the 1950s when private landlords had to sell their holdings to the state in the nationalisation campaign. A low-rent public housing system then replaced the housing market, though a small proportion of owner-occupied private housing remained. In the new system, all urban residents who were registered households were eligible for public housing, normally provided for by state-owned employers, at rents lower than maintenance costs of the buildings. Rents were lower than required by investment return and housing provision was considered a drain on resources. As a result, housing conditions in this second world of real estate either stagnated or deteriorated due to insufficient investment in maintenance and new build (see Section 1.1).

In 1980, the central government proposed an outline policy for housing reform to alleviate the housing burden on the state by requiring greater contribution from welfare housing tenants in the form of higher rents or purchase. A programme of selling public housing was initiated. From 1979 to 1985, attempts were made to sell public housing at or lower than construction costs in 160 cities and 300 counties in 27 provinces, with 10.92 million m^2 of housing space sold (Chai, 2008). However, purchase of housing without right to resell[1] resulted in buyers having tight budgets and unable to benefit from rapidly rising housing standards, which proved unpopular. From 1987 to 1990, attempts were made to raise rents across the board in 12 cities and 30 counties, which proved unpopular due to the vested interest of households in low rents.

In early 1991, a new approach in housing reform was pioneered in Beijing and Shanghai, which involved: step-by-step rent increases without simultaneous pay increases; sales of public housing to sitting tenants at large discounts; and the establishment of compulsory housing provident funds to finance the construction of public housing and the purchase of public housing by individuals.[2] In June 1991, the central government formulated a new national policy on housing reform (State Council, 1991) adopting: step-by-step rent increase; new rents for new public housing; sale of public housing; and housing construction cooperation by employers and employees as a means to reform the provision of housing in urban areas. This flexible and down-to-earth policy overcame the resistance to and stagnation of the housing reform. By the end of 1993, all provinces except Jiangxi and Tibet, had formulated housing reform packages that raised rents from the range of CN¥0.08–0.13 to CN¥0.20–0.50 per m^2. Rent as a proportion of household expenditure increased from 0.87 per cent to 2 per cent (Wang, 1999). Compulsory housing provident funds extended to 60 per cent of all large and medium-sized cities and accumulated CN¥11.0 billion by the end of 1993.

The rapid development of the housing market (see Section 4.3) prompted the central government at the constitutional level in 1994 to link the housing reform to the housing market. The aim of the housing reform was the establishment of a housing system compatible with the socialist market economy, with housing mainly supplied by the market (State Council, 1994a). The tasks of housing reform were identified as increasing coverage of housing provident funds, steady progress in sale of public housing, promotion of housing market transactions, nurturing of market-based housing maintenance and property management, and provision of subsidised housing referred to as Economic and Comfort Housing. These new housing reform decisions by the constitutional level were quickly transformed into operational rules by the local governments at the collective choice level to sell public housing, providing large discounts to buyers and allow them to sell the privatised housing in the market after five years of purchase and holding. By 1995, 30 per cent of public housing had been sold to households and in some provinces, like Guangdong and Zhejiang, the ratio was 60 per cent to 70 per cent (Wang, 1999). By the end of 1997, average rents for public housing were raised to CN¥1.21 per m^2 (Liu, 1998) and the housing provident fund grew to CN¥79.7 billion (Liu, 1999).

The East Asian financial crisis in 1997 and its negative impacts on the Chinese economy prompted the Chinese government to promote the housing industry as a new growth pole and speed up the housing reform to increase demand for housing in the market. In July 1998, the central government enacted the Circular on Further Deepening of Urban Housing Reform. This was to stop welfare housing allocation from the second half of 1998, to require sitting tenants to purchase the public housing they were occupying, and to require new employees of SOEs and the government to purchase housing in the market[3] (State Council, 1998a). Housing allowance on top of wages was provided in localities where average price for a 60 m^2 flat was more than four times the salaries of households with both the husband and wife working. Public housing was sold based at construction costs but with discounts provided for years of service, age of housing and one-off cash payment. With all the discounts, public housing could be bought at less than 30 per cent of the construction costs by sitting tenants with both the husband and wife having over 20 years of service at SOEs or the government. The benefits obtainable from purchasing public housing acted as strong incentives for sitting tenants to comply. As a result, this final phase of the housing reform succeeded in ending the system of welfare housing provision since the 1950s. The provision of social housing equivalent to take care of those left out is discussed in Chapter 13. On the other hand, it awakened a strong demand for commodity housing, contributing to the housing development boom throughout the country from 1998 (see Section 4.3).

The Chinese housing reform, equivalent to a property rights revolution, was successfully implemented from 1980 to 2000 without causing social instability or a reduction of housing standard. On the contrary, the reform helped bring about a dynamic housing market and greatly improved the housing standards of urban residents.

4.3 1979–90: housing market initiation

While initiating the housing reform, the central government also allowed private house building and encouraged the development of the housing market. Allowing private house building, ownership, and exchange was first a tentative measure to solve the housing shortage and relieve the state of the heavy burden of public housing provision. In 1987, building a real estate market which includes the housing market was first proposed in the Thirteenth Congress of the CCP, and housing market creation became a government policy. Housing market development in China can thus be described in four stages according to government policies and market development (Table 4.1).

TABLE 4.1 The four stages of housing market development in China

Stages	*Main developments*	*Time scale*
Stage I: initiation	Embryonic housing market emerging in a few coastal cities; first housing boom and bust as dictated by government fiscal policy	1979–1990
Stage II: expansion	Market coverage extending to most of the country; boom and bust of the housing markets by macro control; property services marketisation and certification started	1990–1998
Stage III: policy boost	Housing market receiving policy support to become a growth engine; tax cuts to boost secondary market transactions; mortgage finance becoming available to the public; property services marketisation and certification expanding rapidly	1998–2002
Stage IV: macro control	Macro control to slow down market growth for financial sector security and to control housing price inflation; short downturn ended by government rescue packages; further control by demand management and monetary tightening	2003 onwards

The first stage witnessed the re-emergence of market activities in housing and was characterised by domestic developers operating in a few cities, with most people and firms unaffected or even unaware of the market provision of real estate. Like the land market, foreign money played a part in kicking off market-oriented housing development. Since 1979, commodity housing had been developed in Shanghai, Guangzhou, and Shenzhen for sale to local residents who were relatives of Hong Kong Chinese and to owner-occupiers from Hong Kong. A number of salient points can be summarised on the first stage when the housing market resumed after more than 20 years of absence. First, private ownership rights in housing were recognised and legalised by the state. The State Council enacted Regulations on Urban Privately Owned Housing Administration in 1983, which set up a system of registration of private housing. By 1990, all private housing in urban areas was registered and 80 per cent of the ownership title deeds had been issued (Zhang, 1991).

Second, private housing construction became a significant part of urban housing construction. Private housing construction grew from only 3 per cent of all housing construction spending in urban areas in 1979, to 18 per cent in 1989 (ibid.). In 1990, 24.2 million m^2 of new housing was built by private funding, equivalent to 24.9 per cent of all housing completion in urban areas.

Third, housing exchange was legalised by the state and enjoyed significant growth during the 1980s. Figure 4.1 shows the sales of new commodity housing from 1986 to 2013, with sales growing from 18.35 million in 1986 to 27.45 million in 1990. In 1990, 27.5 per cent of commodity housing sales were accounted for by individuals. Housing Exchanges where local government departments registered the transaction were set up from 1985. In 1987, the first such Exchange was opened

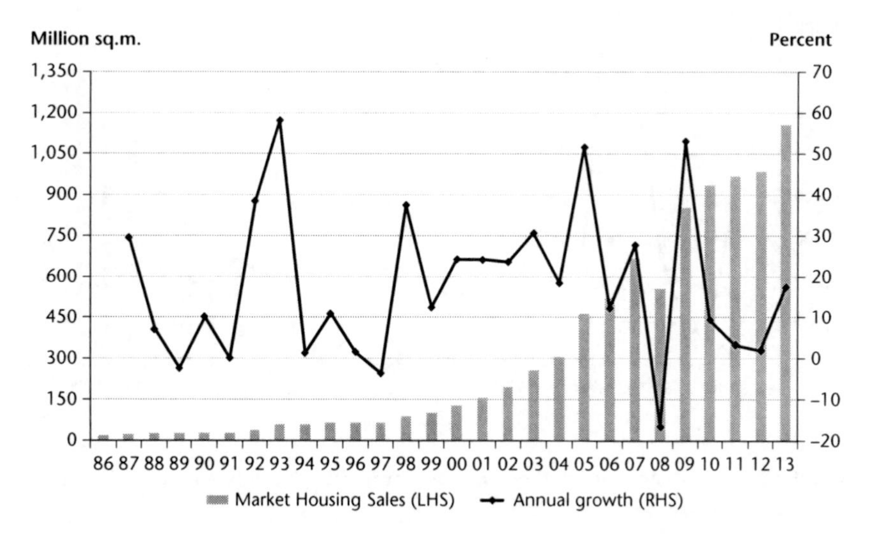

FIGURE 4.1 Commodity housing sales and its annual growth rate, 1986–2013
Sources: NBSC, various years.

in Wuhan, the first such institution administered by a major city's government (Chai, 2008). By 1989, over 1,000 Housing Exchanges in the country were registering 191,700 housing transactions with total area of 19 million m², with transactions between individuals amounting to 5.9 million m² (Zhang, 1991).

Fourth, housing development by developers adopted a model of comprehensive development, in which a developer is responsible for constructing housing and all the municipal and supporting facilities such as nurseries and schools, roads and drainage, greenery and leisure. Developers were also responsible for selling and letting all the market-oriented space in housing estates, and for managing the use of housing and other facilities. This mode of development has the advantage of providing self-contained municipal and supporting facilities in housing estates, to overcome the lack of such facilities in Chinese cities.

This stage of housing market development ended when the government imposed credit control in 1989 to rein in runaway fixed asset investment to counter inflationary pressures. This caused a hard landing, or a significant slowdown in economic growth and housing market growth (see Figure 2.1). In 1989, there was a 2.35 per cent reduction in sales of new housing (Figure 4.1). Market-oriented property development (Figure 4.2), in which housing investment accounted for over 50 per cent, fell by 6.99 per cent in 1989. The downturn resulted in a large amount of newly completed housing space remaining unsold and empty, causing many projects to fail and stand as eyesores in urban areas.

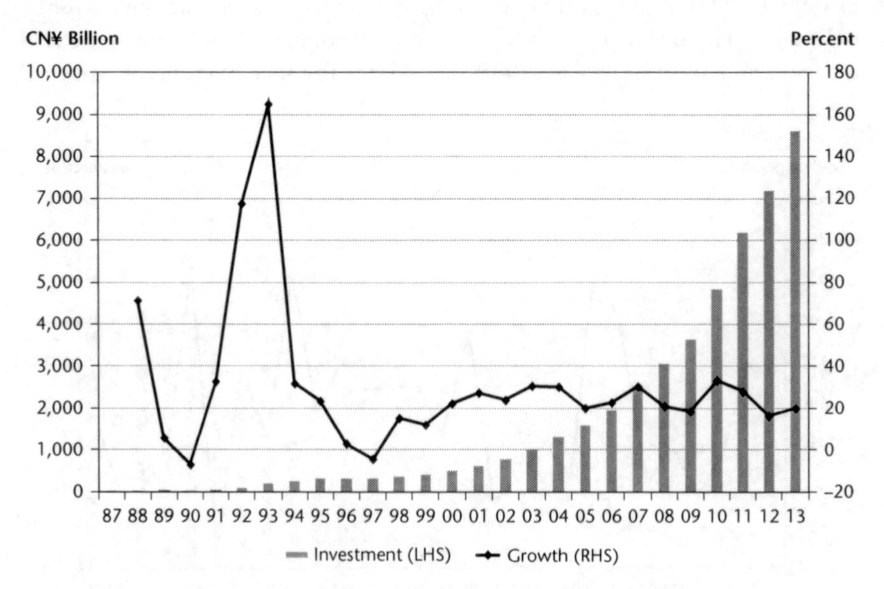

FIGURE 4.2 Investment in market-oriented property development and its annual growth rate, 1987–2013

Sources: NBSC, various years.

4.4 1990–98: housing market expansion

The second stage of the housing market development was characterised by a policy-led boom and bust, expansion of market coverage to the whole country, and the establishment of ancillary markets in mortgage financing and property services. Six important features are discussed as follows:

First, the formal delineation of marketable land tenure, the LUR system (see Section 3.1), by Decree 55 has added to building ownership recognition to complete the property rights reform on real estate in China. It had a major impact on market confidence, especially among home buyers and investors from Hong Kong and other foreign economies such as Taiwan and South East Asia, who contributed to housing market rebound in 1990 and stabilisation in 1991 (Figure 4.1). Prices of commodity housing rose (Figure 4.3). Commodity housing started to become popular among the rich in China due to its investment value.

Second, a property boom and excessive speculation took place due to a favourable policy environment, the deregulation of development control, the relaxation of price administration and lending, and the lack of financial discipline. In Spring 1992, Deng Xiaoping, then the leader, toured South China where reforms and opening up were ahead of the rest of the country, and urged the provincial

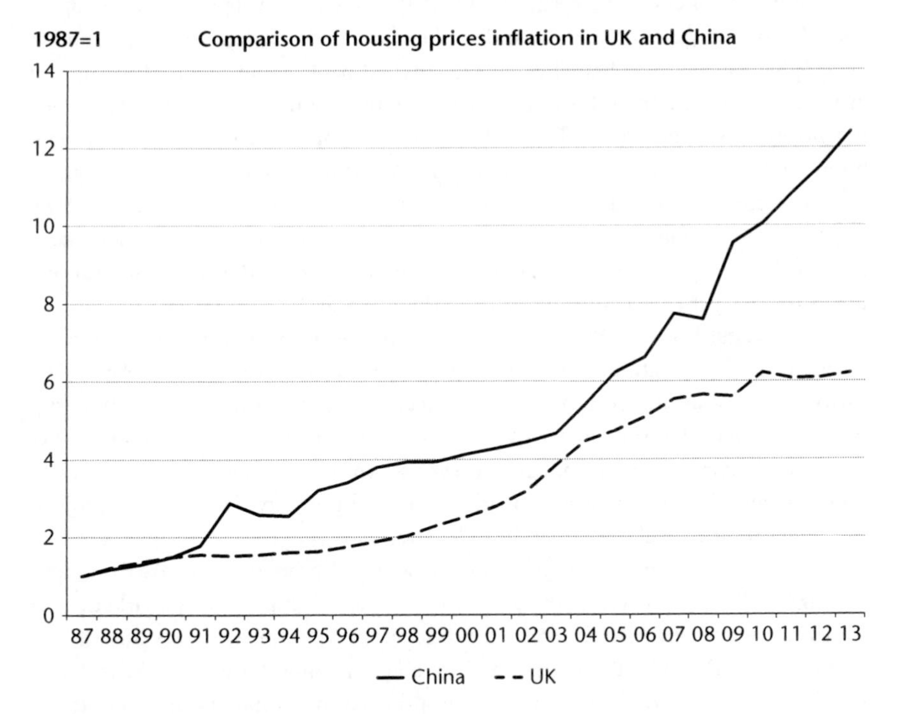

FIGURE 4.3 National average housing price inflation since 1987
Sources: NBSC, various years.

governments to speed up economic development by further reforms and opening up. A favourable policy environment for development formed after Deng's call, and unleashed a wave of reforms and further opening up of the economy.

One of the policy changes following Deng's call was the enactment of the *Decisions on Speeding Up the Development of Tertiary Industry* in November 1992, which encouraged the development of the property industry as a tertiary industry (State Council, 1992). The policy document represented a constitutional-level decision to provide policy guidance to and relax control on local governments and thus push forward reforms and opening up. This, for the first time outlined a blueprint for property market development: a three-tier land market and a system of real estate markets. The document gave the green light to a variety of business models including leasing and mortgages that were still not common at the time, and proposed policy measures to encourage property investment, including foreign direct investment in property.

Deng's call to speed up reforms and open up were generally welcomed by local governments due to the strong incentives brought forward by deregulation. Local governments at the collective choice level enacted new rules for the property market to expand. A number of changes in operational rules followed: land sales were approved by the operational arms of governments, making available sufficient land for development; the delegation of development approval powers to the operational arms of government expanded the scale and accelerated new property developments; the local branches of state-owned banks were required to support property development; and the deregulation of price administration allowed housing prices to be determined by market forces. An investment and development boom in the property sector took place, with investment in market-oriented property development growing 117.56 per cent and 165.12 per cent in 1992 and 1993, respectively (see Figure 4.2). Market-oriented property development new starts reached 114.60 million m^2, 78.1 per cent more than in 1991 (Chai, 2008). Housing prices rose 61 per cent nationally in 1992 (Figure 4.3).

The deregulation in 1992, however, was not accompanied by corresponding changes in the operational rules to require the setting up of the corresponding financial discipline and corporate governance. Rising land and housing prices signalled to society the profitability of real estate investment. A significant amount of the nation's available funds, designated for uses in many other industries, became footloose and found their way to market-oriented property development projects for higher returns. Making use of the delegation of power to attract capital for local economic growth, some local governments adopted property-led growth policies and drastically deregulated operational rules in their local property markets, thus providing venues for such footloose funds to speculate in property investment and development. Beihai City in Guangxi Province, Huizhou City in the Pearl River Delta and Hainan Province soon became property development and investment hot spots. Local governments there set minimal thresholds for investments, sold development sites very cheaply and exercised minimal planning control. Hundreds of development projects went ahead simultaneously. It is worth noting that the two cities and one province were all economically underdeveloped compared to Beijing, Shanghai, Guangzhou and Shenzhen and had no competitive industries. Box 4.1 illustrates the overheated and speculative property market in Beihai before and after central government intervention.

BOX 4.1 PROPERTY SPECULATION AND ITS AFTERMATH IN BEIHAI

Beihai, one of the 14 coastal cities given open-city status by the central government in 1984, was inhabited by 200,000 people in the relatively underdeveloped Guangxi Province in southern China. In 1992, this little known city suddenly became a popular property investment hotspot and its population soared to over half a million by mid-1993.

Inspired by the success of Shenzhen, the Beihai Municipal Government adopted a property-led development strategy and sold LURs cheaply to woo investors. The average land price in the primary land market was only CN¥375,000 (US$68,000) per hectare.[4] Such a low price, coupled with minimal regulation in development and conveyance, attracted numerous investors. From August 1992 to early 1993, 80 sq km of development land, which was seven times of the size of the existing built area of the city, was sold to investors from outside Guanxi Province. The Beihai government received only CN¥3.09 billion (US$560 million), a sum far from enough to provide infrastructure for the land sold. Furthermore, the government substantially reduced the transaction costs of land, thus creating the most active secondary land market in the country, with prices rising ten to 30 times in ten months. Large amounts of cash came from all over the country and the number of property development firms increased from six in early 1992 to 1,056 in early 1993.

However, much of the cash for land purchase and development projects was from funds already allocated for other uses by state-owned banks and state firms. Anecdotal evidence indicated that some branches of state-owned banks in Hunan Province ran out of cash due to the diversion of cash to Beihai. Zhu Rongji, then deputy premier, visited Beihai in January 1993 to inspect this development model and soon started disciplinary action on state-owned banks. In April 1993, those banks involved in unapproved lending to Beihai property investment projects were required to recall the loans. In June, central government's macro control measures imposed a credit squeeze and triggered panic selling in Beihai's property market, with land prices falling from CN¥15 million per hectare to CN¥1.2 million in less than six months. With money and people disappearing almost overnight, the short-lived boom eventually left behind large numbers of derelict sites and semi-completed buildings. Anecdotal evidence shows that some of the semi-completed villas were later let to farmers to raise chickens.

A lot of effort was spent sorting out the mess left behind by the collapse of the local property market. From 1994 to 1995, the municipal government disqualified 2,364 land purchase deals. From 1996 to 1998, 17 km² land planned for development and 1.1 km² land requisitioned were repossessed by the municipal government. Yet a survey in 2000 by the city's real estate administration found that there were still 108 unfinished projects, 1,887 hectares of derelict land, 1.07 million square metres of vacant buildings, and 1.21 million square metres of unfinished buildings (Xinhua, 2003a). Despite all its natural endowments and favourable policies, Beihai has failed to develop quickly and properly in the 1990s, dragging the feet of Guanxi Province in economic development.

With a central government package to exempt Business Tax and Stamp Duty on property transfer in the city, effective from 1 January 2003 to 31 December 2006, Beihai's campaign to save the unfinished projects achieved remarkable success by June 2005, with construction resuming for 92 per cent of the unfinished projects. But anecdotal evidence suggests that unfinished buildings dated back to the 1990s were still to be found in the city centre at the time of writing.

The third feature at this stage of housing market development is the credit tightening imposed by the central government from mid-1993 that led to the property market crash in over-speculated local markets and the soft landing of the property market as a whole. The property boom from 1992 helped start a new round of accelerated economic growth (Figure 2.1) but caused shortages of materials and funds, leading to high inflation (Figure 2.4). In June 1993 the central government attempted to impose macro controls to ease the situation and enacted a decree, the Opinions on Current Economic Circumstances and Strengthening Macro Control, to tighten the credit supply and thereby fight inflation, and safeguard funds and materials for key state projects.

Credit tightening resulted in an immediate property market crash in Huizhou City in the Pearl River Delta, in Beihai City in Guangxi Province, and in Hainan Province, with severe consequences. Box 4.1 details the market crash and the time and effort to absorb the costs of the crash in Beihai. Huizhou City and Hainan Province were also seriously affected by the property market crash. In Huizhou, there were half a million m² unfinished buildings and another half a million m² of vacant buildings left behind. In Hainan Province, there were 18,834 hectares of derelict land and over 16 million m² unfinished buildings (Baidu, 2014). It took nearly a decade for Huizhou and Hainan to regain the momentum in economic growth. In contrast, credit tightening did not have an instant impact on the property markets in cities such as Beijing, Shanghai, Guangzhou and Shenzhen. These cities maintained proper regulation on the land and housing markets and had a more diversified economic structure, domestic and international investment and thus a sound demand for real estate.

Fourth, the housing market slowed down and eventually stagnated from 1994 to 1997 under tight credit conditions from 1993, oversupply from 2005 in all major cities and falling demand caused by the East Asian financial crisis. With inflation running at an all-time high at 24.1 per cent in 1994 (Figure 2.4), the central government in the same year issued a new decree to continue the macro control on fixed asset investment to prevent excessive fixed asset investment, financial ill-discipline and inflation (State Council, 1994b). In 1995 it enacted another decree to restrict new high-end property investment projects (State Council, 1995). With strict control, investment in commodity housing decelerated and declined from 1995 to 1997 (Figure 4.4) and sale of new housing recorded negative growth in 1997 (see Figure 4.1). A market contraction loomed if the macroeconomic environment continued to worsen, which had implications for economic growth.

Fifth, during stage 2, the housing market activities expanded from coastal cities to all important cities during 1991 to 1997, with the last provincial capital, Lhasa, recording sale of new housing of 41,000 m^2 in 1997. Figure 4.5 compares sales of new housing in 1995 to 2012 for the 31 provinces in China. It can be seen in 1995 that the market size was larger in eastern provinces such as Guangdong, Jiangsu, Zhejiang and the provincial cities of Beijing and Shanghai. In the central and western provinces market sizes were larger in Henan and Sichuan, both being the two most populous provinces at the time.

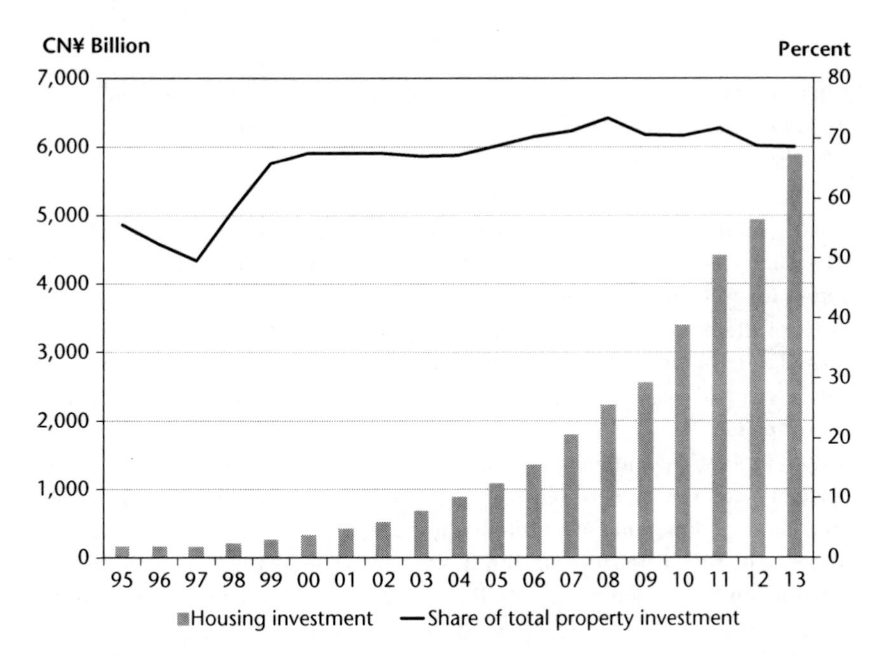

FIGURE 4.4 Investment in commodity housing development and its share of total market-oriented property investment, 1995–2013

Sources: NBSC, various years.

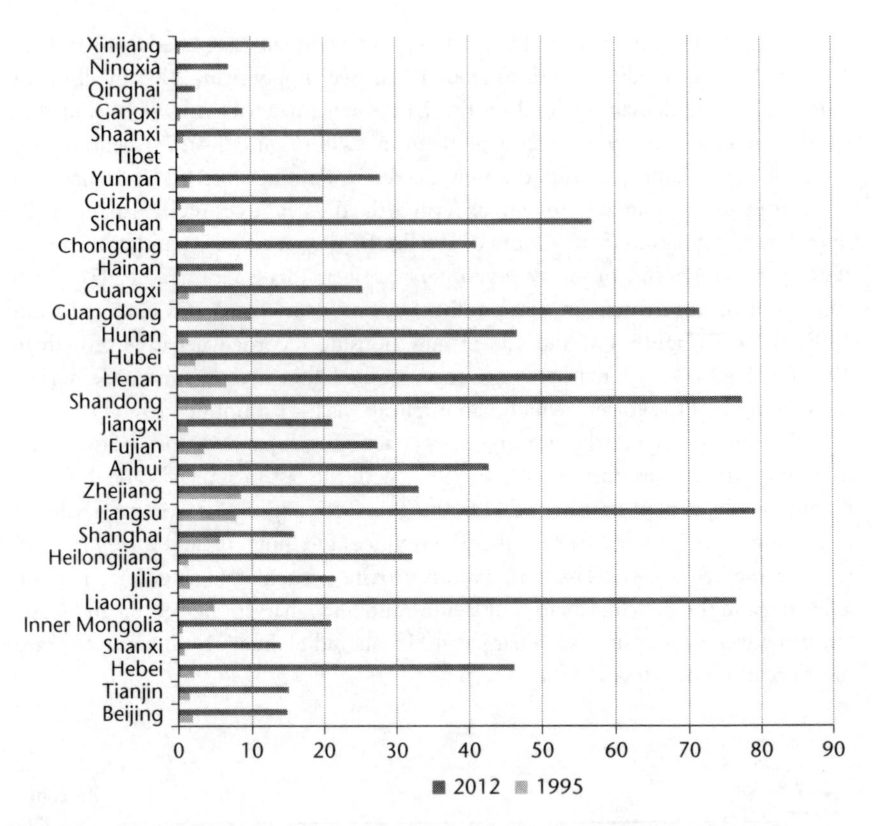

FIGURE 4.5 Sale of new housing in all provinces in 1995 and 2012 (million m²)
Note: In 1995, Tibet had no housing sales, and Chongqing's sales figure is included in Sichuan's data.
Sources: NBSC (2006; 2013).

Sixth, market-based property services such as property management, valuation and agency were established during 1990 to 1998, with systems of national certification for valuation developed in 1994. The marketisation of property services as part of the housing market was first proposed in national housing policy formulated in 1994 (State Council, 1994a), and professionalisation was soon regarded as necessary to ensure the quality of services and facilitate administration by government.

With multi-storey buildings in housing estates becoming the dominant housing choice, a system of owners' self-management and professional property management became important in ensuring the supervision of property management services. The first national regulation was promulgated in 1994 by the Ministry of Construction (succeeded by MOHURD), which set up residential district housing management committees to represent the interests of all owners and to appoint and supervise property management companies. Common disputes at the time between property management service providers and owners were about the details of services and the levels of fees charged. Subsequently, national property management standards and national rules on service fee payment and collection were provided by the Ministry of Construction in 1995 and 1996, respectively.

Quality and reliable valuation services are important when buying and selling, letting, mortgaging and inheriting property. Valuation services were the first property services to receive national certification, with UREAL legalising national certification in 1994. The China Institute of Real Estate Appraisers was established in 1992 with the national open entrance examination conducted in 1994. Meanwhile, the China Real Estate Valuers Association was established in 1994 with the national open entrance examination conducted in 1994.

On the other hand, property agency services, whose quality and standards varied widely, did not receive national regulatory attention. Supervision and standardisation of agency services were mainly undertaken by local initiatives from 1991 to 1998. For example, Guangzhou Municipal Government enacted an interim regulation in 1992 to regulate real estate agents who provided highly unequal services in terms of quality, standards and honesty. The entrance of Hong Kong and international agency firms in 1991 provided benchmarks of services to their local counterparts and regulators.

4.5 1998–2002: policy-boosted rapid growth

The third stage of housing market development from 1998 to 2002 was caused by the need to boost economic growth. The search for a new growth engine started in the period between 1996 and 1997 when the Chinese economy slowed down from its peak from 1992 to 1994. The property industry was deemed a good candidate by Zhu Rongji, then deputy premier, and was referred to as a pillar or key industry. In early 1998, the economy faced strong downward pressure due to the impact of the East Asian financial crisis that caused virtually zero growth in exports. In June 1998, the central government headed by Zhu Rongji, then premier, enacted the Circular on Further Deepening of Urban Housing reform, which put forward the policy of promoting the housing industry as a new growth engine, and pushed forward the last phase of housing reform to stimulate demand in the housing market (see Section 4.2) (State Council, 1998a). Favourable policies provided to the housing market from 1997 to 1998 are as follows.

First, restrictions on home mortgage loans were lifted to release households' demand on housing. China's mortgage loan lending started in 1991 with many restrictions and grew slowly. By 1997, the home mortgage loan balance was CN¥19 billion. In April 1998, the People's Bank of China, the central bank, first called for support for housing construction and consumption by increasing the credit supply and delegated all decision-making on mortgage loans to commercial banks (PBoC, 1998). In 1998, the first home mortgage loan according to the borrower's circumstances was lent by the Construction Bank of China. Compared to the assets of state-owned enterprises, homes whose market value rose quickly were much better collateral so state-owned banks preferred lending home mortgages loans. From 1999 mortgage loans to housing took off (Figure 4.6). Anecdotal evidence suggested that

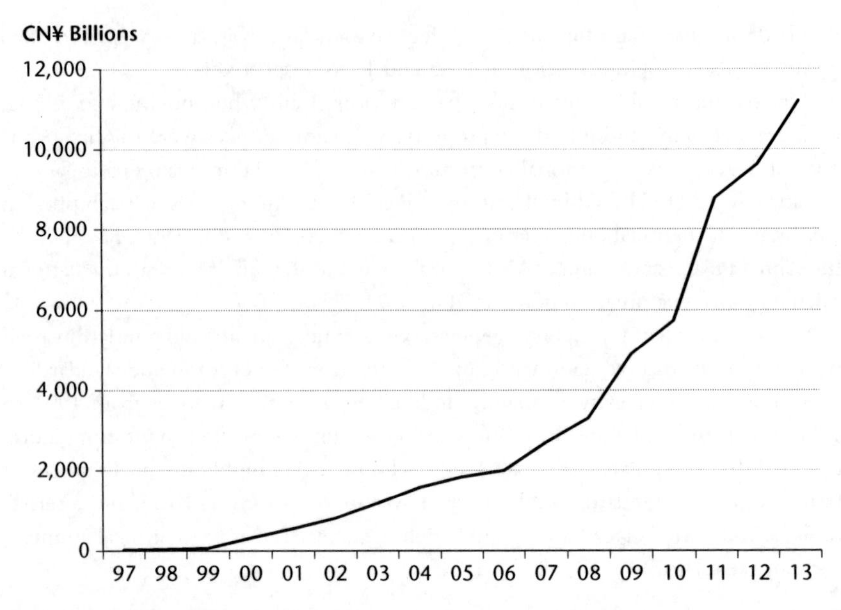

FIGURE 4.6 Mortgage loan balance from 1997
Sources: PBoC, various years.

bitter competition among banks sometimes led to very low thresholds in home mortgage loan lending.

Second, major cuts in taxes levied on property transactions were offered to boost the housing market. In July 1999, the Ministry of Finance and State Administration of Taxation enacted the Circular on Adjusting Some Taxation Policies on the Real Estate Market. This decree waived Business Tax (5 per cent of value) on selling ordinary housing[5] by individuals who had occupied the housing for over a year, and halved the Deed Tax (originally 6 per cent of value) paid by the individual when purchasing ordinary housing (SAT, 1999a). The Land Appreciation Tax was also waived for ordinary housing sold by individuals. In December 1999, the Ministry of Finance, State Administration of Taxation and Ministry of Construction, jointly issued the *Circular on Issues of Income Tax on Housing Sales Revenues by Individuals* (SAT, 1999b), which waived income tax on gains (originally 20 per cent of gains) by an individual when selling ordinary housing if the individual bought the housing again from the market within one year. The tax cuts reduced the high transaction costs in buying and selling housing, incentivising individuals to sell their existing housing (for example, privatised public housing) and buy new houses.

Third, sub-commodity housing for purchase, referred to as Economic and Comfortable Housing (ECH), was provided to developers, encouraging them to build such housing for low and lower-middle income households to buy. ECH was less expensive because the government provided allocated LURs, thus waiving land costs, and offered tax reductions or waivers on construction. More details on ECH are provided in Chapter 13.

As a result of these favourable policies, the housing market achieved rapid growth from 1998 to 2002, with housing investment, new starts, sales and prices all rising significantly. Investment in housing development grew 240 per cent in the five-year period from 1998 to 2002 (see Figure 4.4). Housing new starts, an indication of developer confidence and development activities, doubled in terms of floor areas (Figure 4.7). Meanwhile, sales of new housing in terms of floor areas increased 196 per cent during the same period (Figure 4.1). Although national average prices grew only 16 per cent (Figure 4.3), average prices in Shanghai had grown 32.4 per cent in the same period (Figure 4.8). In the city centre, price inflation was far higher, attracting significant investment from speculators.

Fourth, property services marketisation and certification continued their development. In 2000, the China Property Management Institute had its first National Congress in Beijing to promote professionalisation and professional management in property management, with housing management taking the lion's share. In 2001, the Ministry of Construction enacted a national regulation on real estate agent qualification and entrance examination arrangements to start certification and qualification management of agents. Professionalisation of housing services was necessary because the housing market had become the main source of housing supply.

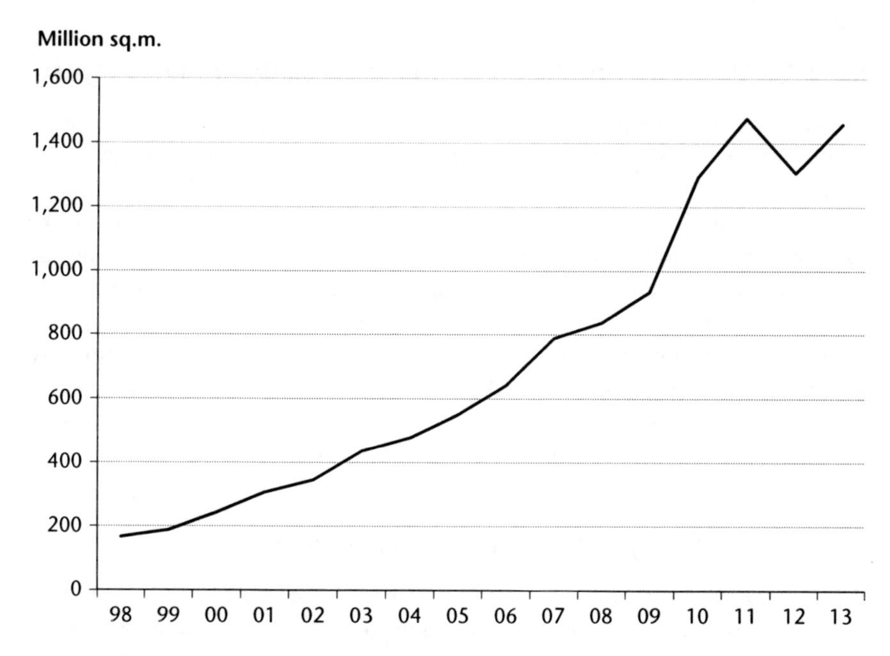

FIGURE 4.7 Housing new starts from 1997
Sources: NBSC, various years.

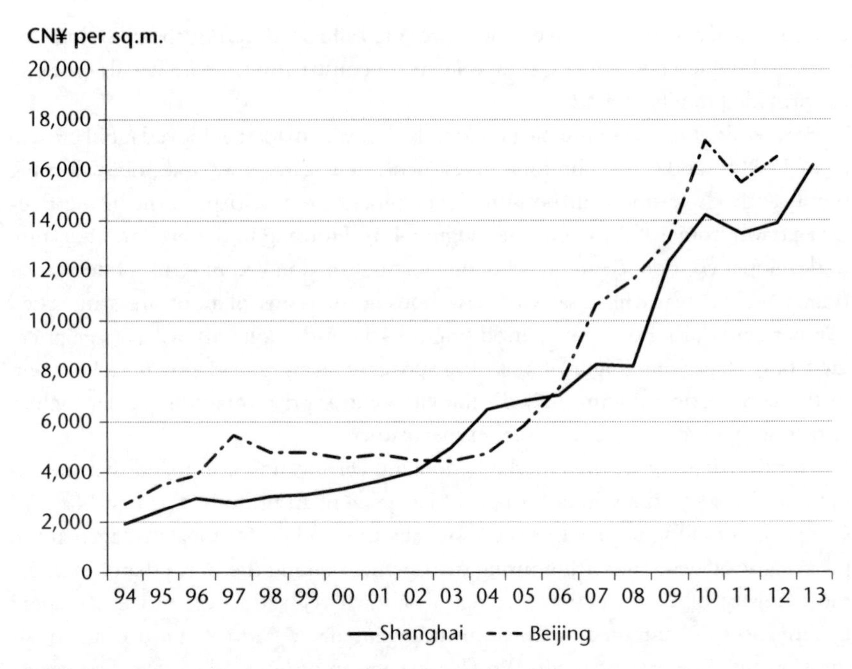

FIGURE 4.8 Housing prices in Beijing and Shanghai from 1994
Sources: *Shanghai Statistical Yearbook*, various years.

4.6 2003 Onwards: growth under the uncertainty of macro control

At stage 4 China's housing market development entered an era of macro control. From 2003 to the present, the real estate market in China has been subject to intense debate on whether the market is overheating and should be cooled down by macro control, and whether short-term correction of the market should be allowed without government rescue. The macro control from 2005 eventually tamed the running away of housing prices, but rescuing the falling market in 2009 created a new round of rapid housing price inflation and intense market activities, forcing the government to resume macro control from 2010. Yet the economic slowdown and tight money heralded a new round of housing market contraction from early 2014, which is causing concern inside and outside China. A number of significant developments at this stage are discussed as follows.

4.6.1 Early warning by the People's Bank of China

First, disagreement between the central bank and the State Council on whether the market was overheating delayed macro control measures which could have been applied earlier. From 2003, investment in housing development, housing new starts and housing prices continued rapid growth over the whole country (see Figures 4.3, 4.4 and 4.7), with housing supply structure biased to high-end market and market transactions becoming increasingly chaotic. In Shanghai, average housing prices grew 24.5 per cent in 2003 and 30.1 per cent in 2004 (see Figure 4.8). To prevent a repeat of the property boom of 1992 to 1993, the central bank took early action to enact the *Circular on Further Strengthening the Management of Real Estate Credit Business* to control the credit supply to the property market (PBoC, 2003) in June 2003. The central bank's caution, however, was rebuffed by the property sector and local governments because the move left some overstretched developers starved of cash. The State Council, to some extent, opposed the Central Bank and supported the property industry by enacting the *Circular to Promote the Sustainable and Healthy Development of the Real Estate Market* in August 2003. This policy document pointed out that the property industry had become a pillar industry for the country and should be supported with more credit, though attention needed to be paid to the rapid increase in investment volume and prices to achieve balanced structure and stable prices (State Council, 2003). In the Circular, the central government emphasised the comprehensive development of the property market and the property industry, without specifically mentioning any effective policy to curb the excessive increases of investment and prices. This reflected the difficult position of the central government: reliant on the property sector to pull economic growth but wary of an excessive investment boom and price hike (Chai, 2008).

Even though the central bank's attempt was suspended, the debate highlighted the concerns over the property market. Thus the central bank's action sent out the first signal of a new round of macro control.

4.6.2 Tightening land administration

Second, strengthening land administration in 2004 effectively squeezed land supply and generated an expectation of land shortage (see Chapter 3), which fuelled housing price inflation. By 2004, illegal occupation and conversion of land, in particular arable land, for property development had greatly alarmed the central government, who acted to tighten land supply. The decision to deepen reforms and strengthen land administration was enacted in October 2004 by the central government, calling for the strictest land approval procedures and advocating the suspension of approval for new construction land (State Council, 2004). A half-year hiatus in approving construction land use ensued, resulting in a sharp slowdown of land conversion to construction land (Figure 4.9).

While effective in protecting arable land, the tightening of land conversion approval had a negative impact on market expectation in land supply. High transaction costs in obtaining new approval and higher land prices produced a sense of land shortage among developers, which was amplified by the media to cause alarm in the housing market (*People*, 2006a). For land already bought by developers, there was a locational mismatch between land ready to develop and housing market demand, and higher transaction costs in demolition and relocation of existing homes to make land ready for development (see Chapter 6 for issues in demolition and relocation). Furthermore, new land acquisition by developers stagnated from 2004

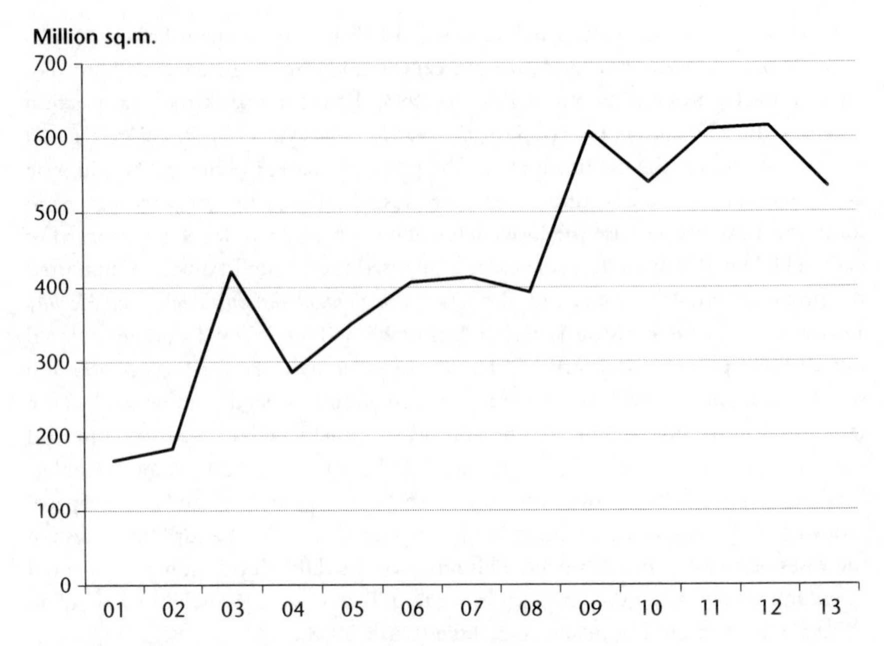

FIGURE 4.9 Approval of land conversion to construction land, 2001–13
Sources: MLR, various years.

(see Figure 3.3) and the growth of new housing starts in 2004 was only 9.34 per cent, an obvious slowdown from 13.71 per cent and 26.31 per cent in 2002 and 2003, respectively (Figure 4.7). The combination of the above developments created inflation expectations in the housing market, with the belief that housing prices would go up amid lower land supply and continuing population growth in urban areas.

Since 2004, land supply policy has become a tool in the macro control of the housing market. From 2005, land conversion has risen in order to increase land supply following the central government's effort to curb housing market price inflation. In 2009, land conversion approval reached a new high as both central and local governments encouraged the property sector to take the lead in generating economic growth (Figure 4.9).

4.6.3 Monetary policy changes

Third, interest rate adjustments and mortgage loan lending conditions were used by China's central bank and the banking regulator to affect the supply and demand of housing. From 2004 to 2007, the operation was to reduce demand and slow down housing price inflation. In 2004, the China Banking Regulatory Committee, the banking regulator, issued a *Guideline on Risks Management of Real Estate Loans by Commercial Banks*, to provide detailed specification on risk management of loans for land reserves (see Chapter 6 for more details), real estate development, individual housing mortgages and commercial property loans. The guidelines tightened lending to property development by raising the minimum equity ratio from 20 per cent to 35 per cent to qualify for development loans, which was higher than the 30 per cent required by the People's Bank in 2003. In the same year, the central bank adjusted the commercial bank lending base rate for loans, which had remained unchanged over 5 years, upwards from 5.76 per cent to 6.12 per cent, and reduced the interest rate discounts commercial banks were allowed to offer (Figure 4.10). From April 2006, the Bank raised the rate frequently and in September 2007 increased the rate to 7.83 per cent, the highest since December 1998. In December 2007, the People's Bank of China and the China Banking Regulatory Committee jointly issued the *Circular on Intensifying Management of Real Estate Lending* (PBoC, 2007), which raised the deposit requirement to 30 per cent for purchasing a first home and 40 per cent for purchasing a second home. Furthermore, the interest rate for loans for second home purchase was required to be 1.1 times the central bank's commercial bank lending base rate. The interest rate hike coupled with the new deposit requirements were widely considered effective in curbing demand for housing, which resulted in a fall in housing prices in early 2008 (Figure 4.3).

From September 2008, the central bank continued its interest rate adjustments to work with the central government's macro control, cutting the commercial bank

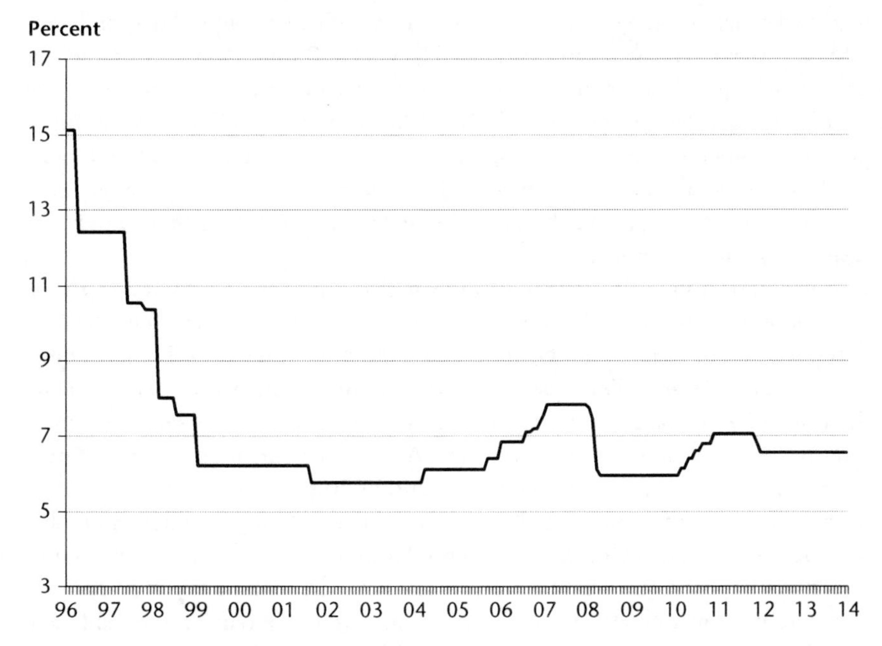

FIGURE 4.10 Commercial bank lending base rates set by the People's Bank of China, 1996–2014

Sources: NBSC, various years.

lending base rate three times in three months and revoking the 40 per cent deposit requirements for second home purchase to help the recovery of housing prices (see Figure 4.3). From October 2010 to July 2011, it raised the commercial bank lending base rates five times to curb rapid housing price inflation. However, the economic slowdown from 2011 prompted the central bank to stop raising interest rate and to cut rates twice in June and July 2012 and hold the interest rate at 6.55 per cent since then, even though housing prices continued to increase. As a result, the interest rate adjustment ceased being used as a macro control tool to fight housing price inflation.

4.6.4 Government intervention

Fourth, the government has directly intervened in the housing market since 2005 to slow down housing price inflation. Under pressure for having the highest housing price inflation, the Shanghai Municipal Government imposed a 5.5 per cent Business Tax on gains of selling housing bought less than one year ago, tabled a series of measures to increase low-end supply of housing, and monitored market behaviour to crack down on speculation and dishonest behaviour on 6 March 2005 (SMG, 2005). The central government promulgated the *Circular to Stabilise Prices of Housing in Earnest* on 26 March 2005 (State Council, 2005a), which outlined a number of policy guidelines. On 9 May 2005, the State Council circulated the *Circular on Opinions for Stabilising the Prices of Housing* (State Council, 2005b) to specify detailed measures. Commodity housing was divided into ordinary and non-ordinary to differentiate tax treatment. A 5.5 per cent Business Tax was imposed on gains of selling ordinary housing bought less than two years ago and on the gains of selling non-ordinary housing bought less than five years ago. The Circular urged local governments to increase the supply of low-end commodity housing and ECH for low- to middle-income households. Local governments were also required to increase market transparency by providing more market information to fight the expectation of rising prices.

Intervention in 2005 saw results in Shanghai but prices continued to rise in the rest of the country, including Beijing (Figure 4.8). On 24 May 2006, the central government enacted the *Opinions on Adjusting the Structure of Housing Supply and Stabilising Housing Prices* (State Council, 2006) to renew attempts to contain housing price inflation. The 5.5 per cent Business Tax was imposed on ordinary housing sold within five years of purchase. A unit size standard, popularly referred to as the 70/90 rule, was imposed by requiring 70 per cent of all new housing projects in terms of floor area to be units of less than 90 square metres. A number of follow-up measures were later enacted as follows: a one-year residency requirement was imposed on foreign buyers to buy housing; income tax at 20 per cent was resumed on capital gains of new housing; and a system of affordable and social housing regime was proposed in 2007 to provide housing to low- and lower-middle income households who had been priced out of the housing market.

The measures introduced by the central government, however, were not very effective. The key issue is that incentives at the collective choice level were against the macro control measures because local governments resented the fall in land sale revenues and the tax incomes caused by macro control. As the gains from price increases far outweighed the taxes, sellers were able to pass on the taxes to buyers with little resistance. Local governments, however, did not take action to tackle such openly practised questionable conduct. The 70/90 rule was bitterly resisted by developers and half-heartedly implemented by local governments. As discussed in Chapter 7, the response of developers to the 70/90 rule was perverse. Yet local governments turned a blind eye to evasive behaviour. Central government at the constitutional level was unable to tighten the rules to force local governments to rule out or penalise such behaviour.

The housing market contracted swiftly in 2008 and early 2009 due to the impact of macro control and the world financial crisis. The national average housing prices fell 2 per cent in 2008, the first fall since 1994 (Figure 4.3). To prevent the collapse of the housing market, both central and local governments tabled rescue packages in 2009 (see Section 2.3), which expanded credit on a substantial scale. Some measures contained in the 2006 macro control policy, such as the 70/90 rule, were not implemented by local governments. Lower prices and plenty of credit resulted in the release of the pent-up demand from 2007 and rapid price inflation returned, with national averaged price rising 26 per cent in 2009 (Figure 4.3).

From the end of 2009, macro control was resumed, with the coverage of the 5.5 per cent Business Tax extended from selling within two years to five years. In April 2010, the central government enacted the *Circular on Resolute Control of Rapid Housing Price Inflation in Some Cities* (State Council, 2010b), which imposed the toughest measures ever seen to increase supply and curb demand. Some of the key measures included that land sale prices should be strictly monitored to avoid rapid inflation, and 70 per cent of land supply should be for small- to medium-sized housing and affordable and social housing. The deposit requirement for second home was adjusted to 50 per cent, and mortgage applications for purchasing a third home were not to be approved.

Local governments at the collective choice level also put forward initiatives to help curb price inflation. A restriction on purchase according to residential status was first attempted by Beijing Municipal Government in May 2010 (BMG, 2011). The restriction, still effective at the time of writing, limits the right of purchase by registered households already owning one housing unit in Beijing to one additional unit. Non-registered households owning no housing units and with five years local tax contributions in Beijing were allowed to buy only one unit. In January 2011, the central government at the constitutional level endorsed the purchase restriction practised in Beijing, and encouraged voluntary adoption to other local governments. By 2012, 45 cities had adopted purchase restrictions, signalling an institutional change in housing market administration. Purchase restriction, though widely disputed by home buyers and developers, successfully lowered housing price inflation. In Beijing and Shanghai, both with purchase restrictions, averaged housing prices fell in 2011 (see Figure 4.8). Nationally, housing price inflation slowed down to 7.5 per cent and 6.5 per cent in 2011 and 2012, respectively.

4.6.5 New trends in 2013–14

In March 2013, a new government, headed by President Xi Jinping and Premier Li Keqiang, took office. It pursued a set of new policies to provide greater freedom to markets and impose less intervention, and no macro control measures were put forward. Housing prices responded by rapid inflation, in particular in the main cities previously affected by purchase restrictions, where pent-up housing demand was very strong. Figure 4.11 shows housing price inflation in four of the main cities, with price inflation over 20 per cent for Beijing, Shanghai and Guangzhou in 2013 alone. Yet housing price inflation in 2013 exhibited divergence. Some cities had rapid price inflation but others had milder inflation or even deflation. Table 4.2 shows housing prices in urban districts of the selected cities, which indicate a divergent growth pattern. Wenzhou, a city in Zhejiang Province, had deflation (see Box 9.2 for more details).

From the second quarter of 2014, a nationwide fall of housing prices is certain. Local governments started to support the housing market by lifting purchase restrictions and tabled other measures to encourage home purchase. More details on the fundamental change of housing market are provided in Chapter 14.

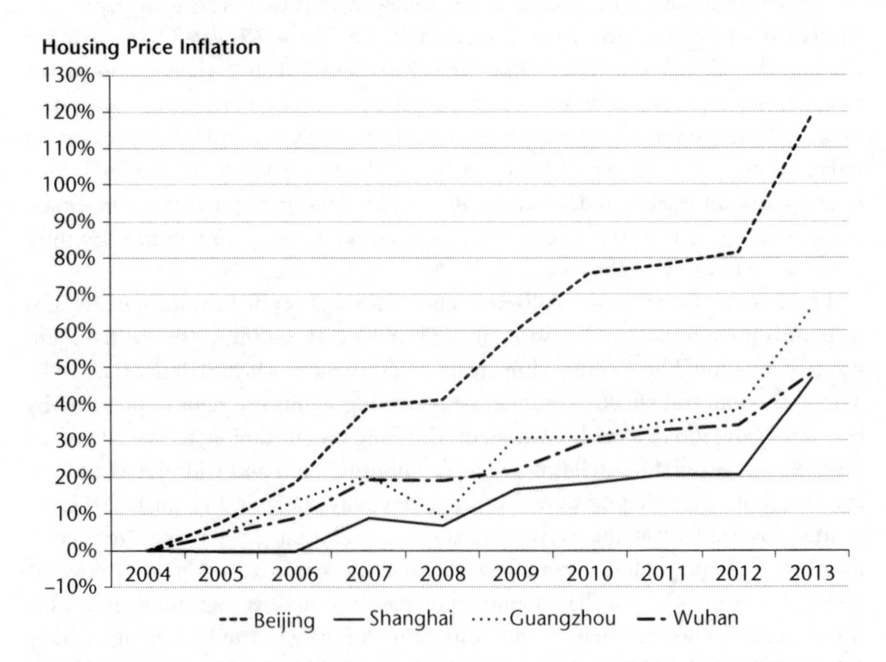

FIGURE 4.11 Housing price inflation in the four main cities in China by 70 Cities Index
Sources: NBSC, various years.

TABLE 4.2 Divergent housing price movements in selected main cities in China

	CN¥ per m² in GFA		*Growth*	
			Annual growth (%)	*Accumulative growth (%)*
	December 2007	*December 2013*		
Guangzhou	7,792	18,163	15	133
Shanghai	18,061	29,433	8	63
Beijing	12,931	42,503	22	229
Chengdu	6,486	8,913	5	37
Xi'an	3,507	6,796	12	94
Shenyang	4,585	8,529	11	86
	August 2009	*December 2013*		
Wuhan	5,011	9,112	15	82
	December 2011	*December 2013*		
Wulumuqi	6,111	7,486	11	23
Wenzhou	29,160	24,571	-8	-16

Source: Fangjia Net, accessed 15 February 2014.

BOX 4.2 CHARACTERISTICS OF THE CHINESE HOUSING MARKET

Qiang Chai

Land in China is publicly owned, with all urban land owned by the state and rural land owned by rural collectives, unless stipulated by law as state-owned. However, buildings are mainly privately owned. Therefore, a tenure consists of private ownership of the buildings and Land Use Rights over the land. LURs are limited in length, with 70 years for those of residential land and 40 years for commercial land. LURs for both residential and commercial land can be renewed. Residential LURs are renewed automatically and thus considered unlimited in length.

The land market in China is the market for LURs, which comprises the market for LUR sales and granted LUR transfers. The market for LUR sales is monopolised by local governments. At present, local governments are reliant on land sale receipts and taxes from housing. In the first quarter of 2014, land sale receipts were CN¥1.08 trillion, up 40.3 per cent from the first quarter of 2013, and total local fiscal revenues were CN¥1.95 trillion.

The Chinese housing market is referred to as a 'policy-based market' because the government is a participant and the regulator of the housing market. It is becoming bigger and bigger and its fluctuation has strong implications to financial security, social stability, and the healthy development of the national economy. Thus the government fosters and controls the development of the housing market. According to the circumstances of the housing market and the macro economy, both central and local governments intervene, either containing the growing market or rescuing the falling market. These interventions have the following objectives: stabilising the housing market; preventing overheating in the economy; or rescuing economic growth. Large-scale and concerted interventions were termed macro control, which includes policy measures in finance, taxation, land supply, affordable and social housing, and others. To some extent, the housing sector is an economic tool used by the government.

To prevent overheating in the economy and to rein in fixed asset investments, the central government in 2003 and 2004 adopted measures such as land supply control, tight credit, higher investment requirements, and stricter project approvals. The upshot was lower housing development investment and reduced new supply, which was one of the main reasons for rapid housing price inflation from 2006 to 2007. In 2008, the central government tabled a series of measures to encourage housing consumption, leading to accelerated growth of housing demand that resulted in rapid housing price inflation in 2009 and 2010.

The development of the Chinese housing market is closely related to macroeconomic policies, in particular, monetary policy that determines the credit available. For example, the money and credit supply from 2009 to 2013 was abundant, resulting in plenty of liquidity. Without alternative investment channels and policy guidance, this liquidity flowed to the housing market, generating huge demand for housing. Prices rose rapidly due to much slower growth in supply.

After decades of development, the Chinese housing market and its surrounding environment have undergone fundamental changes. The changes within the housing market are as follows. First, the relative and absolute levels of housing prices are rather high, with very limited short-term growth potential. The relative level is represented by the housing price to earnings ratio, and the absolute level is based on comparison of housing prices in China and in comparable developed countries. Second, housing shortages in urban areas have been alleviated and demand and supply of housing have been significantly altered. Although differences in housing conditions exist between families and between cities, average per capita housing space has reached 33 m^2 and the average ratio of housing units to households has exceeded 1:1. If housing with small property rights is taken into account, the average housing conditions are even better. Third, economic growth and local fiscal incomes are reliant on the housing sector. Housing demand and housing price inflation have been ahead of the current level of economic growth.

The following are the changes in the surrounding environment of the housing market. The first change occurs in the macro-economy: economic growth has been adjusted from high speed to medium high speed. The second change takes place in the financial sector: funds in commercial banks have been attracted to businesses offering higher yields. The lending market has no longer favoured lending to the housing sector. Thus the credit supply to the housing market declines through the market mechanism, with rising interest rates and longer approval time. Credit tightening lowers the purchasing power for housing and dampens housing demand. The third change happens in the investment market: new investment opportunities arise, offering yields from 5 per cent to 6 per cent, which is far higher than the 2 per cent to 3 per cent from letting housing. Housing investors anticipate limited potential for further price hikes. With low rental yields, housing investment becomes less attractive. As a result, housing as an investment asset becomes less attractive because of alternative higher yielding investment opportunities.

BOX 4.3 AN EVALUATION OF CHINA'S REAL ESTATE MARKET

Hongming Zhang

The development of the real estate market in China can be characterised as occurring in three phases. The first phase of development, from 1986 to 1992, was not smooth. The development of the real estate market was based on the socialist commodity economic theories, which provided an *ad hoc* yet incomplete justification of a market economy. The return of conservatism after the political turmoil in 1989 also slowed down the development. The second phase, from 1992 to 1998, was a development by leaps and bounds, following the establishment of a socialist market economy theory, proposed by Deng Xiaoping. The third phase, from 1998, has been a period of prosperity in the real estate market. The final phase of the housing reforms resulted in a fundamental change in the housing system by ending the welfare allocation of housing in cities and towns, thus clearing the institutional barriers for the development of the real estate market.

The Chinese real estate market can be evaluated from two aspects. First, the Chinese real estate market has achieved a record speed of development, higher than ever seen before in the world. Table 4.3 illustrates this rapid growth. The rapid growth of the real estate market resulted in the continued rise of output by the real estate industry, which accounted for 5.85 per cent of the GDP in 2013, up from 4.18 per cent in 2000. Improvement of housing conditions in urban areas was made possible by such rapid development. From 2000 to 2013, per capita annual increase in new housing space was between 1.38 m^2 and 1.82 m^2; equivalent to an additional room for every person.

TABLE 4.3 Changes in the Chinese real estate market between 2000 and 2013

	2000	*2013*	*Total growth (%)*	*Average annual growth (%)*
Investment in real estate development (CN¥ billion)	498.4	8,601.2	1,726	23.93
Completion of commodity real estate (million m²)	251.1	1,014.4	4,04	9.00
Sale volume of commodity real estate (CN¥ billion)	393.5	8,142.8	2,069	27.34
Sale of commodity housing (thousand units)	2,139.7	9,446.4*	441	10.76

Note: *data in 2012.
Source: NBSC (2013).

Second, housing prices in China grew rapidly, causing worries about housing bubbles. There is no long-term and consistent data on housing prices. From Table 4.3, completion of commodity real estate and sale of commodity housing grew annually at 9 per cent and 10.76 per cent, respectively, from 2000 to 2013. However, the sale volume of commodity real estate grew annually at 27.34 per cent. Housing prices rose significantly even if improvement in housing quality is taken into account. In fact, housing prices rose five to ten times in the last 13 years of the new century, with the fastest growth recorded in the east of the country.

Apart from housing needs in the market, the following factors were important in supporting the rapid housing price inflation: (1) property-led economic growth pursued by local governments; (2) abundant liquidity inside and outside the country; and (3) market expectation supported by (1) and (2).

By the end of 2013, housing price inflation seemed to peak and risks in the housing market loomed. In 2014, the real estate market in China experienced falls in prices and transaction volumes, with local governments busy rescuing the market. A big test for the Chinese government is whether they can tackle the dilemma in slowing economic growth and the falling housing market.

4.6.6 Development of housing services industries

Fifth, property services regulation and practitioner qualification certification made further progress. In 2003, the State Council promulgated the *Regulation on Realty Management* to replace the former national regulation set by the Ministry of Construction to regulate the relationship between owners and property management companies. The regulation was amended to conform to the Property Law in 2007 (State Council, 2007a). In the regulation, the residential district housing management committee was replaced by the Owners' Congress, which represents all owners in the residential district in appointing management companies for property management and manages the maintenance funds for the housing estate. A national regulation on the management of the maintenance funds, which are contributed by owners and managed by the Owners' Congress, was promulgated by the Ministry of Construction and the Ministry of Finance in 2007 to prepare for repairs of the common parts of multi-storey residential buildings and common facilities in residential districts. In addition, a professional qualification for property managers, the Certified Property Manager, was created by the regulation, with the first national examination in 2005. The China Property Management Institute, which had its first National Congress in Beijing in 2000, is the professional body to promote professionalisation and professional management in property management, including training and administration of the Certified Property Manager scheme.

In 2004, the China Institute of Real Estate Appraisers was commissioned by the Ministry of Civil Affairs to conduct the national administration of real estate agents. Renamed the China Institute of Real Estate Appraisers and Agents, the professional body certifies and registers Real Estate Agents nationally. Regulation of the real estate agency industry continued to be pushed by local initiatives, with Guangzhou enacting a local real estate agency regulation in 2003. In 2011, the *Administrative Measures for Real Estate Agents* was enacted by MOHURD, the State Development and Reform Commission and the Ministry of Human Resources and Social Security, as the national regulation for the real estate agency industry.

BOX 4.4 SHANGHAI'S ROLE IN CHINA'S REAL ESTATE MARKET

Xiaobo Dai

In 1949, China adopted a housing rationing system to provide public rental housing, also referred to as welfare housing, to urban residents at minimal standards through two channels: (1) local governments; and (2) employers. However, rents charged on such housing were so low that those public rental housing had to be heavily subsidised by local government and employers. This system of housing provision was difficult to sustain.

After the reforms and the opening-up campaign started in December 1978, housing and transportation became the main problems for Chinese cities and local governments all explored ways to solve the housing problem. In the mid-1980s, Shanghai began to learn from Hong Kong and Shenzhen to lease state-owned land and allow foreign investors to build commodity housing for sale to foreigners. Meanwhile, domestic developers were allowed to build commodity housing for sale to domestic buyers on allocative land. Thus two types of commodity housing were introduced to two groups of buyers with two prices and two title deeds.

In May 1991, Shanghai started its comprehensive housing reforms, which included five features. First, a Housing Provident Fund was set up based on the experience in Singapore as a compulsory saving scheme for housing. The Fund was to help the local government and residents to solve housing funding problems, and is now mainly for mortgage lending and pensions. Second, residents were encouraged to buy the public rental housing they occupied at one third of the market price, and to buy commodity housing in the market. Third, rents for public rental housing were raised 100 per cent to rationalise rents and to alter the price to rent ratio for public rental housing; a measure to encourage occupiers to buy rather than to rent. Fourth, residents who continued to rent were required to buy five-year low interest bonds, which were equivalent to a third of the housing they rented, to raise funds for further housing construction. Fifth, marketised property management was introduced to sold housing to break the link between sold public rental housing and their former management by local government or employers.

In 1998, Shanghai followed the national drive to end welfare housing allocation and embraced the full market allocation of housing. First, allocation of the public rental housing was cancelled, ending the era of welfare housing. Second, the government built low-price housing for low-income households and households living in housing poverty. Third, the land and housing markets were opened to both domestic and foreign developers for sale to domestic as well as foreign buyers.

The marketisation of housing in Shanghai has major implications for urbanisation, general economic growth, the development of many industries, housing conditions and the wealth of the urban residents. From 1990 to 2013, the number of urban residents in Shanghai increased 80 per cent to 24.2 million. Housing investment grew from one-eighth of total fixed asset investment to one-third. Total housing space in Shanghai grew from about 100 million m² to 650 million m², and households owning housing grew from 12 per cent of total households to 80 per cent.

Nationally, the real estate industry has made a huge contribution to economic growth. It directly and indirectly accounted for one-sixth of GDP and one-third of GDP growth in the country.

4.7 Discussion: ghost towns or vacant to be sold?

In 2013 the Chinese government took on the problems of high local government debts (see Section 2.2) and high corporate debts accumulated from the last five years by the stimulus package. The policy-induced deleveraging has led to more cautious lending by banks with implications for the property market. Furthermore, interviews in China in 2014 reveal that commercial banks no longer regard housing as the best collateral, they have become less willing to lend mortgage loans while oversupply of housing has become widespread in many cities. This shift in financial support has resulted in falling housing sales. An analysis of the first 11 months of 2014 indicated that new housing sales shrank with a 10.0 per cent fall in terms of floor area and a 9.7 per cent fall in terms of sales revenues; and housing new starts fell by 13.1 per cent (NBSC, 2014c). Such a reduction in sales is the precursor to a fall in prices, and evidence of falling housing prices has already emerged in Beijing, traditionally the housing market with the strongest demand. Such a fall gave ammunition to the pessimists who have been claiming housing has become oversupplied in many cities and towns and that more ghost towns will appear and a resultant sharp fall in housing prices will materialise. On the other hand, government officials and optimists have been arguing that only a temporarily slowdown will occur, because further urbanisation will take up the slack.

A key point in such debates is whether urban housing as a whole has been oversupplied in the absolute sense, i.e. there is more urban housing than there is urban population. Interviews in China in 2013 to 2014 revealed that the number of housing units in urban China was enough to house all the urban families. However, there is a lack of authoritative data on how much housing exists in urban China. The latest release of total urban housing floor area data by the Ministry of Construction was in 2006, which indicates that in 2005 total housing floor area in cities and towns was 10.77 billion m^2 (MoC, 2006b).

There are a number of ways to estimate the total urban housing floor area. One way is adding the 2005 total floor area to the total urban housing completions reported by the National Bureau of Statistics from 2006 to 2012. Total new completion is 5.87 billion m^2 suggesting that the total housing floor area by the end of 2012 was 16.64 billion m^2. The problems of using this method are that a certain amount of housing has been demolished since 2006 and some urban housing construction is not accounted for.

Another means of calculation is to use the housing floor area data estimate from the sixth population census, which indicates the total urban housing floor area was 17.9 billion m^2 and the number of units was 220 million in 2010. The number of units is roughly equal to the number of households in that year. However, among the 220 million units, only 72 per cent are self-contained (Wang, 2013). Residential property, such as dormitories in schools and factories, is likely to be included, so the real total urban housing floor area suitable for family occupation should be less than the estimate.

The third way of calculating the total housing floor area in urban China is to multiply the per capita housing space by the urban population. The former is 32.9 m^2 and the latter 711.82 million (NBSC, 2013). Then the total floor area suggested is 23.42 billion m^2. However, only the registered household population, which is approximately two-thirds of the urban population, is used to calculate the floor space per capita, so this method generates an overestimation. For example, the total housing floor area in Guangzhou urban districts was 217.69 million m^2 in 2012 and the population of registered households in the urban districts was

6.78 million (GBS, 2013). The calculated per capita housing floor area of 32.10 m^2 is in line with the figure quoted for per capita housing floor area. The same is true for Shanghai.

From the above estimation it can be concluded that urban housing in China has not yet reached the absolute oversupply. With natural population growth in the urban areas and migration in the hundreds of millions from the countryside, more housing is needed. Nevertheless, there is a regional mismatch between population and housing space. East China and all major cities throughout the country will benefit from migration, while small cities and towns in Central and West China will see slower population growth or even decline. It is true that some cities have reached absolute oversupply in housing space and low occupancy is real. In particular, some of these cities are not attractive to migrants so it will be difficult to digest the surplus housing space. Nevertheless, many of the dozens of so-called ghost towns are located in economically active areas so the problem is just poorly conceived housing projects. Such projects are vacant but will be occupied in the future. The speed of occupation depends on the speed of local job creation and economic growth. Yet such projects tend to depress local economic growth due to a contraction of the construction industry. Thus market discipline matters. The two decades to sort out oversupply in Beihai City are a reminder of the patience needed.

4.8 Conclusion

In less than 30 years, China has made huge strides in converting the system of welfare housing allocation to market allocation of housing. The housing reforms were initiated to ease the state burden on housing provision and to generate economic growth by encouraging market provision and mobilising non-state resources. With the establishment of private property rights in housing, the housing market became established and grew. Contextual factors such as foreign remittance and foreign housing investments were important in igniting the housing market and activating it after a period of depression in the late 1980s and early 1990s. With the help of an expanding market, the housing reforms could be pushed ahead to end the welfare housing allocation. Policy support to reduce transaction costs in housing transactions was instrumental for the housing market to enjoy rapid growth and become the main means of housing provision. However, concerns over housing price inflation by the central bank were rejected by the central government and the opportunity to act earlier was missed. Furthermore, land supply was tightened to cool down the supply side of the housing market without any measures to cool down the demand side. Rapid housing price inflation resulted in the use of monetary policy by the central bank and direct government intervention to arrest the runaway prices. New measures to increase supply, including the 70/90 rule, took effect to bring down housing prices. However, the international financial crisis slowed down economic growth and caused the housing market to contract. Rescue measures were overdone and housing price inflation resumed, leading to a new round of macro control. Purchase restrictions were imposed to achieve a temporary effect in slowing down the price hikes, but the change of government and economic policy released the pent-up demand, leading to a new round of price inflation.

The slowdown of the housing market in 2014 caused concerns about a possible sharp fall of housing prices due to oversupply, leading to the lifting of purchase restrictions and the return of some local support for the housing market. Analysing data released by the various channels of the government reveals that there is no absolute housing oversupply on a national scale and new housing supply is needed for China to continue its urbanisation campaign and improve the housing conditions of urban residents. However, a mismatch of supply and demand exists and some cities, particularly those in Central and Western China, have excessive oversupply. Apart from generating further economic growth to increase jobs to attract migrants, these cities need to spend years or decades to digest the existing supply.

Notes

1 There was virtually no black market in housing at the time in those cities.
2 The new thinking behind this approach was that the pay packages to state employees after 1985 gradually included housing benefits as wage levels grew substantially. Thus the government could raise rents without paying compensation.
3 Those who could not afford it will be able to remain in the public housing. Due to family support and the prospect of capital gain, very few people did not take up the offer.
4 In 1992 the exchange rate between CN¥ and the US $ was 5.5149 to 1.
5 Ordinary housing means medium-sized housing. However, it was not properly defined until 2005. The definition of ordinary housing is discussed in detail in Section 7.3.

5

THE COMMERCIAL PROPERTY MARKET

This chapter explores the rise of the commercial real estate market and its expansion. It starts by examining property rights in commercial real estate. The concept of tiers in city classification, widely used to classify the potential of cities for property investment, is introduced and a modified version of a popular classification system is provided. The chapter then discusses the emergence of the commercial real estate market in the 1980s and the expansion in the 1990s, and illustrates the problem of failed projects in the 1990s. It goes on to explore the development of the office, retail, and industrial real estate markets after China's accession to the World Trade Organisation in 1999 and points out that a new round of oversupply looms large in the 2010s. The chapter argues that oversupply has its internal logic in China but warns of the resultant high risks. Market data on stock, rental trends and vacancy rates are provided to illustrate the office market cycles that epitomise the development of the Chinese commercial real estate markets.

5.1 Property rights in commercial property

The property rights system in commercial property in China consists of building ownership rights and LURs on state-owned land. Building ownership is permanent but LURs have maximum terms according to the uses determined at the time the LUR is sold by the state. Maximum terms range from 40 years for commercial use and 50 years for industrial, mixed uses and other uses. According to Decree 55, LURs for commercial uses can be extended upon application by LUR holders at least one year before the expiry of the LURs, if the sites involved are not required for public use and premiums for new terms are paid (see Chapter 3 for more details of the legal specifications for LURs).

There have been calls from legal professionals for greater clarity on LUR extension, in particular how much new premium should be paid and how many years in advance the LUR owners can apply. However, interviews with real estate administration officials and professionals in 2000, 2005 and 2010 by the author indicated that this will not put off investors. A few practitioners pointed out that LURs for commercial buildings usually carry the longer 50-year terms. This is because local governments, as LUR sellers, benefit from higher land prices and show greater acceptance when selling LURs for 50 years on mixed uses. Likewise, developers, as LUR buyers, maximise their profits when obtaining 50 years for the LURs. Because the LUR system was created in 1990, most LURs of commercial buildings still have more than 30 years remaining so there is no problem with investors.

Although usually selling LURs for the maximum length of time, local governments do have the discretion to sell LURs for shorter lengths of time. To make better use of its limited land stock for industrial use, the Shanghai Municipal Government in March 2014 promulgated *Opinions on Further Increasing the Intensity of Land Use in Shanghai*, which stipulates the length of time for LURs for industrial land to be 20–50 years, with 20 years being the norm. Extension will be decided by an assessment to determine if the industrial processes built on site are viable for an extended period. If not, the municipal government will repossess the site without compensation.

With the exception of industrial real estate, commercial real estate in China is often built in the form of large buildings for multiple users, such as offices and shopping centres. The units of these buildings are often sold individually like apartments in residential blocks. Article 70 of Property Law stipulates that an owner of units in buildings of multiple ownership should have ownership over the exclusive parts within the buildings, and shall have common ownership and the right of common management over the common parts. The management of the common parts is normally conducted by an Owners' Congress through appointing a management company. This arrangement is similar to the strata title schemes first devised in Australia, which are composed of individual lots and common property, such as lobby, common stairwells, driveways, roofs, gardens and so on. Accordingly, strata titles are used by real estate professionals in China to refer to the arrangement of multiple ownership.

5.2 The tier system of Chinese cities

China has a tradition of dividing up its vast number of cities into four or five tiers. Classifying cities is an effective way to slice this big country into specific local categories for the sake of administration and economic planning by the government. The same approach can also be used to understand consumer behaviour, income level, and local trends by businesses; to enable firms to tune their strategies to suit national and local conditions. An earlier classification, backed by the Urban Planning Law 1989 (now repealed), divided cities into four tiers: (1) small cities with a population less than 200,000; (2) medium-sized cities with a population between 200,000 and 500,000; (3) large cities with a population between 500,000 and 1,000,000; and (4) super-large cities with a population over 1,000,000. This system became impractical due to the rapid urban population growth from 1979 and fell into disuse. This classification was not updated by the subsequent Urban Planning Law and became obsolete. In 2010, a new system was developed by the Commission of Economic Development of Small and Medium-sized Cities under the China Academy of Social Sciences to classify cities into five tiers: (1) small cities with a population less than 500,000; (2) medium-sized cities with a population between 500,000 and 1,000,000; (3) large cities with a population between 1,000,000 and 3,000,000; (4) super-large cities with a population between 3,000,000 and 10,000,000; and (5) megacities with a population over 10,000,000 (CEDSMC, 2010).

However, many economists, consultants and businesses use different methods for classification, including parameters such as GDP, population and retail sales, etc. As China's extraordinary pace of urban development has led to massive changes across China, city classifications are often rearranged to reflect the changes.

This book adapts the classification system by Jones Lang LaSalle (JLL, 2012a) that analysed economic strengths and property market development of cities and divided them into four tiers. The adapted system adds a fifth tier to capture many other cities not included in the JLL classification (Table 5.1). It agrees with the JLL system on tier 1 cities but regards the other tiers of cities as slightly more mature and developed than the JLL system. Such a difference could be due to the fact that the system is based on more up-to-date data. The author assessed the issue of tiers using data from fieldwork that covered 13 major cities and three medium-sized ones in South, Southwest, East, Central, North, Northeast and Northwest of the country in August and October of 2013, and June 2014.

TABLE 5.1 The tier system of Chinese cities

Tiers (in this book)	*Tiers (JLL)*	*Cities*
1 (Core)	1 (Core)	Beijing, Shanghai, Guangzhou, Shenzhen
1.5 (Mature)	1.5 (Transitional)	Chengdu, Chongqing, Shenyang, Hangzhou, Tianjin, Dalian, Wuhan, Suzhou, Nanjing
2 (Transitional)	2 (Growth)	Qingdao, Xiamen, Xi'an, Ningbo, Changsha, Hefei, Zhengzhou, Wuxi, Dongguan, Jinan
3 (Growth)	3 (Emerging)	Fuzhou, Kunming, Changchun, Harbin, Foshan, Shijiazhuang, Nanning, Changzhou, Nanchang, Hohhot, Wenzhou, Yantai, Nantong
3 (Emerging)	3 (Early adopter)	Zhuhai, Guiyang, Taiyuan, Urumqi, Shaoxing, Zhongshan, Jiaxing, Weifang, Tangshan, Xuzhou, Jinhua, Quanzhou, Luoyang, Lanzhou, Haikou, Jilin, Xiangyang, Shantou and cities in the Pearl River Delta and Yangtze River Delta not named in this table
4 (Early adopter)		All other provincial capital cities, including Yinchuan, Xining, Lhasa, most prefecture-level cities not named in this table and many county-level cities not named in this table

Note: adapted from JLL (2012a).

A brief description of the characteristics of the tier system is as follows (Table 5.1). Tier 1 cities are national and international business hubs, which have well-developed international investment grade stock, and the most liquid and transparent commercial real estate markets in the country. They serve as the benchmark for cities in other tiers. Tier 1.5 cities boast large and mature commercial real estate markets that are slightly riskier than those of tier 1 cities due to smaller international investment grade stock, slightly lower liquidity and less international and national tenants. Tier 2 cities have maturing markets and robust infrastructure but very small international investment grade stock, lower liquidity and mainly local tenants. Tier 3 cities have expanding markets and rapidly improving infrastructure but with little or no international investment grade stock, rather low liquidity, and mainly local tenants. The difference between tier 3 (growth) and tier 3 (emerging) is in market size and the quality of professional services on commercial real estate. Tier 4 cities have some market transactions in commercial real estate but need further commercial real estate development and infrastructure improvement.

5.3 1980–91: initiation in BSGS

Before 1980, no commercial real estate market existed in China's second world of real estate (see Section 1.1). Purpose-built office buildings were owner-occupied by government agencies and large state-owned firms. The small and informal private sector had no demand for offices. Retailers and manufacturers requiring higher-specification premises were all owner-occupiers. Furthermore, commercial real estate was administratively allocated by the government.

In the early 1980s, the commercial real estate market was created in Shenzhen when foreign investors, mainly from Hong Kong, needed reasonably specified space as offices and factories, but did not want to commit capital to become owner-occupiers. Following Hong Kong's commercial real estate market model, state-owned firms in Shenzhen built modern office buildings in Luohu to be as close as possible to the train station at the border between Shenzhen and Hong Kong, creating the first office market in China. Retail space was available in the new buildings for leasing. Further away from urban Shenzhen, Hong Kong investors leased hundreds of industrial buildings for manufacturing purposes.

Since its inception in Shenzhen, the commercial real estate market has undergone three stages of development (Table 5.2). The first stage covers the first market cycle in offices in Beijing, Shanghai, Guangzhou and Shenzhen (BSGS). To provide modern business accommodation to foreign business people, BSGS embarked on the development of modern office buildings with both foreign and domestic funding in the 1980s, creating the first batch of foreign investments in Chinese real estate. The first purpose-built office building in Shenzhen was completed in 1982, in Beijing in 1985[1] and in Shanghai in 1985,[2] providing alternative office accommodation to adapted starred hotel rooms in those cities. The completion of the 53-storey Shenzhen International Trade Tower in 1984 was sensational, setting a record for construction in terms of speed (three days per storey) and in height. In Beijing, construction of the

TABLE 5.2 The three stages in China's office market development

Stages	*Main developments*	*Time scale*
Stage 1: Initiation	Office markets emerged in Shenzhen, Beijing, Shanghai and Guangzhou and an office development cycle occurred. An industrial real estate market emerged in Shenzhen and the Pearl River Delta.	1981–1990
Stage 2: expansion	Office market had major expansion in BSGS and emerged in many other major cities. Retail and leisure real estate markets emerged and expanded in BSGC and other major cities. Industrial real estate market expanded to Yangtze River Delta, but controlled by local governments.	1991–1998
Stage 3: maturing	Office, retail and leisure real estate development boom took place in all main cities; Commercial real estate in market matures in Beijing and Shanghai; Industrial real estate market emerged due to land shortage and the rise of modern logistics industry	1999 onwards

China World Trade Centre started on 1 September 1985 by a state-owned firm and a Malaysian investor. Construction of the 100,000 m^2 Guangzhou World Trade Centre began in 1986. In 1987 and 1988, limited supply of quality office space resulted in near zero vacancy rates. In 1989, however, economic slowdown hit, when the projects started in the boom time entered the market, causing high vacancy rates in Beijing (Figure 5.1), Guangzhou, Shanghai and Shenzhen.

During the period from 1981 to 1990, the retail real estate market was still in its infancy. Letting of prime shop space first appeared in Shenzhen when foreign retailers opened outlets. In the rest of the country, however, leasing only occurred for tertiary shops for retailers from the non-state sector. Prime and most of the secondary shops were all owner-occupied by state-owned retailers.

The industrial letting market developed during this period in Shenzhen and neighbouring cities in the Pearl River Delta. As local governments drove down rents to attract foreign manufacturing investment, the industrial letting market was dominated by policy-based low rents.

It is worth noting that the emergence of the commercial real estate market was largely driven by the demand of foreign occupiers and investors at this stage. Both the central government at the constitutional level and local governments at the collective choice level, played a passive role in this process. The lack of policy attention sowed the seeds of chaos for the future development of the commercial real estate market.

5.4 1991–98: expansion and downturn

The second stage of commercial real estate development witnessed the rapid expansion of the office market, the emergence and significant development of a retail and

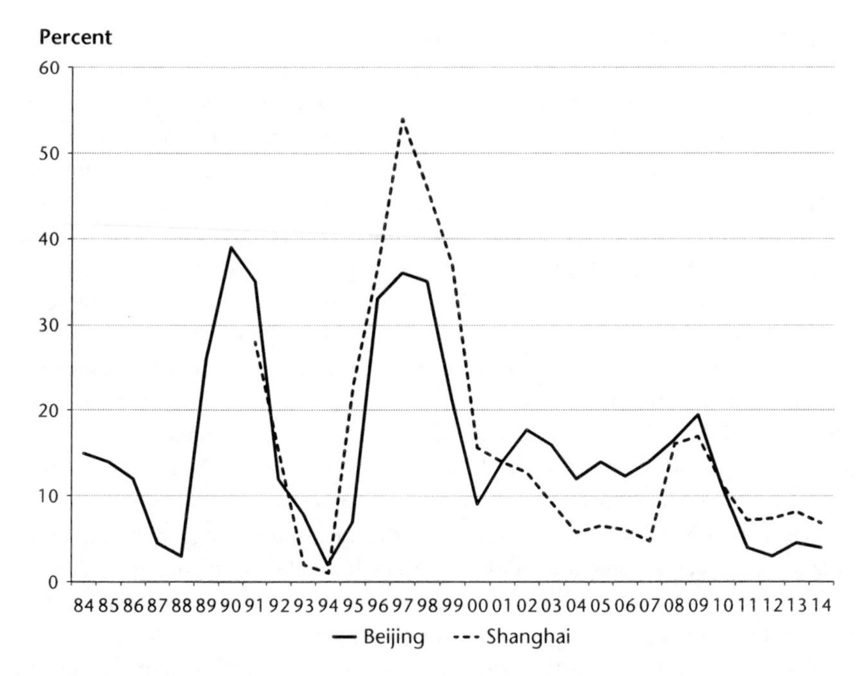

FIGURE 5.1 Prime office vacancy rates in Beijing and Shanghai, 1984–2014
Sources: adapted from Savills (2002a; 2002b; 2008; 2014a; 2014b).

leisure real estate market, and expansion of an industrial real estate market. Important changes in the office, retail, and industrial markets are discussed below, followed by a summary of failed projects.

5.4.1 Office markets

Implementation of strategic urban development targets by major cities from 1992 brought unprecedented levels of office development and generated an office development boom from 1991 to 1995. Inflow of foreign direct investment (see Figure 2.6) generated increased demand for office space from 1991. To implement their ambitious urban development plan to become business centres, major cities led by BSGS initiated a large number of infrastructure and office projects. BSGS also tabled plans to build high standard central business districts (CBD) in their cities. It is worth noting that the local governments obtained central government approval for such ambitious development. Unlike its attention to housing, the constitutional level did not produce policy guidance on commercial real estate, for example, planning policy guidance on commercial and industrial developments and city centre developments, leaving local governments at the collective choice level to set their own plans and rules for market operation and delivery.

Nevertheless, delivery of such a large number of expensive projects was beyond the capacity of BSGS. Outside help became indispensable. In 1992, Beijing sent a delegation to Hong Kong to attract investment into urban regeneration projects, and signed contracts with many Hong Kong developers and investors such as Cheung Kong, New World Development and Sun Hung Kai Properties to develop office, retail and residential projects. Shanghai designated an area the Pudong Central Business District and obtained central government support to require all ministries and provinces to construct a new office building in the largely low-density residential and rural district. Such assistance, however, is not a modern-day example of 'robbing the poor to pay the rich'. Rather it was a form of payback to Shanghai whose urban development was almost frozen after 1949 because most of the surplus produced in the city was used to fund urban development elsewhere. Many major Hong Kong developers were also invited to join the massive rebuilding of Shanghai. Guangzhou also designated a large tract of farmland as a site at Tianhe District for the Pearl River New City CBD. Shenzhen expanded its urban area and built a second business district as Luohu, the original CBD, was overcrowded. Other cities such as Chengdu, Chongqing, and Wuhan also signed contracts with foreign investors to participate in urban regeneration projects. With rental values rising rapidly and office vacancy rates at an all-time low in Beijing, Shanghai and Guangzhou, many domestic development firms were established to participate in the unprecedented urban regeneration, turning these cities into giant construction sites.

There were issues associated with the office market development during this period. The scale of development for all major cities was simply too large. For example, occupiers of the newly developed modern offices were mostly foreign firms who could afford high rents. Without many local tenants, absorption of the huge volume of space being developed was impossible. This was particularly true for tier 2 cities like Chengdu, Chongqing and Wuhan where foreign investors were few in number. Starting at the same time, these new projects hit the market at more or less the same time. Furthermore, the number of projects outweighed the number of capable developers in Hong Kong and domestically. Many projects were given to developers with limited expertise and shaky financial standing. In particular, some projects were simply poorly conceived. For example, a 54-storey office building project was approved on a site in Guangzhou's Haizhu District, which was remote from the city's central business areas. Construction started in 1994 but was then suspended in 1996 for a decade. Box 5.1 tells the story of one of Guangzhou's flagship projects that took over one and a half decades to complete.

In 1995, rental values started to decline in Beijing and Shanghai when the first round of new projects was completed and entered the market. More projects were completed in 1996 and 1997 when the economy slowed down significantly. The East Asia financial crisis hit the market hard in the second half of 1997 by wiping out the demand of foreign occupiers. By 1998, vacancy rates were 35–55 per cent in Beijing and Shanghai office markets (see Figure 5.1) and rents were depressed in Beijing, Guangzhou and Shanghai (Figures 5.3 and 5.4). Many office projects throughout the country slowed down or were suspended. Many failed, with semi-completed buildings becoming eyesores. In Guangzhou, some office development projects in the Pearl River New City CBD were allowed to be converted into residential properties

BOX 5.1 ODYSSEY OF CHINA'S LARGEST FAILED REAL ESTATE PROJECT

The fate of Zhongcheng Square (now renamed Sinopec Tower), China's largest failed real estate project, epitomises the problems in office development in the 1990s when local governments pushed hard for property-led urban economic growth.

With a state-of-the-art design, the office and retail complex is located in Guangzhou's CBD, only 200 metres from CITIC Plaza, the country's tallest building then (now the city's second tallest office building and the twentieth tallest building in the world). With a GFA of 232,766 m^2, Zhongcheng Square offers a 38,809 m^2 shopping mall and two 50-storey office tower blocks with 3.45 ceiling height and 2,000 m^2 floorplate (Figure 5.2). Furthermore, it has an 18-metre-high lobby, 56 lifts and 600 parking spaces in its underground car park.

Construction started in 1992. The developer, Guangzhou Pengcheng Real Estate Company Ltd, was a joint venture between Guangzhou's best developer at the time and a Hong Kong investor. A significant part of the building was pre-sold in 1993 and fetched HK$28,000 per m^2 for office space. In 1995, the partnership broke up because of major disagreement and the Guangzhou

FIGURE 5.2 Zhongcheng Square under construction in Guangzhou in December 2006

Source: photo taken by Albert Cao (2006).

developer walked away. Construction was slow but the main structure was completed in 1997. In 1998, construction was suspended because the property could not be let after the Southeast Asia financial crisis in 1997, leaving 104 creditors suing Pengcheng for compensation.

To prevent the project becoming an eyesore, the municipal government orchestrated a fundraising activity to install glass walls in 2001 for the 9th National Games, which was held in the sports stadium on the other side of the street. However, only 80 per cent of the glass wall project was completed because the contractor had not been fully paid and became a big creditor of the building complex.

Pengcheng finally sold the project on 22 November 2002 when the Supreme People's Court approved a deal between Pengcheng and a consortium formed by Beijing Jingmao and Guangzhou Junpeng. After the first instalment of RMB350 million was paid in February 2004, work resumed in May of that year. Nevertheless, Junpeng was soon found to have generated RMB300 million fraudulently. Work was suspended again in October 2004. Junpeng's chairman ran away and Beijing Jingmao became the sole owner by order of the Guangzhou Intermediate People's Court.

In December 2005, Shenzhen Jiezhaoye bought Beijing Jingmao and renamed the company Guangzhou Jingmao, which then sold one office tower to China Petrochemical Corporation (Sinopec), one of the largest firms in the world, for CN¥1 billion. Work resumed and the project was finished in March 2008 and occupied afterwards. All the debts were settled in January 2012, 18 years after the first presale. In the same year, the deputy head of the Supreme People's Court was deposed because he had taken bribes in handling the sale of Zhongcheng Square.

A number of lessons can be learnt from this extraordinary development. First, Guangzhou allowed too many large-scale property developments to proceed during 1992, without realistic checks on demand, the capacity of the local construction industry and the track record of the developers. Second, corruption stood in the way of the court's handling of this case, causing further complication and delay. Although two judges were brought to justice, there were still many unanswered questions. Third, the practice of presale, currently the common practice in China, should be strictly reviewed and modified to discourage excessive risk-taking by developers and to minimise the risks to buyers.

so that land could be sold to pay for the underground railway under construction, and thereby this undermined the future capacity of the CBD project.

5.4.2 Retail markets

The retail leasing and investment market emerged at this stage. From 1992 to 1999, retail development schemes in main cities and a large number of housing and office

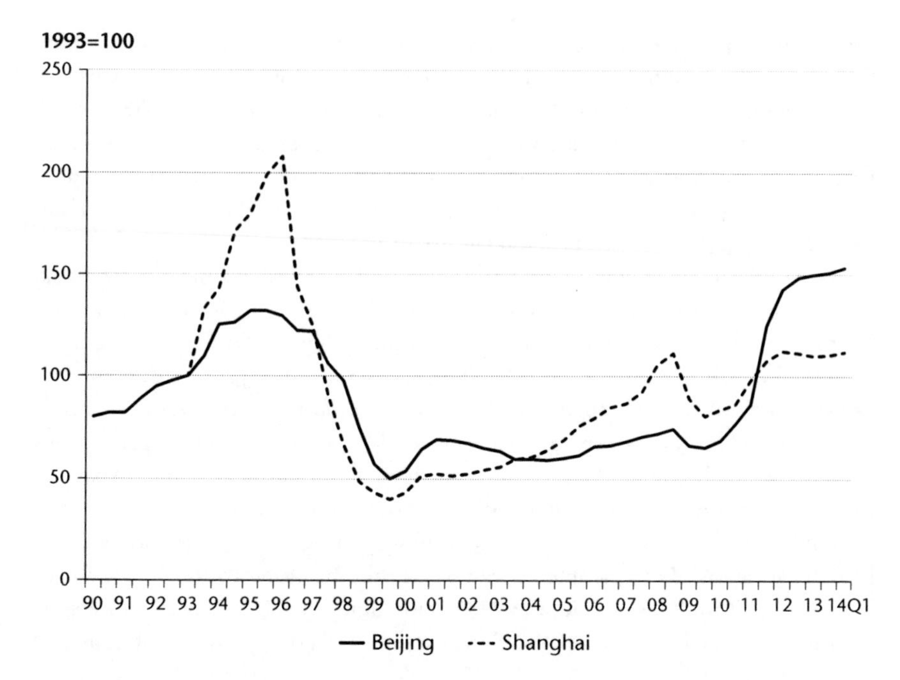

FIGURE 5.3 Grade A office rental indexes in Beijing and Shanghai
Sources: adapted from CBRE rental indexes (CBRE, 2005; 2014a).

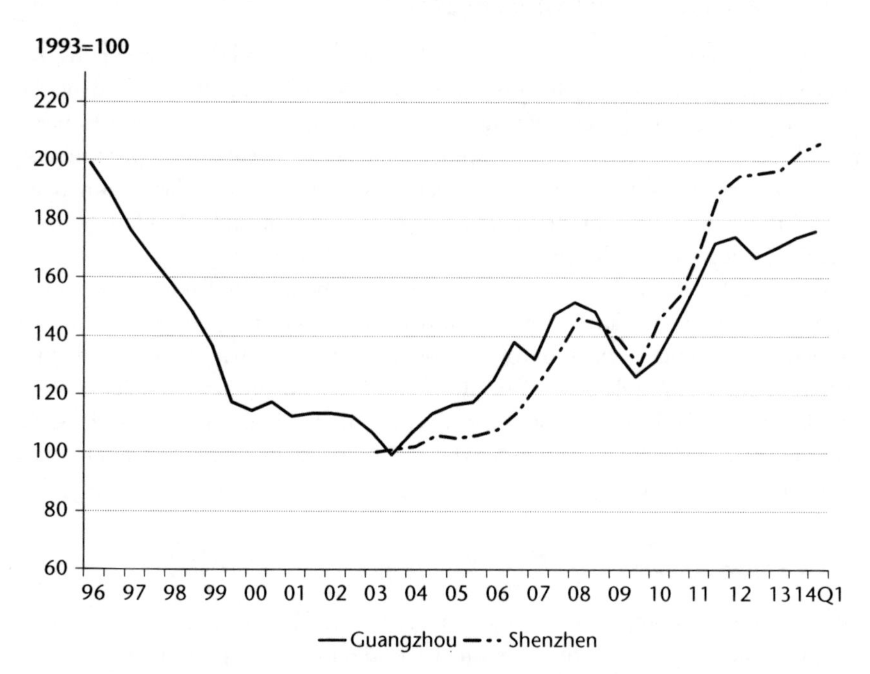

FIGURE 5.4 Grade A office rental indexes in Guangzhou and Shenzhen, 1996–2014 (Q1)
Sources: adapted from CBRE rental indexes (CBRE, 2005; 2014a).

development projects provided a large amount of prime and secondary space throughout the country. Shanghai regenerated its four major shopping streets and one shopping centre, and built three new shopping centres, providing 2,400,000 m^2 of new and renovated shopping space.[3] Guangzhou regenerated its two major shopping streets and built several shopping centres. For example, two major retail projects, Liwan Square and Teemall, were completed in Guangzhou in 1996, adding 140,000 m^2 and 100,000 m^2 prime retail space.

Retail sales grew rapidly during this stage and many newcomers entered the retail market. In particular, foreign retailers, mainly from Hong Kong, Taiwan and South-East Asia, set up their stores in the prime shopping districts. In Beijing, a 22,000 m^2 department store built in 1992 became China's first joint venture in retail in 1993, and was renamed the Lufthansa Centre. Compared to the inefficient state-owned owner-occupiers, foreign retailers were able to extract much bigger margins and thus pay higher rents. Many state-owned retailers in prime locations expanded their business and let some of their space to private and foreign retailers.

Similar to office development, many retail projects were too large and in unsatisfactory locations. For example, Liwan Square in Guangzhou was too big for its location and it took years to let the 2,000 shop units. Grandview Mall and neighbouring Teemall failed to be completed because of poor pre-commitment and financing problems.

5.4.3 The industrial real estate market

The industrial real estate market was marked by oversupply and below-cost rents and prices by local governments who took a holistic view on the impact of inward investment on city image, jobs, and tax revenues. To be competitive, some local governments provided two to five years of tax exemption and three more years of 50 per cent discounts on corporation taxes for projects over ten years in duration, plus lengthy rent-free periods for industrial property. The designation of thousands of industrial parks (development zones) resulted in the oversupply of industrial land in many localities.

On the other hand, the tightening up of national policies on land administration to preserve cultivated land eventually had its impact on the supply of industrial property. The implementation of the open market mechanism on land sales increased the costs of land for non-industrial uses and prompted developers to purchase industrial land in expectation of converting it into other uses, legally or illegally. As a result, interest in industrial land increased, bidding up the prices of industrial land and industrial property.

5.4.4 Failed projects

A large number of projects failed due to three factors: (1) poor judgement by inexperienced developers and investors on the feasibility of the projects; (2) massive oversupply resulting from the building boom; and (3) the slippage of demand caused by the East Asian financial crisis in 1997. Failed projects were found in all the major cities, including BSGS. A total of 182 failed projects was identified by the municipal government in Guangzhou, one-third of which were still dormant by the end of 2005 (Qianlong, 2005). In Shanghai, the number of failed projects reached 624 by the end of 2002. All of those failed projects in Shanghai were started before 1997; they froze about CN¥100 billion invested in those projects (Xinhua, 2003b).

The institutional arrangements for urban development at the time were responsible for the emergence of large numbers of failed projects in many major cities from 1997. As there was no policy guidance from the constitutional level, local governments at the collective choice level were allowed to have full discretion on urban renewal and the development of commercial real estate. Thus there were no checks and balances on local government decisions on the large amount of capital intensive projects. Local governments in fact knew what the consequences would be if all the development projects went ahead. The author visited the Pearl River New City CBD showroom with some British real estate professionals in 1997 and was briefed on the number and scale of the building projects. From the briefing and the exchange of views, the author found the officials were clear that the scale of developments in the CBD, which was close to the total prime office floor space in London, was very risky. Interviewees explained that selling the sites at the CBD was important in order to obtain funds to construct the underground railway and was crucial for the rapid urban development that was required by the top local officials to improve their performance in governance. What would happen next, however, was not the concern of the top officials, who would either have finished their terms of office by then or been promoted.

5.5 1999 Onwards: further expansion and maturing

The third stage of the commercial real estate market development began after the landmark agreements between China and the USA and between China and the EU on China's accession to the World Trade Organisation in 1999. News of the forthcoming conclusion of this 18-year-long negotiation strengthened market confidence that had hitherto been depressed by the East Asian financial crisis. The optimism on future exports, FDI and economic growth encouraged multinational companies (MNCs) to expand office occupation, resulting in rapidly falling office vacancy rates and rising rents in both Beijing and Shanghai (Figures 5.1 and 5.3). Suspicion of the development model in the 1990s was cleared and ambitious building programmes were then vindicated. As a result, many cities were keen to imitate the success of Beijing and Shanghai and become international metropolises. A new round of urban expansion and construction started alongside the rapid expansion of the housing market following the completion of the housing reform (see Section 4.5).

5.5.1 Office markets

The lack of policy guidance, for example, in planning, on commercial real estate development from the constitutional level enabled local governments to set unrealistic targets and mobilise resources for these. In the 1990s, Beijing and Shanghai's rise in real estate infrastructure, i.e. in offices and other business facilities, had led over 50 cities in 1995 and 75 cities in 1996 to set urban development targets to become international metropolises (Yang, 2006). Following the recovery of the office markets in Beijing and Shanghai, 182 cities in China set urban development targets to become international metropolises by 2003, which was regarded as unrealistic by the central government (*Dalian Evening Post*, 2003). By 2013, the number of cities targeting international metropolises exceeded 200 (Zhao *et al.*, 2013). Under the banner of building an international metropolis, ambitious urban development plans to build CBDs and state-of-the-art offices was born. A new office development boom has swept through China's large cities since 2000, which could not be stopped by the central government. Section 6.3 will explore this planning failure, which is the outcome of a lack of constitutional-level guidance on commercial real estate development.

From 1999 to 2014, the office market enjoyed rapid growth: fully established in BSGS; maturing in tier 2 cities; and expanding in tier 3 cities. The expansion of the office market has partly been the upshot of changes in demand and supply of offices in Chinese cities. On the one hand, the growth of financial and business services requires more office space and Chinese companies are willing and able to pay for more efficient office accommodation. The expansion of the national operations of many companies generated demand for quality office space in many nontraditional business hubs. Furthermore, the purchase of office space for owner occupation and investment was active (Figure 5.5). On the other hand, the success of BSGS in becoming national business hubs has influenced thought and practices in urban development in China. A large and modern office stock has been considered by local governments at the collective choice level as a key infrastructure for cities in attracting domestic and international companies to invest and set up operations such as regional and national headquarters, thereby enabling cities to become regional, national and international business hubs. Developing CBDs has been high on the agenda of provincial capitals and large cities to define their cities' roles and to stand out in inter-city competition. As a result, CBD development in Chinese cities has been driving office supply in China. Table 5.3 provides data on Grade A office stock in China's major business hubs. Nationally the sales volume of new office space is illustrated by Figure 5.5.

Throughout this stage, the rapid economic growth of the national economy (Figure 2.1) had generated demand for office space in many cities. The international financial crisis hit BSGS hard in raising vacancy rates (Figure 5.1) and caused falling rents in 2008 (Figures 5.3 and 5.4). Nevertheless, the rescue package of the central government and local stimulus packages quickly arrested the downturn of the office markets, strengthening optimism in both local governments and the market, leading to further expansion of office supply in major cities, in particular in cities with a shorter history and experience of office markets. Fieldwork in 2013 found massive urban redevelopment and expansion in Kunming and Nanning, which included large-scale office development. Nanning,[4] the capital city of Guangxi Province and

TABLE 5.3 Grade A office stock in China's major business hubs in 2013/2014

Cities	*Status*	*Grade A office stock (million m²)*
Beijing	National business hub	9.34
Shanghai	National business hub	5.40
Guangzhou	Sub-national business hub	3.29
Shenzhen	Sub-national business hub	2.94
Hangzhou	Provincial business hub	2.00
Chengdu	Provincial business hub	1.78
Qingdao	Sub-provincial business hub	1.39
Dalian	Sub-provincial business hub	1.10
Chongqing	Provincial business hub	0.83
Shenyang	Provincial business hub	0.67
Xian	Provincial business hub	0.56
Tianjin	Provincial business hub	0.52
Hong Kong	International business hub	7.40

Sources: DTZ (for Guangzhou, Wuhan and Xi'an); Savills (for all other cities).

a tier 3 city, has been selected as the city to host the annual China-ASEAN[5] conferences and boasts a CBD in its Wuxiang (literally meaning five elephants) New District (Figure 5.6). Nevertheless, the city is building a brand new CBD in yet another new district, which was still mostly undeveloped. While Nanning can argue

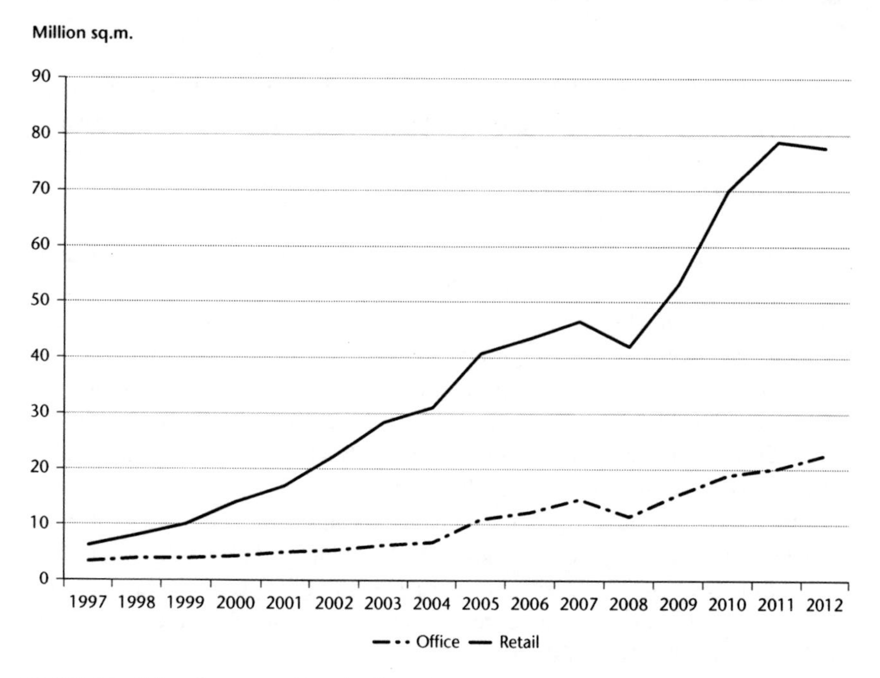

FIGURE 5.5 Sale of new retail and office space
Source: NBSC (2013).

FIGURE 5.6 Wuxiang CBD in Nanning, Guangxi Province
Source: photo by Albert Cao (2013).

for more office development due to its strategic role in China-ASEAN connection, Kunming,[6] the capital city of Yunnan Province, has little to justify its major expansion of office space. Further afield, Urumqi, the most inland city of the world, has also developed its own CBD, with a new 59-storey, 301 metre office building altering the skyline of the city (Figure 5.7).

Expansion in office development has brought unprecedented supply in a period of national economic slowdown. From 2011 to 2013, 8.2 million m^2 of new, quality office space, equivalent to an annualised supply of 2.7 m^2, entered the office market, driving up vacancy rates in Guangzhou and Shenzhen (Figure 5.8), Hangzhou and Dalian (Figure 5.9), and Chongqing and Qingdao (Figure 5.10). Furthermore, 25.6 million m^2 of quality office space was planned for completion between 2014 and 2016. Of this, 12.4 million m^2 or 48 per cent is in BSGS, with over 6 million m^2 for Shanghai, over 4 million m^2 for Shenzhen and more than 2.2 million m^2 for Beijing and Guangzhou. In particular, new completions will double the space in Shanghai and Shenzhen. For tier 2 cities, Chongqing's 2.4 million m^2 planned space, if delivered on time, will quadruple its existing space. If all of the planned space were finished on schedule, the total office stock in the country would increase by 87 per cent by the end of 2016 (DTZ, 2014).

Chronic oversupply of office space has implications for the future use of completed office space. From an investment value point of view, office projects are better

FIGURE 5.7 CBD of Urumqi, China's most inland city
Source: photo by Yuanze Gao (2014).

held by the developer/investor or sold en bloc so that there is a centralised management to lower transaction costs and maximise management potential. However, high vacancy rates, depressed rents and little prospect of capital appreciation will deter large institutional investors, in particular, from buying new office projects for investment purposes. Thus developers may have to opt for less efficient outcomes by dividing up large office buildings for strata-title sales, impacting on the future performance as strata-titled buildings are less likely to be let at higher rents than identical buildings under simpler ownership.

On the demand side, new supply enables potential occupiers to upgrade and consolidate their operations, which accounted for much of the new take-up. Depressed and slow rental growth has been useful to help occupiers to control costs. The ready availability of space in different forms is useful for nationally operated firms to strategically locate their operations, increasing efficiency, growth and profits.

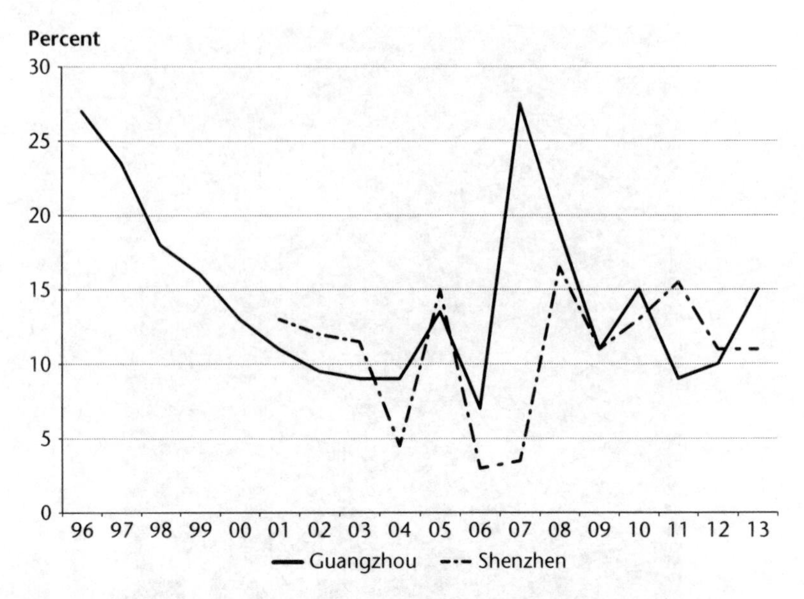

FIGURE 5.8 Prime office vacancy rates in Guangzhou and Shenzhen, 1996–2013
Sources: adapted from Savills (2013; 2014f).

On the financial side, developers and investors have difficulty balancing their books due to slow sales and letting, low prices and rents achieved, and extra marketing and management costs. Lenders bear the risks of non-performing loans, foreclosures and disposal costs.

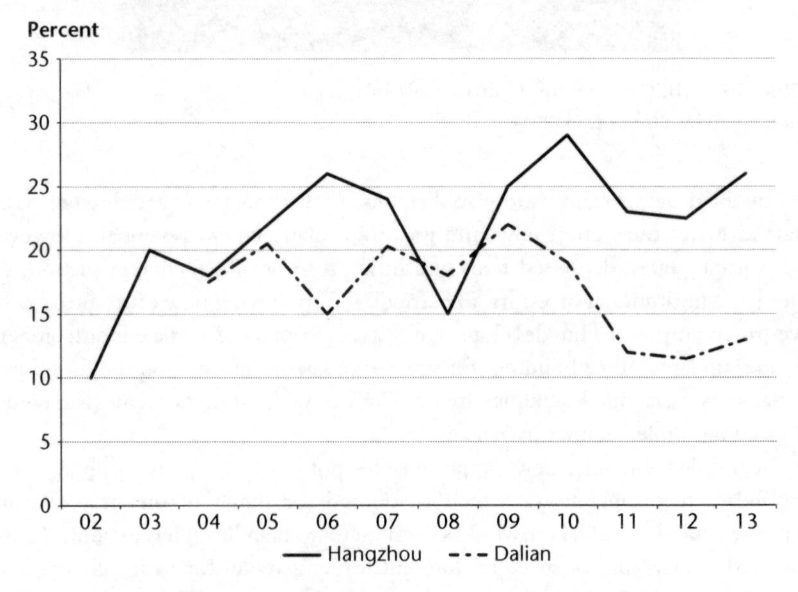

FIGURE 5.9 Prime office vacancy rates in Hangzhou and Dalian, 2002–13
Sources: adapted from Savills (2014d; 2014e).

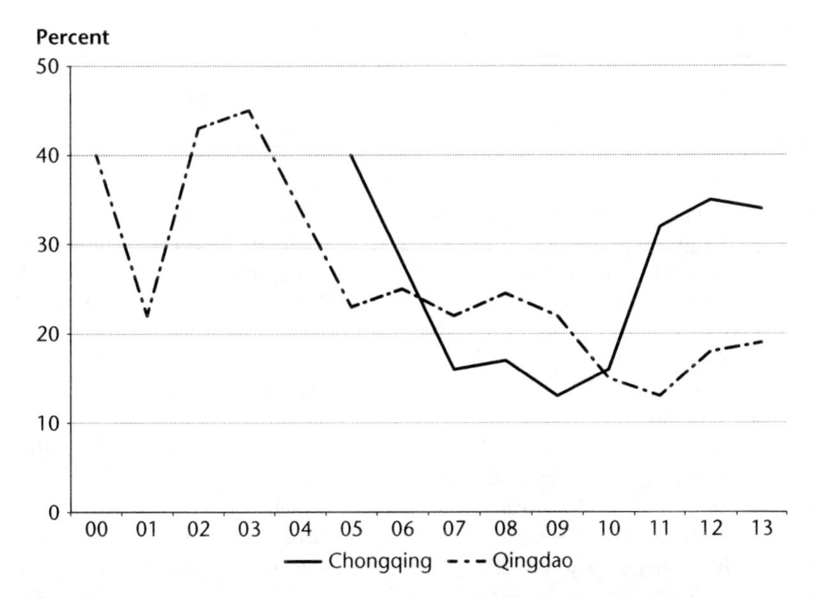

FIGURE 5.10 Prime office vacancy rates in Chongqing and Qingdao, 2000–13
Sources: adapted from Savills (2013; 2014f).

The reason for local governments' enthusiasm for office development is that modern tall buildings promote a city's modern image. For many years, competition for the tallest building in China was among Shanghai, Guangzhou and Shenzhen. Guangzhou led the country until 1999 and Shanghai now leads. Other cities are also actively engaged in constructing very tall buildings to improve their images. Table 5.4 lists the 25 buildings in Chinese cities which are among the world's 100 tallest buildings.

5.5.2 Retail markets

Rapid economic growth in China in particular, rising disposable incomes, has provided great opportunities for the retail trade and retail real estate. In 2012, total retail sales in China were on par with those of the USA (CBRE, 2014b). In 2013, China became the largest retail market in the world.

From 1999 onwards, the retail real estate market has been maturing in BSGS, expanding in tier 2 and tier 3 cities and emerging in tier 4 cities. The expansion of the retail real estate market has been the outcome of changes in demand and supply of shops in Chinese cities. On one hand, fast growth of retail sales (Figure 5.11) generated demand for more shop space and both domestic and foreign companies compete to take space to raise their market shares. Rents were bid up in this process. Retailers not only expanded into tiers 1 and 2 cities, but also into tier 3 cities. In 2002, UNIQLO opened its first store in Shanghai, marking the entry of internationally well-known brands into China. It had 211 stores in tiers 1, 2 and 3 cities by the middle of 2013 (CBRE, 2013). Many other international retailers followed suit.

TABLE 5.4 China's tall buildings and their place in the world

World Ranking	*Building*	*City*	*Height (m)*	*Floors*	*Built*
2	Shanghai Tower	Shanghai	632	121	2014
6	Shanghai World Financial Centre	Shanghai	492	101	2008
12	Kingkey 100	Shenzhen	442	100	2011
13	Guangzhou International Finance Centre	Guangzhou	440	103	2010
15	Jin Mao Tower	Shanghai	421	88	1999
20	CITIC Plaza	Guangzhou	391	80	1997
21	Shun Hing Square	Shenzhen	384	69	1996
30	The Pinnacle	Guangzhou	360	60	2012
31	SEG Plaza	Shenzhen	356	70	2000
40	Hefei Feicui TV Tower	Hefei	339	75	2012
42	Tianjin World Financial Centre	Tianjin	337	79	2010
44	Shimao International Plaza	Shanghai	333	61	2005
46	Modern Media Centre	Changzhou	332	57	2013
47	Minsheng Bank Building	Wuhan	331	68	2007
48	China World Trade Centre Tower 3	Beijing	330	74	2009
50	Longxi International Hotel	Huaxi Village	328	72	2011
53	Deji Plaza Phase 2	Nanjing	324	62	2013
56	Wenzhou World Trade Centre	Wenzhou	322	68	2010
63	Moi Centre	Shenyang	311	75	2012
68	Pearl River Tower	Guangzhou	310	71	2012
74	East Pacific Centre Tower A	Shenzhen	306	85	2012
83	Moi City	Wuxi	303	72	2012
84	Leatop Plaza	Guangzhou	303	65	2011
87	Gate of the Orient	Suzhou	302	68	2012
95	Shanghai Wheelock Square	Shanghai	298	58	2009

Source: Wikipedia (2014b).

Meanwhile, domestic retailers also vastly expanded their operations nationwide. Investment in retail real estate has also grown rapidly and nationally the sales volume of new retail space is illustrated by Figure 5.5.

Throughout this stage local governments had been promoting major retail development as part of urban regeneration and urban expansion to provide quality shopping space and so establish their cities as regional retail hubs. Some cities proposed building Central Living Districts, which include large shopping areas. With local governments' support, a retail development boom took place and fundamentally changed the retail landscape in Chinese cities.

Development of large shopping malls has been part of the retail development boom since 2002. At that time, shopping facilities in many Chinese cities were inadequate and dilapidated, leading to poor shopping experiences. Inspired by the success of earlier modern shopping malls such as Teemall in Guangzhou, many cities tabled shopping malls development proposals or approved large shopping mall

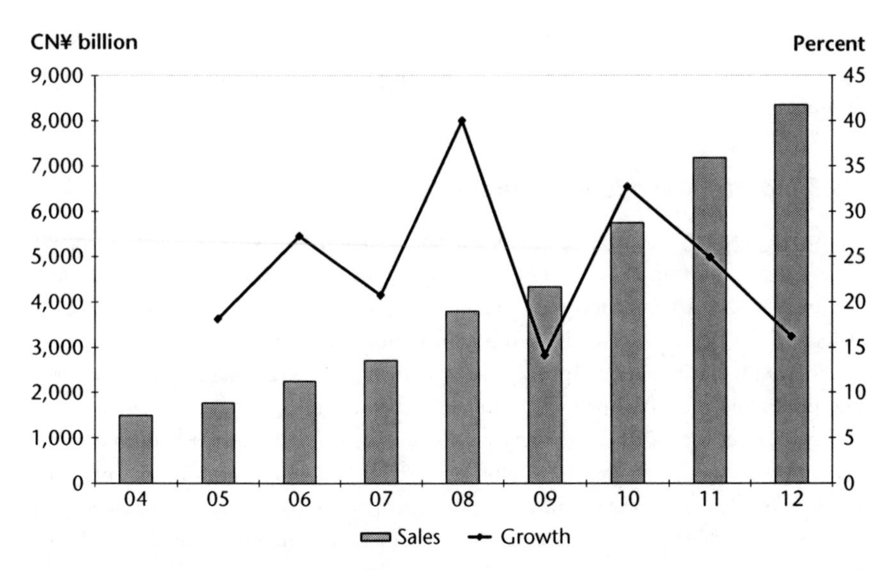

FIGURE 5.11 Retail sales in China, 2004–12
Sources: NBSC, various years.

development by developers. Specialist developers emerged and built shopping malls in many cities, raising the standards in specification and management. In particular, Wanda, a developer originating from Dalian, has successfully developed 72 Wanda Plazas in 25 provinces.

Meanwhile, old shopping districts have been partly or fully pedestrianized to provide better shopping environments and experiences to shoppers. These projects combine history, traditional local shops and modern stores to compete with the shopping malls. For example, the 1.2 kilometre Nanjing Road Pedestrian Shopping District in Shanghai boasts many preserved old buildings showing the history of Shanghai. In Guangzhou, the 1.2 kilometre Beijing Road Pedestrian Street exhibits 1,000-year-old road pavements and other historical sites with old and new shops. In Beijing, Wangfujing Pedestrian Street also combines traditional and modern shopping to give shoppers the Beijing-style shopping experience.

The significant increase of retail space has given rise to a rapid increase in the purchase of retail space for investment and owner occupation (Figure 5.5). Owner occupation has been increasingly adopted by retailers to reduce expenditure on rising rents. The increase in retail space availability has made possible rapid changes in store types (Table 5.5). It is worth noting that traditional department stores have been declining rapidly. The greater availability of cars has given rise to a rapidly increasing number of hypermarkets.

Internet sales in China are rising rapidly. The share of online sales in total retail shares rose from 1.1 per cent in 2008 to 7.8 per cent in 2013 (CBRE, 2014b). The sensational online sales on 11 November 2013, which exceeded CN¥35 billion, heralded the shift in retail sales in China. As a result, traditional in-store retail sales need to face the challenge by enriching the shopping experience of shoppers.

5.5.3 Industrial and logistics property

Industrial property, in particular, logistics properties, has witnessed rapid growth since 1999. Soaring exports and domestic consumption have generated huge demand for industrial property in the manufacturing heartland of the Yangtze River Delta and the Pearl River Delta and beyond. New industrial land supply, however, is increasingly limited and is approaching exhaustion in BSGS. As a result, industrial land prices and industrial property rents have been rising.

The market for logistics property has been developing rapidly. Modern logistics has gained recognition since 1990s and was established by foreign logistics property developers. Although many new logistics properties have been developed, the nationwide undersupply situation for logistics markets remained unchanged. Rents have been rising in all tier 1 and tier 2 cities for logistics property. The prime logistics market in most tier 2 cities is still in the early stage of development as the total stock of modern logistics facilities has yet to grow to scale.

5.5.4 Property services

There has been significant growth in professional services in commercial real estate. Major international consultancy firms such as Jones Lang LaSalle, CB Richard Ellis, Savills, DTZ, Colliers International and others have been increasingly penetrating Chinese cities. These firms mainly provide consultancy and agency services on commercial property. Domestic agency firms and their counterparts from Hong Kong, traditionally focusing on residential properties, are competing with their international giants to provide consultancy and agency services on commercial properties. Property investment funds, both local and international, have also established in the country, providing investment services for a host of clients.

TABLE 5.5 Change in retail store types in China

Store type	*2002 (%)*	*2012 (%)*
Specialty stores	44.40	46.30
Supermarkets	37.50	16.10
Exclusive stores	10.60	15.00
Department stores	5.70	2.30
Others	1.90	7.30
Convenience stores		6.90
Hypermarkets		6.30

Source: CBRE (2014b).

BOX 5.2 REAL ESTATE VALUATION IN CHINA

Min Liao

The valuation profession in China is full of Chinese characters – a product of the institutional arrangements in China. There are three valuation professional bodies, providing valuation services under different purposes at different stages. Table 5.6 lists the three bodies with their supporting ministries.

There are no legal or regulatory limits on the scope of valuation work for the three types of valuation professionals. In reality, each of the three types of valuation professionals is protected by the administrative jurisdiction of the supporting ministries (Table 5.6). Land valuations that the Ministry of Land and Resources (through their local branches) checks or requires information on need to be conducted by REVs. Such valuations include valuations of land for sale by the local government or transfer among land users. Real estate valuations that the Ministry of Housing and Urban and Rural Development checks or needs information on can only be conducted by CREAs. Such valuations comprise mortgage valuations and valuation for demolition and compensation. Valuations of company real estate assets and those valuations that the State Asset Administration Bureau checks or requires information on are conducted by CPA. It is worth noting that a CPA is trained to value all types of real property and moveable properties.

The institutional setting for valuation requires a valuer to obtain at least two qualifications: Real Estate Valuer and Certified Real Estate Appraiser. Some valuers also obtain the qualification of Certified Public Appraiser. Professional valuers normally work in specialist valuation firms, though some such firms also conduct agency services. They normally provide valuation and related services on a provincial basis, though some cover multiple provinces. The firm I work in covers valuations in Guangdong Province. In 2013, it completed 2,758 valuation reports, with 2,343 reports for banks and 415 reports for courts, asset management companies, local government demolition and relocation administrations, and other clients. Total floor area covered was 1.57 million m^2 and total land area was 8.74 million m^2.

TABLE 5.6 Valuation professional titles and their supporting professional bodies

Professional titles	*Professional Body*	*Supporting ministries*
Real Estate Valuer (REV)	China Real Estate Valuers Association (CREVA)	Ministry of Land and Resources
Certified Real Estate Appraiser (CREA)	China Institute of Real Estate Appraisers and Agents (CIREA)	Ministry of Housing and Urban and Rural Development
Certified Public Appraiser (CPA)	China Appraisal Society (CAS)	Ministry of Finance

The central government has tried to reconcile the segmentation of and competition among the three valuation professional bodies. In 2011, a law on valuation was drafted, with first and second readings conducted in 2012 and 2013. Public consultation resulted in nearly 200,000 returns being filed, mainly by the valuers of CREVA and CIREA. A joint seminar was held in April 2014 by CREVA and CIREA on international asset valuation legislation, the bias of the draft law to CAS, and the nature of asset valuation. Before the law on asset valuation could be enacted, the State Council cancelled the administrative power to approve the qualification of Certified Public Appraiser by the Ministry of Finance and the Ministry of Manpower and Social Security in July 2014. This is a major step by the central government to reduce government intervention in the market, leaving market qualifications to be approved by professional bodies operating in the market. The approval of CPA now becomes an issue for the CAS, levelling the playing field with CREVA, which has never received support from the Ministry of Manpower and Social Security. CREA, however, is enshrined in real estate administration law (REAL) and can continue to enjoy the status of being approved by the Ministry of Manpower and Social Security. However, CIREA also claims it has been carrying out self-management of CREA and real estate agents.

Although confusion and scepticism remain, the delegation of power to professional bodies to self-manage professional qualifications represents a step forward to limit the government's power to intervene in the market.

5.6 Discussion: the economics of oversupply

In 2014, a massive oversupply in office space in Chinese cities looms large. Yet it causes little concern to local governments at the collective choice level, which are responsible for the oversupply. Nor does it concern the central government at the constitutional level, which has yet to enact policies for the commercial real estate market. Why this lack of policy attention?

There are several reasons to this lack of policy attention. First, the bitter experience of under-capacity before the mid-1990s and the long history of rapid economic growth in the last 35 years have nurtured a unique attitude among local officials and developers to oversupply. When there is fierce inter-city competition, local governments adopt strategies to vie for more resources, which include funds, firms and talents, through real estate development, occupation and management. Buildings, even though empty, embody the successful capture of resources by the locality. Fast economic growth will generate demand for the empty buildings and bring more resources. In the last 30 years, overbuilding in Chinese cities only left short periods of oversupply. On a city scale, availability of new accommodation at low rents allows the city to bid for inward investment.

Second, the costs of erecting and maintaining those buildings are low. Sites for key office projects could be bought with preferential treatment for low land prices, low finance costs and tax breaks. Construction costs are low because the costs of labour and materials are low. Loans from state-owned banks are low in interest rates and normally come with the possibility of policy support should the projects face temporary difficulty. Thus, part of the costs brought on by vacancy is borne by the city's tax payers as a whole and by state-owned banks.

Third, investment values and psychological values matter. As Chinese cities expand rapidly, sites at central locations quickly run out. Thus it is rational to buy the sites and start construction in the expectation that the values of the buildings are rising. For many developers, construction and ownership of a state-of-the-art office building improve the company image and bring in valuable assets. For a city, a cluster of modern high-rise office buildings conveys the message that this modern city is open for business.

Fourth, development of new office space generates land sale receipts and tax revenues, which can be invested in infrastructure to make the city more attractive to inward investment. Selling sites for office development alone brings in CN¥ billions for the city to build roads, airports and underground railways and upgrade the city's infrastructure. For example, Kunming has opened its first underground railway line, becoming the fourth city[7] in Western China to operate underground railway lines.

Nevertheless, the unprecedented level of oversupply coupled with economic slowdown imposes high risks on many tier 2 and tier 3 cities and causes a great waste of resources. It is time for the central government on the constitutional level to design a policy framework to rein in the limitless ambition of local governments in their city-scale speculative development. For example, tighter control can be exercised in approving designation of CBDs and mega projects. The central government has been reducing the scale of administrative approvals to give the market more say on matters at the operational level. Yet office development at the current scale is a collective choice-level decision. The central government must provide a framework of rules on local governments for rational decisions to be made on large-scale office development.

5.7 Conclusion

Commercial real estate markets have been established in China and are playing important roles in China's economic growth. Occupational demand from foreign investors, and foreign capital and knowhow, were important in kicking off the office market and subsequent market expansion. Local governments in China, using their power in planning control and resources mobilisation, have played the key roles in the supply of offices, retail, leisure and industrial real estate. For major cities in China, the office market has become a key infrastructure for those cities to attract inward investment. However, the lack of policy guidance at the constitutional level caused a severe oversupply in offices in BSGS and other cities in China in the 1990s, causing huge waste in hundreds of failed projects. After 30 years of development the office markets in Beijing and Shanghai have matured with dynamic market activities and low vacancy rates. However, a new round of office oversupply on a bigger scale looms large in 2014 in many tier 2 and tier 3 cities. It will take much longer for those cities to generate sufficient demand to absorb the vacant space from this wave of oversupply. Policy guidance, in particular in planning commercial property development, from the central government is imperative in order to avoid future city-scale speculative developments in the country.

Notes

1 Beijing International Tower, a 31-storey office building, was started in 1982 and completed on 1 March 1985. It was developed by CITIC, one of the earliest developers with international experience.
2 Union Friendship Tower, a 29-storey office building, was completed in May 1985. It was the first full glass wall office building in China. With a height of 106.1 metres, it remained the tallest building in Shanghai from 1985 until 1987.
3 The four shopping streets are: Nanjing East Road, Nanjing West Road, Huaihai Middle Road and Sichuan North Road. The four shopping centres are New Shanghai Shopping Centre, Yuyuan Tourist and Shopping Centre, Xujiahui Shopping Centre and New Train Station Shopping Centre.
4 Nanning had a population (registered households) of 7.13 million and a per capita GDP of CN¥35,138 (US$6,396) in 2013.
5 ASEAN: Association of South East Asian Nations.
6 Kunming had a population (resident population) of 6.58 million and a per capita GDP of CN¥52,094 (US$8,544) in 2013.
7 The other cities are Chongqing, Chengdu and Xi'an.

PART III

Market regulation

6

LAND SUPPLY, PLANNING AND DEVELOPMENT CONTROL

This chapter investigates the system of land supply and administration, the regulation of development by the planning authorities and the distortion of planning control due to the zealous pursuit of local economic development. It starts by explaining the unique land administration system that determines the supply of land to the real estate market. The chapter then explores the widespread adoption of property-led growth by local governments and their fiscal dependence on the real estate market. It goes on to provide a brief review of China's urban planning system and problems in urban planning in the country. Finally, the chapter argues that the constitutional level should rein in local urban development by controlling the power of local top leaders at the collective choice level.

6.1 Supply of land for construction

Land ownership in China includes: collective ownership of rural land; state ownership of urban land; and state ownership of some rural land and other land (see Chapter 3 for details). Collectively owned land, mainly for agricultural uses, can be converted into construction land for the purpose of housing members of the collective that own the land, and for accommodating economic activities of the collective. However, ownership of such land and buildings developed upon it cannot be legally transferred accordingly to current statutes.

At present, real estate development must be on state-owned land with granted Land Use Rights (LURs) (see Chapter 3 for details). Laws and regulations in China stipulate that land administration agencies sell LURs in the primary land market for profitable land uses, and allocate LURs for free for non-profitable land uses. The Ministry of Land and Resources is the national level land administrative organ and oversees Departments of Land and Resources at provincial levels, which in turn

TABLE 6.1 Structure of land administration on land supply

Levels	*Organisations*	*Duties*
National	Ministry of Land and Resources (MLR)	Formulating national policies of land supply, making national land supply plan and supervising local implementation of the national land supply plan
Provincial	Department of Land and Resources, Bureau of Land and Resources*	Formulating provincial policies of land supply, making provincial land supply plan and supervising local implementation of the provincial land supply plan
Local – prefecture (municipal) level	Municipal Bureaux of Land and Resources	Formulating local land supply policies, making local land supply plan and supervising implementation of the local land supply plan at prefecture and county levels
Local – district, county and municipal level	District, county and county-level municipal Bureaux of Land and Resources	Making local land supply plan under their prefecture-level counterparts and implementing local land supply plan
Local – township level	Office of Land and Resources	Assisting the implementation of county and municipal land supply plan

Note: *for the four provincial-level municipalities: Beijing, Chongqing, Shanghai and Tianjin

supervise Bureaux of Land and Resources at municipal and county levels to perform land administration functions on behalf of the state (Table 6.1).

Supply of land for construction purposes in China is increasingly under strict state control to prevent unplanned loss of arable land, and from 2004 as a tool of macro control to rein in housing price inflation. Mechanisms of such control include the Master Land Use Plan (MLUP), the Annual Land Use Plan (ALUP) to set the quota for local land sales, and an open market land sale mechanism.

The MLR orchestrates the production of a national MLUP, which sets out the lay-out and arrangements of land uses in a 15-year period and is the basis of land use regulation by the state. Land Administration Law sets out the five principles of formulating MLUPs:

1. Strictly protect arable land and control the occupation of arable land by construction.
2. Increase land use efficiency.
3. Coordinate land uses in different regions.
4. Protect the ecological environment to ensure sustainable land uses.
5. Balance arable land conversion into construction and reclamation of arable land.

The MLUP designates three types of land uses: construction, agriculture and other uses. It is formulated to facilitate the central government to strictly control the conversion of land uses from agriculture to construction to protect the country's limited arable land. The formulation of MLUPs adopts both top-down and bottom-up approaches and is a negotiation process among different levels of governments.

The arable land protection system based on the MLUP and ALUP has increasingly been tightened to arrest the rapid loss of arable land due to conversion from rural land to construction land. China's economic development is through urban development and infrastructure development, both requiring land, a lot of land. For example, the total land sold for market-oriented property development from 1991 to 2013 was 32,664 km^2 (Figure 3.1). In fact, much more land was converted for non-market-oriented property and infrastructure development. From 1991 to 2013, a great deal of arable land was saved from this system of MLUP and ALUP and various crackdowns on illegal land uses. Even with an active land reclamation programme, net arable land loss in 2013 was 802 km^2, which amounts to 0.06 per cent of the country's total arable land stock (MLR, 2014b). Thus it is important to uphold this system of land administration to 'strictly protect arable land and control the occupation of arable land by construction', an institutional arrangement key to China's self-sufficiency in food supply. However, the tightening of land supply has serious implications for land prices.

ALUP is formulated based on MULP. Since 2005, it has been part of the macro control toolkit of the central government to manipulate land supply to combat housing price inflation. ALUP sets out the arrangements for various land uses in the coming year, which includes the amount of new construction land supplied. It underlines decisions by land administration agencies at various levels to approve the conversion of agricultural land to construction land, the land use requirement for new development projects, and land reclamation to increase agricultural land. The formulation of ALUPs also adopts both top-down and bottom-up approaches and is a negotiation process among different levels of governments. Upper levels of land administration agencies, such as the MLR and provincial Departments of Land and Resources, have the right to adjust local ALUPs to incorporate new national or provincial policies that affect local land uses, and tackle local violations.

A land sale mechanism was first created by MLR (2002) to require land for market-oriented uses to be sold by public auctions, public tender and public listing. If there is only one party interested in the land being sold, a private contract can be used to complete the sale. In most instances, the private contract only applies to land used for major infrastructure projects and large industrial projects with the resale of land unlikely in the near future.

Table 6.2 records total land supply for construction purposes to meet the needs of economic and social development throughout the country, including land supplied for housing development. Total land supply declined in 2004 and 2005 in response to tougher control on land conversion to protect loss of arable land and to cool down the housing market. The resultant tight supply, however, led to reduced residential land supply and an expectation of a shortage of residential land

TABLE 6.2 Total land supply and land supplied for housing development from 2003 (thousand hectares)

Years	*Land supply*	*Residential land supply*	*Land for ordinary housing*	*Land for Economic and Comfortable Housing*	*Land for Low Rent Housing*	*Land for luxury housing*
2003	286.437	63.752	36.251	3.555		1.003
2004	257.919	59.689	42.755	4.645		0.201
2005	244.269	55.144	37.993	5.821		0.026
2006	306.805	65.153	47.975	5.278		0.024
2007	341.913	80.174	59.630	5.084		0.068
2008	234.184	62.030	46.362	5.107		0.007
2009	361.468	81.548	69.089	10.343	1.394	0.720
2010	432.561	115.272	97.887	13.294	3.380	0.711
2011	593.285	126.453	103.696	16.238	6.122	0.396
2012	615.200	110.800				
2013	730.500	138.100				

Sources: MLR, various years.

for development in 2005. From 2010, total land supply increased significantly to provide land for accelerated economic growth and the construction of major infrastructure, such as high-speed railway lines, and to combat housing price inflation. It is worth noting that residential land supply declined in 2012 following the easing of housing price inflation due to housing purchase restrictions (see Chapter 4). In 2013, a record level of residential land was supplied to combat rapid housing price inflation in that year. It is worth noting that the supply of residential land for affordable and social housing has increased significantly since 2009.

The mechanism for land supply has been untouched in the past decade because of its impact on availability of land supply and land prices (see Figure 3.4 for national land price inflation). The challenge is to the unequal rights of the two land ownerships, i.e. state-owned land and collectively owned land; the former is without the right to issue LURs and thus is not tradable. Opponents of this restriction argue that giving equal rights to collectively owned land can increase the income of members of collectives, narrowing the income gap between the peasants and the urbanites. Easing the restriction can also increase the land supply for housing to mitigate housing price inflation. The opponents point to the more spontaneous provision of housing on collectively owned land, referred to as 'small property rights' housing (see Section 4.1 for more details), as proof that such housing solves the needs of many low-income households. This challenge led the constitutional level to consider reforms to make collectively owned land legally tradable in the Seventeenth Congress of the CCP in 2007. However, supporters of the status quo argued that arable land protection will be compromised if the restriction is lifted. Furthermore, land sale revenues received by local governments will be reduced, undermining local governments' ability to supply infrastructure. The MLR has been adamant that 'small property rights' housing is illegal and will not be given the same rights as housing on state-owned land. As a result, the reforms were suspended.

The new government led by President Xi Jinping revisited this challenge and has proposed reforms to equate the rights of state-owned and collective land in the Third Plenary Session of the Eighteenth Congress of the CCP (see Chapter 2 for details). However, this reform initiative involves a systematic change in China's land administration and ownership, and requires reforms on many fronts. To prevent disruption of land administration, the government has to obtain agreement from all stakeholders, which is unlikely to be a quick process for such a controversial issue. For example, the task of amending the current Land Administration Law has been stalled for ten years, because there has been no consensus on the amendment (see Section 2.6 for more details on slow legislation).

6.2 Property-led development

Local governments in China are responsible for developing the local economy, with local top leaders being assessed on their performance in generating local economic growth for their job security and promotion (see Section 2.3). With limited natural endowments, low productive capacity and limited human capital to start with, most local governments adopted property-led urban development as the growth model. Through provision of land, infrastructure and other incentives, local governments actively capture footloose capital, technology and talents produced by globalisation. Land sale revenues are important in financing infrastructure development to enhance urban competitiveness.

Property-led regeneration has been powering urban redevelopment in China's cities. In the 1990s and the 2000s, Chinese cities were ripe for urban redevelopment because the values of many urban sites, suitable for proposed high-rise development, were much higher than the low-rise, dilapidated buildings on site and the costs of relocating existing residents and occupiers. In most cases, affected residents were happy to leave their congested old residences and occupy spacious and modern new homes. The sale of the sites generated revenues for local governments to improve the infrastructure. Subsequent construction, sales and the occupation of completed new housing or offices generate tax revenues. As a result, local governments have incurred limited or no expenditure to initiate redevelopment. This model of using 'land to breed development' has been adopted by the majority of local governments. Figure 6.1 shows urban redevelopment in process in Shanghai, with blocks of low-rise and poorly maintained buildings being demolished to make way for the metro and high-rise buildings.

The economic case for urban expansion is even more convincing. Acquiring rural land involves lower costs due to the lower amount of compensation due, and construction on such land is cheaper due to fewer restrictions on site and the economies of scale. Owners of rural collectively owned land, village committees and rural economic organisations (see Section 3.1) are extensions of local governments. They are normally cooperative in allowing local governments to acquire their land,

FIGURE 6.1 Urban redevelopment in Shanghai's Hong Kou District
Source: photo taken by Albert Cao (2007).

typically for a large sum of money. As a result, urban expansion took place at an alarming pace. For example, built-up urban areas in cities grew from 10,161.3 km^2 in 1986 to 45,565.8 km^2 in 2012 (NBSC, 2013), a 348.42 per cent overall increase or 5.72 per cent year-on-year growth. In Guangzhou, the built-up area grew from 79 km^2 in 1978 to 1,010 km^2 in 2011, equivalent to 1,178.48 per cent overall growth or 7.78 per cent year-on-year growth (NBSC, 2012). Urban growth for key cities between 1990 and 2012 is illustrated in Table 6.3.

Urban expansion on such a scale has consumed large amounts of the country's best arable land, which is often located in suburban areas. To individual cities and counties, however, having more urban land means more economic growth and higher living standards. To protect it, the grip by the MLR on arable land has been tightened and local governments have to stick to ALUPs. By initiating major urban developments that are of strategic importance to local, regional or even national economies, however, local governments can apply for approval from higher levels of

TABLE 6.3 Urban expansion of selected cities

Regions	*Provinces*	*Cities*	*Built-up areas (km²)*		
			1990	*2012*	*Annual growth (%)*
North	Beijing	Beijing	397	1,261	5.39
	Tianjin	Tianjin	335	722	3.55
	Henan	Zhengzhou	112	373	5.62
	Shandong	Qingdao	94	375	6.49
Northeast	Liaoning	Shenyang	164	455	4.75
	Liaoning	Dalian	131	395	5.14
Northwest	Shaanxi	Xi'an	138	375	4.65
	Xinjiang	Urumqi	64	384	8.49
	Gansu	Lanzhou	163	199	0.91
East	Shanghai	Shanghai	250	886	5.92
	Jiangsu	Suzhou	37	437	11.88
	Zhejiang	Hangzhou	69	453	8.93
	Zhejiang	Wenzhou	27	204	9.63
South	Hubei	Wuhan	189	520	4.71
	Changsha	Hunan	101	326	5.47
	Guangdong	Guangzhou	182	1,010	8.10
	Guangdong	Shenzhen	69	863	12.17
	Guangdong	Zhaoqing	15	94	8.70
	Hainan	Sanya	16	47	5.02
	Guangxi	Nanning	70	242	5.80
Southwest	Yunnan	Kunming	96	298	5.28
	Chongqing	Chongqing	87	1,052	12.00
	Sichuan	Chengdu	87	516	8.43

Sources: Urban and Social Survey (1991; 2013).

government to alter MLUPs so that they can deliver such urban developments. These higher levels of government include the National Development and Reform Commission (NDRC) and the State Council, which have higher authority than the MLR. Once approval from the NDRC or the State Council is obtained, local governments can have more land to sell, which means more economic activities, more resources, more taxes, higher income and ultimately a higher standard of living.

Guangzhou presents the most successful example of property-led development. Guangzhou used to be a city specialising in commerce and was weak in manufacturing and technologies. After the reforms and the opening-up campaign started, Guangzhou took advantage of its early development of the property market to practise property-led development. Table 6.4 provides a list of major urban developments that allowed the city to expand its construction land stock legitimately. It is worth noting that by annexing the four cities and converting them into urban districts, Guangzhou is in a better position to propose new urban developments in those new districts to convert more rural land into construction land.

TABLE 6.4 List of new urban expansion projects in Guangzhou approved by the State Council

Year	*Designation*	*Justification*	*Area (km²)*
1984	Guangzhou Economic and Technological Development Zone	Attract foreign high-tech investment	9.6
1992	Pearl River New City CBD	New CBD	6.4
1993	Nansha Economic and Technological Development Zone (NETDZ)	Attract foreign high-tech investment and build deep water ports	54.0
1998	Guangzhou Science City	Attract foreign high-tech investment for R&D	22.7
2000	Panyu District (by annexing Panyu City)	Urban expansion	530.0
2000	Huadu District (by annexing Huadu City)	Urban expansion	907.0
2003	University City	Higher education development	34.4
2007	Guangzhou Asian Games City	International sports events	2.7
2012	Nansha District (by annexing NETDZ)	Industrial and urban expansion	803.0
2014	Zengcheng District (by annexing Zengcheng City)	Urban expansion	1,614.0
2014	Conghua District (by annexing Conghua City)	Urban expansion	1,974.5

The addition of five new urban districts provides land for Guangzhou to initiate more urban development proposals. A new proposal, if favoured by the housing industry, will significantly raise land and housing prices in the area designated for the proposed urban development. Some of the proposals promoted by the Guangzhou Municipal Government in 2013 and 2014 include the following:

- Guangzhou International Financial City
- Haizhu (District) Ecological City
- Tianhe (District) Wisdom City
- Guangzhou International Health Industry City
- International Airport City
- South (Train) Terminal Business District
- International Innovation City
- Huadi Ecological City
- Huangpu Port Business District.

Delivering the above proposed developments requires heavy investment in infrastructure, which is usually partly or mostly financed by land sale revenues. Without land sales, urban growth in most Chinese cities will face a shortage of funds and will slow down, and the large debts already incurred in supporting previous urban growth

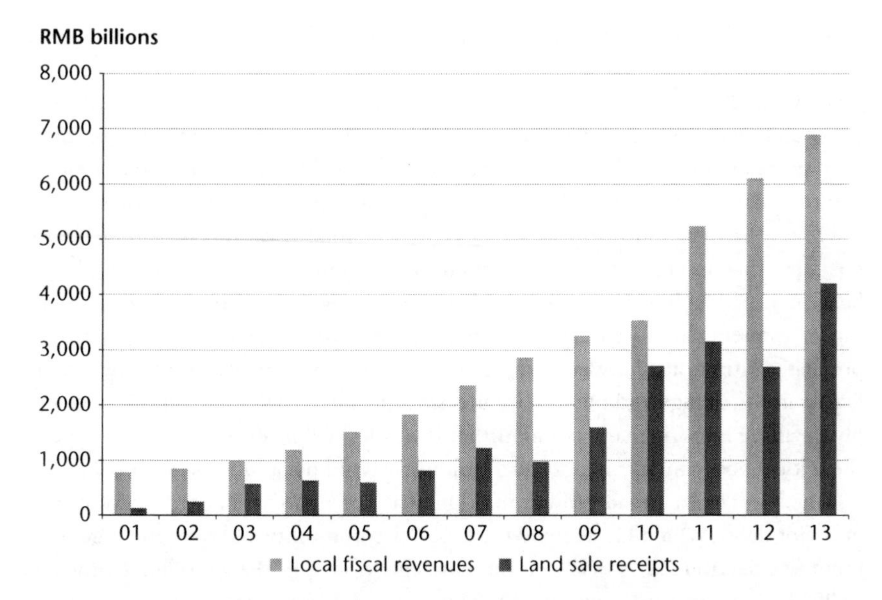

FIGURE 6.2 Local budgetary fiscal revenues and land sale receipts for all local governments in China, 2001–13

Sources: NBSC, various years; MLR, various years.

will be difficult to repay. Figure 6.2 illustrates the increasing importance of land sale revenues, an extra-budgetary income, to local government's financial capacity. It can be seen that in 2013 total land sale revenues were over half of the budgetary fiscal revenues. As extra-budgetary incomes, land sale revenues can be spent on projects at the discretion of local governments. After land sales, subsequent real estate development and transaction also bring in tax revenues. Table 6.5 shows the importance of land sale revenues, land-based taxes and land-related taxes to local governments' fiscal income. All land-related revenues accounted for 45.05 per cent of the local government revenues in 2012. It is worth noting that net land sale revenues was

TABLE 6.5 Land sale revenues and land-based and related taxes in local fiscal capacity in 2012

Items	*Local*	*National*
Net land sale revenues	1,375.92	
Land-based taxes	856.80	
Land-related taxes	518.60	
Other taxes and non-tax revenues	3,356.51	
Total local fiscal revenues	6,107.83	
Total national fiscal revenues		11,725.35
Share of land-related revenues in total local fiscal revenues	45.05%	
Share of land-related revenues in total national fiscal revenues		23.46%

Source: NBSC (2013).

CN¥1,375.92 billion, equivalent to 30.37 per cent of the budgetary revenues (see Section 14.3 for more details).

Nevertheless, some local governments have become dependent on land sale revenues and taxes from the property market. Once the property market slumps, these local governments will be hit hard and may not be able to honour their commitments to debts if no financial assistance is provided from the central government. Table 6.6 lists the ratio of land sale revenues to total fiscal revenues in 2013 for 45 cities that imposed purchase restrictions on housing in 2010 and 2011 (see Section 4.6). The housing markets of these 45 cities were hotspots of national housing investment due to its capital appreciation prospects, hence the imposition of purchase restrictions. However, there are no restrictions on sales. Once the market expectations change and out-of-city investors decide to sell, there will be oversupply in falling housing markets and further land sales will be difficult. If that happens, these local governments may need a bailout to avoid defaulting on their debts.

The spectacular urban growth in China has been partly financed by local government debts, which has alarmed the business community and the central government (see Section 2.2). Local government debts were CN¥19.94 trillion by the end of 2012 (Economic Information, 2014), which is 3.36 times the local governments' total revenues in 2012 and 14.49 times their total net land sales revenues (Table 6.5). The high indebtedness of local governments indicates the unsustainability of the current growth model, especially when further across-the-board housing price inflation looks unlikely (see Chapter 7 for more details), with negative implications for land sale revenues.

6.3 Planning and development control

China has a sophisticated land planning system, consisting of territorial plans, regional plans, urban system plans, urban master plans and urban detailed plans. Territorial plans are broad-based spatial plans that influence the distribution of people and activities according to natural resources and environment in spaces of various scales: national, regional and provincial. Regional plans are arrangements to achieve development objectives in regions that may cross provincial boundaries and include a number of cities. For example, the Yangtze River Delta Regional Plan identifies the overall objectives of economic and social development, and sets out the layout of development patterns and coordinates transport and other development projects in the region. Another example is the proposed decentralisation of Beijing's administrative functions to neighbouring cities, which calls for a new plan for the region encompassing Beijing, Tianjin and part of Hebei Province. Urban system plans define an urban hierarchy within a region in order to coordinate the infrastructure, the development and the environment. Examples of existing urban systems are city clusters in the Pearl River Delta, the Yangtze River Delta and Beijing, Tianjin and part of Hebei cities. Emerging urban systems include seven city clusters with the core cities of Jinan and Qingdao, Shenyang and Dalian, Zhengzhou and Luoyang, Wuhan and Huangshi, Xiamen and Fuzhou, Chengdu and Chongqing,

TABLE 6.6 Land sale revenues dependency ranking for local governments in 2013

Rank	*Cities*	*Land sale revenues*	*Fiscal revenues*	*Ratio (%)*
1	Hangzhou	147.8	94.5	156
2	Foshan	64.6	43.8	148
3	Nanjing	91.4	83.1	110
4	Changsha	56.5	53.7	105
5	Sanya	6.5	6.8	96
6	Hefei	41.9	43.9	96
7	Fuzhou	40.9	45.4	90
8	Kunming	40.4	45.1	90
9	Jinan	43.1	48.2	89
10	Xuzhou	37.3	42.3	88
11	Ningbo	67.4	79.3	85
12	Wenzhou	27.0	32.4	83
13	Chengdu	71.3	89.9	79
14	Nanning	20.3	25.6	79
15	Shaoxing	23.2	29.3	79
16	Holhot	14.3	18.2	79
17	Taizhou	19.4	24.8	78
18	Suzhou	103.3	133.1	78
19	Nanchang	22.1	29.2	76
20	Wuhan	74.0	97.9	76
21	Lanzhou	8.8	12.5	71
22	Changchun	26.4	38.2	69
23	Xining	4.5	6.7	67
24	Shengyang	52.8	80.1	66
25	Taiyuan	15.6	24.7	63
26	Guizhou	17.0	27.7	61
27	Guangzhou	69.1	114.2	61
28	Shijiazhuang	18.4	31.5	59
29	Xiamen	27.3	49.1	56
30	Shanghai	226.2	411.0	55
31	Xi'an	26.7	50.2	53
32	Zhoushan	4.8	9.3	51
33	Wuxi	36.4	71.1	51
34	Beijing	182.2	366.1	50
35	Jinghua	11.6	24.2	48
36	Dalian	39.8	85.0	47
37	Qingdao	35.6	78.9	45
38	Tianjin	80.5	207.8	39
39	Zhengzhou	26.0	72.4	36
40	Quzhou	2.3	7.3	31
41	Haikou	2.7	8.7	31
42	Urumqi	9.0	30.2	30
43	Yinchuan	3.7	13.5	27
44	Shenzhen	46.7	173.1	27
45	Harbin	15.2	63.2	24

Source: Topspur (2014).

and Xi'an and Xianyang. Urban master plans and urban detailed plans are made with reference to the above plans.

6.3.1 China's urban planning system

Urban planning in China involves the making of an urban master plan and urban detailed plans, and their implementation. It plays a key role in coordinating a very large number of developments to achieve the objectives of urban development for the city concerned, and thus has vital importance in urban growth and property development.

Urban planning in China was set up to restore ruined cities and cater for rapid urban growth. Planning the development of cities started in the period from 1949 to 1951 to restore cities and towns damaged in the Japanese invasion and subsequent civil war. From 1952 to 1960 the Soviet system of urban planning, which featured industries providing housing to their workers, was influential in the steady growth of towns and cities during this period. Urban planning was weakened due to de-urbanisation from 1960 to 1965 and abandoned from 1966 to 1976 due to domestic instability. This led to uncontrolled urban development during this period. In 1978, urban planning was restored, to order urban development that had become chaotic. In 1985, the first round of urban master plans was formulated to cover 98 per cent of cities and 85 per cent of capital towns of counties.

Pressure for economic and urban development in China has shaped the role of urban planning as a means to achieve economic growth. Urban areas are considered growth poles, thus plan-making is dominated by economic considerations at the expense of social and environmental concerns. Urban planning thus becomes a convenient tool to rationalise property-led development by ensuring standards of development projects, rather than a process to coordinate developments according to the wishes and needs of the public. For example, Article 26 of the Urban and Rural Planning Law 2008 stipulates that before sending an urban or rural plan for examination and approval to the upper level of government, the local government department responsible for making the plan shall announce the draft plan and collect opinions from experts and the general public by way of argumentation, hearing or other. The draft shall be made public for at least 30 days. In reality, public hearings are just a public relations exercise.

The nature of urban planning as a delivery mechanism for economic growth legitimises its frequent changes as adjustments to rapid economic and population growth, and the pressure on urban competitiveness. The mockery that master plans are for exhibition purposes but not delivery is revealed in the fact that a change of leadership often results in a change in the urban master plan, as those leaders impose changes to urban development to leave their footprints on the city as their achievement in office. Frequent adjustments to the urban master plan to include more urban development projects are an important factor that results in such plans quickly becoming obsolete.

Examples of urban master plans rapidly becoming out of date abound. Take Beijing as an example. The city produced its master plan in 1983, which aimed at controlling its population within 10 million in 2000. In 1986, the city's population reached 10 million. The next master plan for 1991 to 2010 aimed at controlling its population within 12.5 million in 2010. Yet the fifth population census showed that in 2000 the city's population was 13.82 million. The following master plan for 2004 to 2020 stipulated a population control target of 18 million by 2020. Nevertheless, the sixth population census revealed the city's population was 19.61 million in 2010 (Lu, 2013). Another example is Chongqing, whose urban master plan 2005–20 indicates that by 2010 the city's built-up area will be 580 km^2 (CMG, 2005). Yet by 2010 the city's built-up area had already reached 870.2 km^2 (NBSC, 2011).

The Chinese urban planning process consists of plan formulation and planning control. In the plan formulation phase, the planning executor, the Municipal Bureau of Urban Planning or Municipal Urban Planning Committee, produces a draft master urban plan, which is a strategic plan for a period of up to 20 years. The system of approval for master urban development plans is illustrated in Table 6.7. It is designed to ensure that proper checks are made to new development plans and projects so that enough attention is paid to all impacts, rather than just economic benefits, brought by the development plans and projects. Nevertheless, approval by an upper level of government is the norm rather than the exception for a number of reasons: there is asymmetry of information, with upper-level governments lacking information; bigger, more grandiose cities are the physical representation of the achievement of the government in general; and there is also bribery and corruption.

Once the master plan has been approved, municipal urban planning bureaux develop a set of detailed plans, which are short-term plans based on the master urban plan, and implement planning control according to the detailed plans. All developments require planning consents to go ahead. A planning consent contains three certificates issued at different stages of property development. The first certificate, the Site Planning Consent, specifies use, plot ratio, building density and greenery ratio of the site. The second certificate, the Building Planning Consent, covers the details of the design of the buildings and other improvements. The third certificate, the Building Construction Consent, authorises the construction work on the site according to the

TABLE 6.7 The system of approval for urban development plans and projects

Nature of developments	*Level of approval*	*Approval by*
Small development	District/County	District/County Bureau of Urban Planning
Medium to large development	Municipal level	Municipal Bureau of Urban Planning or Municipal Urban Planning Committee
Very large development and municipal urban master plan	Provincial level	Department of Housing and Urban Rural Development
Major development and urban master plan of very large cities	Central government	MOHURD and State Council

approved design. When the construction work is completed, a Building Completion Certificate is issued if the new buildings pass the completion inspection.

6.3.2 Problems in urban planning

In 2014, a deputy minister in MOHURD from 2001, Qiu Baoxing, retired due to old age. Being a senior urban planner himself, Qiu was the most outspoken senior official who criticised the problems in urban planning. He mounted a major critique of the country's urban planning system in 2005, which is still regarded as valid today (China Plan Net, 2011).

Qiu identified many problems in China's urban planning, which are related to local leaders. These problems are: wasteful prestige projects; planning changes by seeking shortcuts in urban growth; unsustainable use of land; overexploitation of limited resources; blind lay-out of infrastructure; chaotic use of suburban land; distortion of urban and rural planning at will; and unstoppable illegal construction. Qiu acknowledged that the causes of these problems were complex, but explained that local leaders bore special responsibilities for these problems:

- Unrealistically upgrading the position of cities in the urban hierarchy. It is unrealistic that all cities want to become regional centres or international metropolises.
- Overexpansion of population sizes. Higher population sizes justify larger urban areas, which in turn bring in more land sale revenues.
- Blind large-scale urban rebuilding. The old urban area carries the root of local civilisation and urban forms. Large-scale demolition of old urban areas destroyed many historical buildings and city scapes.
- Thoughtless urban expansion, modernisation and Westernisation. There was a wave of building urban landmarks, such as large urban squares, luxury office towers and housing with special characteristics, throughout the country.
- Thoughtless transportation development to cater for cars. To prevent traffic jams, Chinese cities built boulevards, flyovers and junctions, and widened old streets. However, these cities discriminate against cycling and walking.
- Excessive geographical division according to functions. Working and living were increasingly separated and development zones were built far away in the suburbs, generating more and more commuter traffic.
- Wrong understanding of authority to change urban plans. Some local leaders thought they had absolute power to alter urban plans because they were responsible for implementing the plans.

Qiu went further to critcise urban planning in China on the following two points. First, urban plans have lost some of their guiding and controlling function on urban development. On one hand, urban construction exceeded the arrangements provided in the urban plans, with many 20-year urban plans being completed in 5 years; on the other hand, urban construction lagged behind the stipulation in the urban plans, with many infrastructures not completed in time. Second, urban planning and rural planning were separated, causing chaos at the urban and rural boundaries. Urban planning authorities and land administration authorities did not coordinate their administration, resulting in poor management on development at urban and rural boundaries. For example, urban villages are without building regulation control, with overbuilding violating minimum health and safety standards. Figure 6.3 shows an urban village close to Guangzhou's new television transmission tower, a skyscraper, with very high building density, poor access roads and no open space.

Qiu's critique, however, did not go as far as explaining why the State Council and its ministry at the constitutional level could not make effective rules for the local governments at the collective choice level to follow. In fact, the approval mechanism described in Table 6.7 represents an institutional arrangement to scrutinise local development plans. The intriguing issue is why such a mechanism is not effective to contain unrealistic local ambitions.

Frequent changes in urban master plans and detailed plans place property development and investment at risk. One problem is the difficulty forecasting the supply of commercial property, which makes forecasting future cash flows difficult. As a result, developers and property services providers have to spend time and money to gather information on future urban plans. Home buyers and investors often suffered 'shocks' because they did not expect new buildings to crop up and congestion to develop in the vicinity of their properties at such a speed.

FIGURE 6.3 An urban village in Guangzhou's urban core, seen from the TV tower through the fog

Source: photo taken by Albert Cao (2013).

BOX 6.1 PROPERTY-LED DEVELOPMENT: THE CASE OF SUZHOU

Yunqing Xu

Suzhou is one of the oldest cities in China, with a recorded history of over 4,000 years and a historical city site over 2,500 years old. It was the second largest city in China and one of the top 10 cities in the world in 100 AD,[1] and one of the four business centres in China from the fourteenth to the nineteenth century. The city declined due to civil wars and the Japanese invasion from the 1860s, and the designation of Suzhou City was abolished in the first half of twentieth century. In 1949, the new government restored the designation of Suzhou City, whose built-up area was 19.2 km^2. Before the reforms and the opening-up campaign, Suzhou had slow growth, with the built-up area growing to 26 km^2 in 1978. During the 1980s, the city was still locked in slow growth, with the built-up area growing to 32.2 km^2 by 1984 and 38 km^2 in 1991.

In the 1990s, two major property-led growth schemes transformed the city. The first scheme was the designation of the 52 km^2 Suzhou National High-Tech Industries Development Zone in the west of the city. In the following year, the 70 km^2 Sino-Singapore Suzhou Industrial Park (SIP) was established

FIGURE 6.4 Modern Suzhou: boulevards and high-rise buildings
Source: photo taken by Lynn Xu (2014).

in the east of the city to attract FDI in manufacturing. The city's economy, in particular, real estate, manufacturing and foreign trade, has taken off since then. GDP grew from CN¥20.2 billion in 1990 to CN¥154.1 billion in 2000 and to CN¥1.3 trillion in 2013. FDI grew from US$0.07 billion in 1990 to US$2.83 billion in 2000 and to US$8.7 billion in 2013. Exports grew from US$0.16 billion in 1990 to US$10.48 billion in 2000 and to US$175.7 billion in 2013. More significantly, investment in market-oriented real estate development grew from CN¥0.17 billion in 1990 to CN¥6.12 billion in 2000 and to CN¥147.58 billion in 2013 (SSB, 2010; 2013). By 2012, the built-up area of Suzhou was 437 km^2 (Table 6.3).

Through more than two decades of rapid development, the city restored its former glory, with GDP ranked sixth in the country in 2013, only behind BSGS and Tianjin. At present, Suzhou is the second largest industrial city in China (Figure 6.4). By end of 2011, over 140 Fortune 500 companies had invested in Suzhou, mostly in SIP.

Land and property-related revenues have been instrumental in bringing about such remarkable growth. Excluding the share of the real estate sector in Corporation Tax and Business Tax, the Suzhou Municipal Government derived 31.74 per cent of its total income from land sales (Table 6.8), and

TABLE 6.8 The importance of land sale revenues and real estate taxes in Suzhou

	CN¥ billion	*CN¥ billion*	*(%)*
Total local fiscal revenues		91.33	
Urban Construction and Maintenance Tax	2.44		
Real Estate Tax	2.75		
Stamp Tax	0.97		
Urban Land Use Tax	2.62		
Land Appreciation Tax	2.51		
Occupancy of Cultivated Land Tax	0.71		
Deed Tax	5.84		
Total real estate-related fiscal revenues	17.84		
Land fund incomes		42.47	
Total local government incomes		133.80	
Land fund incomes as a proportion of total local government incomes			31.74
Total real estate-related incomes as a proportion of total local government incomes			45.07

Source: SSB (2010).

45.07 per cent from land sales and real estate-related taxes in 2009. The scale of urban land sale has peaked at 16.5 million m^2 in 2003, corresponding to the fastest growth of GDP at 18.0 per cent in the same year.

The property-led growth model that contributes to Suzhou's rise, however, has become unsustainable in 2014. With falling housing prices and an office vacancy rate of 22.7 per cent, the property market in Suzhou is likely to generate less land sale revenues and real estate related taxes. This poses a challenge to the city that depends a lot on land sale revenues (Table 6.6). Suzhou needs to change into a new, and more sustainable, growth model.

6.4 Discussion: the costs of planning failures

On 11 July 2003, two new buildings in Chengdu were blown up to demolish them. One of the demolished buildings was the residents club of Hongfeng Peninsula Garden, resulting in a loss of CN¥19.29 million. The other was part of Jincheng Garden, resulting in a loss of CN¥12.59 million. The reason for demolition was that the two buildings intruded on the green belt reserved for the Yingbin Boulevard by 11 and 10 metres, respectively. Such intrusion was approved by the deputy head of the Planning Bureau, who had received bribes from the developers involved in these cases, against the planning conditions (*Tianfu Morning Post*, 2004).

Planning failure in this case was corrected, albeit at a high cost. In this case the perpetrators were put behind bars. However, planning failure at a larger scale is very difficult to put right and the losses can be astronomical. For example, the hundreds of failed projects in Guangzhou and Shanghai alone originated in the two city's urban development plans that ignored market conditions and the stage of development in the two cities (see Section 5.4 for more details). Although most of those failed projects have now been revived, the financial costs were hundreds of times more than those for the two demolished projects. The looming oversupply of office space (see Section 5.4 for more details) is likely to drive up office vacancy rates and result in failed projects, causing a huge waste of resources and financial loss.

Interviews with researchers in international consultancy firms indicated that many of the tier 2 cities have limited office demand, which do not justify the large-scale building activities in the CBDs of those cities. Why did urban planners in those cities produce such urban development plans that are well ahead of the development stages of their cities? Interviews with planners, some at senior levels, revealed that the problems were from the top leaders who made decisions on those cities' development strategies and mega projects. Therefore, the solutions to the problems in urban planning in China lie in the constitutional-level decisions to make local leaders accountable for their decisions. A new institution of *ex ante* prevention of planning failures should be put in place to minimise the costs incurred by planning failures.

6.5 Conclusion

China has built an administrative mechanism for selling urban land for property development to provide land for urbanisation while protecting the country's limited arable land, environmental and ecological resources. This mechanism protects the country's self-sufficiency in food supply but has negative implication on land supply for construction and prices. Sales of housing land have been linked to macro control of the housing market in an attempt to influence housing prices. Land conservation is an uphill struggle against local development initiatives that found building on greenfield sites cheaper and more convenient. The property-led development model widely adopted by Chinese cities has resulted in built up areas in urban China becoming 3.5 times larger in the last 27 years. Expanding the built-up area allows local government to sell more land for more revenues to invest in infrastructure and city image projects. At the time of high local government debts, the ability to sell more land is very important to keep infrastructure projects going. As a result, local top leaders are found to interfere in urban planning to upgrade their cities in the urban hierarchy, promote new developments so that more land can be sold, and regenerate old urban areas, leading to destruction of historical buildings. Problems in China's urban planning system can be minimised if constitutional decisions are available to limit the interference of local leaders on the operation of the urban planning system.

Note

1 From: http://geography.about.com/library/weekly/aa011201b.htm.

7

HOUSING MARKET ADMINISTRATION AND MACRO CONTROL

This chapter examines the government's housing market administration and control on housing price inflation since 2005 in the context of rapidly rising housing prices and investment demand. Market administration conducted by the government includes administration of development firms, intermediary firms and professional services firms. The chapter then goes into the details of the macro control measures to rein in housing prices from 2005 to 2013, which brought down housing prices twice. A full analysis on the central government's effort to impose a size standard is provided. Later the chapter details the imposition of purchase restrictions and the behaviour of different levels of government and the market. The chapter finally tackles the issue of the housing bubble in China and argues that the housing market correction started in 2014 will not drag on the economy: instead, it provides the opportunity for the economic and the real estate sector to implement the long-awaited restructuring.

7.1 Market administration

Real estate market administration by the Ministry of Housing and Urban-Rural Development (MOHURD) and its provincial and local counterparts (Table 7.1) includes the management of development firms' qualifications, presale management, real estate professional services and market transaction management.

7.1.1 Development firms' qualification management

China's housing market has been dominated by the sale of new housing at its inception in 1980 in terms of floor area sold and sales volume. Sales of old housing have been growing and in some cities are close to that of new housing in terms of floor area sold and sales volume. For example, in Shanghai, the sale of old housing was

TABLE 7.1 Structure of real estate administration

Levels	*Main administrators*	*Auxiliary administrators*
National	Ministry of Housing and Urban-Rural Development	National Development and Reform Commission; Ministry of Land and Resources
Provincial-level	Department of Housing and Urban-Rural Development	Department of Land and Resources
Prefecture-level (municipal)	Bureau of Housing Administration or Bureau of Land and Housing Administration	Bureau of Land and Resources
County-level (including municipal & district)	Bureau of Housing Administration or Bureau of Land and Housing Administration	Bureau of Land and Resources

11.36 million m^2 in 2012, close to sales of new housing at 15.93 million m^2. As a result, the administration of the housing development sector, i.e. the housing development firms, is of particular importance.

The management of the housing development firms is through a qualification management scheme. China's house builders are classified into five types according to ownership: private; state-owned; collectively owned; Hong Kong, Macao and Taiwan; and foreign. State-owned developers outnumbered other ownership types until 1999 when private developers became the largest ownership type, reaching over 80,000 in 2012 (Figure 7.1).

Different types of development firms entered the development sector at different time periods. Older state-owned development firms were mainly former house-building teams administered by the local real estate administration bureaux. Mass entry by state-owned, collectively owned (rural and township government), Hong Kong, Macao and Taiwanese development firms occurred from 1992 to 1994, when the property market was booming in the coastal areas. Since then, their number has gradually declined, though the number of Hong Kong and Taiwanese firms increased again in 1998. The number of foreign firms, initially ones run by ethnic Chinese from South-East Asia, peaked in 1995 and has since fallen. However, the number of American and European firms has grown. The number of private development firms has been growing since 1992 and now dominates the development sector.

Administration of real estate development is through management of developers' qualifications. Developers are divided into four classes of qualifications according to: (1) their registered capital; (2) years of experience; (3) accumulative completions in floor area; (4) pass rates at quality checks; (5) number of technical staff; (6) the quality warranty system with the provision of Housing Warranty Certificates and Housing Use Manuals; and (7) the records of significant construction quality incidents. Developers holding Class 1 qualification have no limits in carrying out real

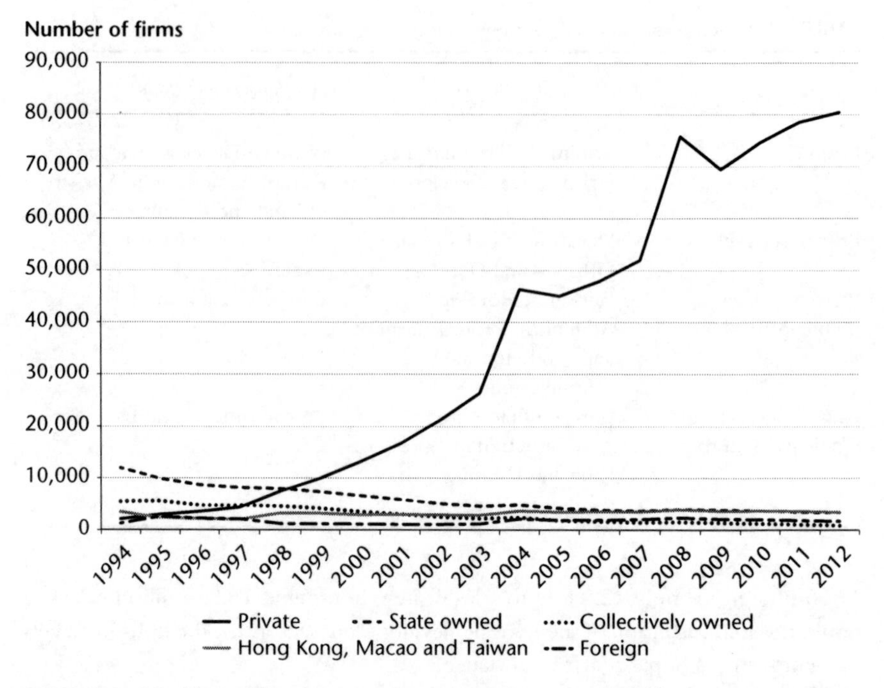

FIGURE 7.1 Number of development firms according to ownership types
Sources: NBSC, various years.

estate development throughout the country. Those holding Class 2–4 qualifications can only carry out projects with less than 250,000 m^2 in total floor area per year in their province of registration. Developers whose projects are low in quality or involve accidents could have their qualification lowered or revoked. If a developer's qualification is revoked, its licence for property development will be cancelled. Developers, on the other hand, are keen to upgrade their qualification. For example, real estate trust products, a type of loan finance, are available only to developers with grade 2 or higher qualifications (see Section 11.2). It is worth noting that the qualification scheme is effective in policing construction quality and reducing quality accidents.

7.1.2 Management of real estate intermediaries

Real estate market transactions include purchase and sales, lease and mortgage, all of which are conducted through the real estate intermediaries, i.e. the real estate intermediary firms. Similar to development companies, real estate intermediary firms consist of private, state-owned, collectively owned, Hong Kong, Macao and Taiwanese, and foreign firms. However, private, Hong Kong and foreign firms dominate the real estate intermediary market.

Although assuming a very important role in the increasing dynamic market, the real estate intermediary is an area of increasingly intensified and repeated management effort due to a number of issues outlined by the *Circular on Intensified Special Administration of Real Estate Intermediary* jointly issued by MOHURD and the State Administration for Industry and Commerce (MOHURD, 2013). Those issues are serious and at the core of complaints by buyers, sellers and renters, including the following:

- Disseminating false information about the availability of real estate stock. Generating and disseminating rumours about market demand and supply to mislead consumers.
- Inducing and helping buyers to forge documentation to obtain qualification to buy housing under purchase restriction, and to cheat lenders.
- Generating false demand for particular real estate projects by employing people to pretend to be buyers.
- Assisting parties to transactions to sign dual contracts for tax evasion.
- Violating provisions in Administrative Measures on Commodity Real Estate Leasing on the following: (1) subdividing units to let by unauthorised structural alterations; (2) providing agency services to buildings that are unsafe and with fire hazards; and (3) cheating on market rents data.
- Embezzling clients' money for transactions.
- Compulsory provision of services such as mortgage advisory and guarantee services for fees without prior information.
- Disclosing, for profit, clients' information without prior approval by clients.
- Providing agency services without licences.
- Signing agency contracts without an agency qualification.
- Lending agency qualification to unqualified others to provide agency services.

To tackle the problems in the real estate intermediary field, MOHURD and other ministries have relied on *ad hoc* campaigns, with several nationwide campaigns being carried out. For example, in the 2007 campaign, MOHURD orchestrated a market order administration campaign that covered 31,577 real estate development projects in 286 cities at prefecture level or higher. The campaign imposed a self-check according to MOHURD's criteria on over 30,000 development firms, over 20,000 agency firms, over 3,000 valuation firms, over 20,000 property management firms, over 2,000 demolition and relocation units, and over 5,000 local government branches, all of which were related to the 31,577 projects. The self-check revealed that 13 per cent of the 31,577 projects had various problems and further investigation followed. In particular, over 2,000 cases of wrongdoing by intermediary firms were handled (Qi, 2007).

Nevertheless, those *ad hoc* campaigns were more educational than punitive, and were expensive to carry out. As a result, market disorder regularly recurred. Presently, MOHURD is working on establishing long-term mechanisms to manage real estate intermediaries by building online signing of contracts, online management of intermediary firms and a service provision platform to provide real estate availability enquiry and credit records. However, the key to such long-term solution is setting rules and implementing them. MOHURD, or perhaps a higher-level government organ, should enact constitutional rules to empower local governments at the collective choice level to enact operational rules to penalise wrongdoings.

7.1.3 Other administrative services

Real estate administrative agencies also provide services, such as standard sale contracts, leasing contracts, best service examples such as standard property management residential districts and promotion of corporate social responsibility among developers and intermediary firms.

Professional services under administration include valuation, agency, and property management. All those service providers need to obtain licences from real estate administration bureaus. There are professional qualification requirements on valuers, agents and property managers (see Box 5.2 for details).

BOX 7.1 REAL ESTATE AGENCY IN CHINA

Junping Liao

In China, real estate agents are intermediaries in property transactions, including lettings and sales. A real estate agent (REA) is professionally qualified through examination and certification, both organised by the China Institute of Real Estate Appraisers and Agents (CIREA). In July 2014, the State Council abolished administrative approval for the REA, which is now fully under the administration of CIREA, a semi-governmental professional body. In 2013, there were 52,436 qualified agents, 28,848 of them registered to carry out agency work in 23,090 agency firms (see Figure 7.2). In the same year, the agency industry employed over one million practitioners in over 50,000 agency firms and their branches. The largest agent, Centraline from Hong Kong, had 2,340 branches in 38 cities and 52,625 staff. The second largest agent, B.A. Consulting & 5i5j of Beijing, had 1,500 branches in 12 cities and 28,000 staff.

The agency industry is monitored by the government regarding the levels of fees charged. Leasing commission is normally half to a month's rent. Sales commission is normally 0.5 per cent to 2.5 per cent of the price. Exclusive agency commission for developers is normally less than 3 per cent of the price.

The agency sector faces two major issues in its continued development. First, there are still a significant number of cases involving violation of regulations and codes of conduct, and frauds. The government carried out a crackdown on the problems in mid-2013 (MOHURD, 2013). Second, the sector has faced changes of business models and intensified competition from the internet. In 2013, CIREA chaired a conference on the development of the agency sector in the era of big data. In the same year, some agency firms adopted internet business models to open internet branches and internet unions. New websites were set up to provide short-term leasing information and complete transactions. Some internet websites provide leasing information and transaction services, posing a challenge to traditional agency businesses. The widespread use of mobile devices such as smartphones prompted the introduction of apps on home search and

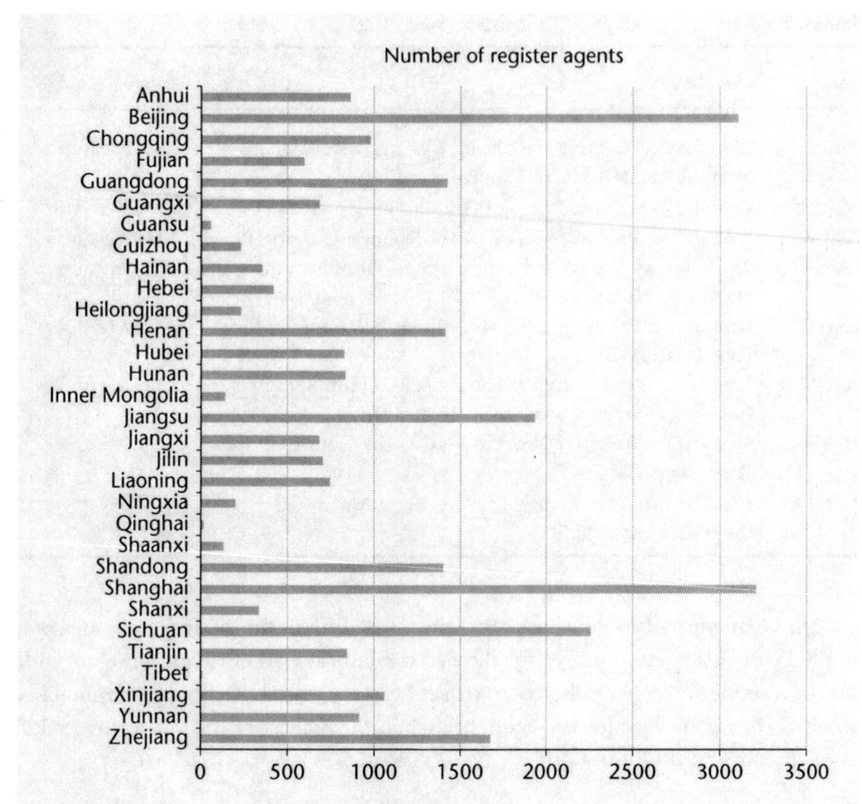

FIGURE 7.2 Distribution of registered agents
Source: China Institute of Real Estate Appraisers and Agents.

leasing, and the use of WeChat, a social media for mobile phones, to provide housing information and conduct leasing transactions. The agency sector has been adapting to the changes of business models and competition.

7.2 Macro control

Macroeconomic regulation and control, abbreviated to 'macro control', is the use of direct intervention by the central government to cool down the overheated economy or a sector of the economy to achieve the government's economic development objectives. It is a unique economic adjustment measure exercised in China, given the Chinese government's power and institutional support for its actions. The instruments used in a macro control include fiscal policy instruments, monetary policy instruments and directives. Macro control was first exercised in 1993 to cool down the overheated economy (Table 7.2), in particular the real estate market, and bring down inflation (see Section 4.4). Although it caused the collapse of the real estate market in Hainan Province, Beihai City in Guangxi Province and Huizhou

TABLE 7.2 Macro control policy documents since 1993

Dates	*Policy documents*	*Abbreviations*
Jun-93	Opinions on Current Economic Circumstances and Strengthening of Macro Control	1993 Measures
Mar-05	Circular to Stabilise Prices of Housing in Earnest	2005 Principles
May-05	Circular on Effectively Working to Stabilise Housing Prices	2005 Measures
May-06	Opinions on Adjusting the Structure of Housing Supply and Stabilising Housing Prices	2006 Measures
Dec-09	Circular on Promoting the Healthy Development of the Real Estate Market	2009 Measures
Apr-09	Circular on Steadfastly Preventing Rapid Housing Price Inflation in Some Cities	2010 Measures
Jan-11	Circular on Relevant Issues to Further the Control of the Real Estate Market	2011 Measures
Feb-13	Circular on Continuously Implementing Macro Control of the Real Estate Market	2013 Measures

City in Guangdong Province, the 1993 measures did not hit the real estate markets in BSGS and other cities badly, but allowed those local markets to grow in line with the local economy. As a result, macro control was regarded by central government as an effective way to balance the economy when the real estate market was regarded as overheating and housing prices rose too fast in 2005.

7.2.1 2005: setting the tone of macro control

The central theme of macro control from 2005 to 2013 was to control housing price inflation. The first macro control policy document, 2005 Principles (State Council, 2005a), set out the principles in macro control in the housing market (Table 7.3). It reflects the start of a search for new housing policy by the central government at the constitutional level when housing provision shifted from public dominance to market dominance, with supply and prices being determined in the market rather than by public sector. The enactment of 2005 Principles ended the debate from 2003 (see Section 3.5) on whether the housing market was overheating, and reaffirmed that the responsibility of providing affordable housing rested on the government rather than on the market. The 2005 Principles foresaw that housing issues could be transformed from a local and sectorial issue to a national and holistic one. It sounded the alarm bell to many officials who no longer were paying attention to housing affordability by reminding them that housing was a political issue.

The 2005 Principles and its follow-up the 2005 Measures announced a number of policy initiatives on issues that were important but did not receive due attention at the national policy level. Such initiatives became the centre of attention and contention during the whole period of macro control. First, the 2005 Measures tabled a size standard on 'ordinary housing', i.e. commodity housing that, according to the

TABLE 7.3 The eight principles contained in the 2005 Principles

Number	*Principles*
1	Taking the task to stabilise housing prices very seriously
2	Earnestly taking responsibility to stabilise housing prices
3	Making great efforts to adjust and improve the structure of housing supply
4	Strictly controlling passive housing demand*
5	Properly guiding housing consumption expectations
6	Monitoring the operations of the property market fully
7	Proactively implementing the various policy measures to adjust and control housing demand and supply
8	Seriously organising inspections on the task to stabilise housing prices

Note: *Passive housing demand means demand generated by involuntary relocation.

2003 national housing policy document (State Council, 2003), should be mass-built for mass occupation and ownership. Details of the size standard and the reactions and contentions surrounding it are discussed in Section 7.3. Second, the 2005 Measures imposed additional taxes on housing transaction and ownership to penalise speculation and tax capital gains. Details of the new tax regime and the reactions and developments surrounding it are discussed in Chapter 8.

Third, the 2005 Principles and the 2005 Measures emphasised the importance of increasing the provision of ordinary housing, Economic and Comfortable Housing (ECH: a sub-market of housing for purchase by qualified applicants), and Low Rent Housing (LRH) for the poor to rent. This was a deviation from the 2003 housing policy (State Council, 2003) that had marginalised affordable and social housing. It heralded the establishment of a new affordable and social housing system in China to run parallel to commodity housing. Details of the affordable and social housing regime are discussed in Chapter 13.

Fourth, the 2005 Principles and the 2005 Measures called for a combination of measures, including land supply, taxation, and monetary measures to affect market expectations on future prices and guide housing consumption behaviour. It joined the attempts by different government agencies and broke the inter-ministerial separation in administration. In particular, the central bank became a stakeholder in macro control. Details of the monetary operations on the housing market and the reactions and developments surrounding it are discussed in Chapter 8.

The 2005 Principles and the 2005 Measures also highlighted the importance of market administration to tackle excessive speculative investment, to check violations in housing sales and to combat activities that drove up housing prices. It called for information disclosure and market transparency, and required the real estate administration agencies to monitor market movements and develop warning systems on prices and supply. In this sense the 2005 Principles started a long-term attempt to build a market in a country where the housing market had only operated as a main provider of urban housing for less than seven years.

7.2.2 2006–07: boosting supply and curbing demand

Nevertheless, the 2005 Principles and the 2005 Measures only achieved partial results in terms of housing price control, with housing prices in Shanghai stagnant for a short time (see Figures 4.8 and 4.11). However, housing prices in the rest of the country continued to soar, with prices in Beijing overtaking Shanghai and becoming the national barometer of housing prices. To alleviate rising housing price inflation, the 2006 Measures, sponsored by nine ministries (State Council, 2006), were promulgated in May 2006. They were followed by a series of policy documents and market intervention by the ministries, forming a complete set of housing policies which founded the modern housing policies of the country. In addition to further definition of size standards, taxation, monetary intervention, affordable and social housing provision and market building, they required that local governments produce Housing Construction Plans to specify the targets for delivering ordinary housing, ECH and LRH for the next two years and next five years. These plans had to be made public in September 2006 and filed with higher-level governments. With some delay, local Housing Construction Plans were made and publicised in 2007. In addition, purchase restriction was imposed on foreign capital by a follow-up measure (MoC, 2006a) to restrict housing purchase by foreigners (see Chapter 12 for details). In 2007, the central bank and the China Banking Regulatory Committee jointly issued a circular to tighten lending to home buyers (PBoC, 2007), which was effective in reducing demand. This circular was considered the last straw to fall on the weakened growth momentum of housing prices and caused housing prices to fall.

Macro control substantially reduced banks' exposure to the housing market and brought down housing price inflation before the waves of international financial crisis reached China. Without the burden of non-performing loans, Chinese banks emerged as the best capitalised banks in the world. Nevertheless, much of the liquidity in Chinese banks was then put into the property market as part of the rescue packages tabled by both central and local governments, fuelling another round of housing price inflation from 2009 to 2010.

7.2.3 2008–09: rescuing the housing market

Continuous implementation of the 2006 Measures and its follow-up measures, coupled with the 2008 Global Financial Crisis, brought down housing price inflation throughout the country (see Figures 4.3 and 4.8). Panic selling occurred in some Chinese cities and prices fell sharply for many housing projects. Some home buyers who had bought at higher prices in 2007 and earlier in 2008 became angry that similar units in the same housing estate were being sold at much lower prices. They smashed the sales office and threatened to sue the developer. Land sales were brought to a halt, hitting local governments' infrastructure development projects. To rescue the housing market, many local governments at the collective choice level tabled their versions of macro control through tax cuts, subsidies, provision of household registration status and other means to attract home buyers to part with their money to buy homes. Many of the measures formulated from 2005 to 2007 were suspended (see Sections 7.3, 7.4 and Chapter 8 for more details), and the central government was silent on local governments' market boosters and violations of the 2006–07 macro control measures. Following the central government's CN¥4 trillion rescue package, local governments rolled out their own versions of much larger rescue packages. Excessive credit supply, the result of those rescue packages, restored purchasing power in the market and resulted in rapid housing price inflation in 2009.

With hindsight, the opportunity to make housing market adjustment could have been better used to achieve structural change in the housing market. However, the constitutional level was too afraid of economic slowdown and adopted a quick fix to pump liquidity into the economy. Much of the new credit ended up in the housing market. With that, the housing market recovered quickly and prices rose sharply in 2009.

Rapid housing price inflation in 2009 resulted in the resumption of macro control from late 2009. In response to rapid housing price inflation, the 2009 Measures (State Council, 2010a) were promulgated in December of that year to increase the supply of affordable housing and ordinary commodity housing; to reduce speculative housing demand; to strengthen risks control and market administration; and to speed up the construction of affordable and social housing.

7.2.4 2010–13: the 'toughest' macro control

Faced with rapid housing price inflation, the central government enacted the 2010 Measures (State Council, 2010b) in April 2010, which was referred to as the 'toughest' macro control measures on record. The 2010 Measures reimposed the taxes from previous macro control measures, conducted monetary intervention, speeded up affordable and social housing provision, and worked to build market order (see Sections 7.3, 7.4 and Chapter 8 for more details). Furthermore, it stipulated that top leaders at the provincial level were responsible for housing prices stability and affordable and social housing provision. It also required developers to make public the housing stock they held. The 2009 Measures, and in particular the 2010 Measures, were referred to as the 'second round' of macro control in China to differentiate them from the 'first round' of macro control that had started in 2005.

In this round of macro control, initiatives were attempted at the collective choice level. On 30 April 2010, Beijing Municipal Government tabled the country's first purchase restriction to curb housing demand by households with or without registered resident status (see Section 7.4 for more details). Purchase restriction was quickly adopted by another 45 cities, which represented the toughest administrative intervention in the housing market and resulted in milder increases or stagnation of housing prices throughout the cities with purchase restrictions (see Figures 4.3 and 4.8). The 2011 Measures were promulgated to verify the purchase restriction adopted by local governments. To enforce local responsibility on price inflation, it required local governments to make and publicise housing price inflation targets (see Section 7.4 for details).

The last major policy document in the 'second round' of macro control was the 2013 Measures, tabled when the premier, Wen Jiabao, was stepping down in March. The 2013 Measures imposed tougher taxation on resale incomes, which triggered unexpected market behaviour such as bogus divorce by couples who wanted to avoid the tax (see Chapter 8 for details). The new premier, Li Keqiang, has not been keen to add to the existing macro control. Market response to the change of government policy towards the housing market resulted in rapid housing price inflation in 2013 as pent-up demand from 2010 was released, particularly in BSGS (see Figure 4.9). On the other hand, differentiation of housing price inflation from the second half of 2013 occurred, with housing prices in tier 3 and tier 4 cities showing stagnation or mild decline.

Because the new government has not tabled any new macro policies, macro control as practised from 2005 to 2013 ended in 2013. The emerging new regime of housing market administration is explored in Chapter 14.

BOX 7.2 HOUSING POLICIES AND THEIR IMPACTS IN CHINA

Lin Liu

Since the reforms and the opening-up campaign, the State Council has enacted 15 documents on important housing policy issues. Seven of them were on building and improving the urban housing system with the promotion of housing marketisation as the main policy objective. The other eight were on short-term housing adjustment and control policies to intervene in the operation of the housing market, with controlling rapid housing price inflation as the main policy objective.

The essence of policies on building and improving the urban housing system, which was pushed forward from 1978, was the clarification of the responsibilities of the state and the market in urban housing provision after the housing market was established. At present, consensus has been reached on the responsibilities of the state and the market to meet the housing needs of different social groups. There are disagreements on the scope of the responsibilities of the state and the means for the state to take up its responsibility. Overall, policies on building and improving the urban housing system have produced conspicuous positive impacts.

First, a market-focused urban housing provision system was established by the urban housing reform from 1978. This system allocated housing by monetary means rather than administrative means, replacing the vicious cycle of housing funding shortage by a virtuous cycle of housing funding accumulation. It greatly released the productivity in urban housing production and fostered the private housing development sector as the core of China's housing industry. State housing management has been taken over by marketised professional housing management. The establishment of the housing provident fund and the development of housing finance and insurance led to the coexistence of policy-based and market-oriented housing credit systems.

Second, urban housing conditions have been greatly improved, especially during the rapid expansion of the housing market. Since the housing reforms were initiated, both per capita housing space and owner occupation in cities and towns have risen substantially. In 1978, per capita housing space measured by gross floor area was 6.7 m^2; in 2012, it had increased to 32.9 m^2. Owner occupation increased from 57 per cent in 1995 to 75 per cent in 2010. In particular, per capita housing space grew more than 1 m^2 every year from 1998 to 2007, when the housing market had its fastest expansion.

Compared to policies on building and improving the urban housing system, policies on short-term housing adjustment and control had unsatisfactory impacts. From a macroeconomic perspective, economic cycles, wide fluctuation in housing prices and an imbalance in the housing market are all necessary conditions for government intervention. From 2004, the Chinese housing market exhibited a structural imbalance and rapid price inflation, and the central government initiated an intervention that was the largest in scale and longest lasting since the last phase of housing reforms in 1998. However, most policies failed to achieve their design goals for the following three reasons.

First, housing policies were not implemented independently as important measures in macro control for the whole economy for most of the time after 1998. Because of the importance of housing investment to economic growth, control in housing was regarded as an important measure for macro control for the whole economy in 2004 to 2010. To slow down the growth in fixed asset investment in 2004, investment in housing was controlled, leading to lower level of supply. This contradicted the control on housing prices that required a higher level of supply. Since 2010, adjustment and control policies on housing have gained relative independence from macro control on the whole economy, and maintained consistency in increasing supply, limiting speculation, promoting the provision of affordable and social housing, and strengthening monitoring. Such consistency is conducive to stabilised market expectations.

Second, housing price inflation is closely linked to economic growth and ease of liquidity. Taking into account the contextual factors such as economic growth and liquidity, policies to contain housing price inflation, enacted frequently from 2004, were not effective. From 2004 to 2007, annual economic growth was over 10 per cent and there was plenty of liquidity, and housing prices continued to grow rapidly. In 2008, economic growth slowed down sharply due to the international financial crisis, and housing prices fell for ten months. In 2009, housing prices rebounded rapidly following economic stimulus policies that brought plenty of liquidity. In 2011, the economic stimulus policies were withdrawn due to rising inflation, resulting in tightened credit supply and a slowing down of economic growth. In that year, growth in housing demand fell to a historically low level, with falling housing prices and slowing housing supply. Hence, housing price inflation occurs simultaneously to an easing of liquidity. Policies to control housing price inflation when there is plenty of liquidity are not effective.

Third, 'one-size-fits-all' -style housing policies were enacted without taking into account regional differences. Housing adjustment and control in the last ten years since 2003 had been centred on the initiatives from the central government. Although providing national consistency, such policies did not take into account that housing problems were local in nature. There was a wide regional difference in housing conditions and economic growth in China, which required regional differentiation in housing policies. However, the short-term adjustments and control in housing were based on considerations of the central government without giving sufficient discretion to local governments.

7.3 Housing size standards: contention and resistance

Housing sizes vary according to availability of land, construction technology and financial constraints. Before 1979, China was under severe financial constraints so conducted very limited housing construction. Most of the housing built then was not self-contained.

7.3.1 Usable area and Gross Floor Area

In 1986, the National Planning Commission formulated the Residential Buildings Design Standards, which required a residential unit to contain a living room, a number of bedrooms, a kitchen and a lavatory. The minimum sizes of residential units were 34 m^2, 45 m^2, 56 m^2 and 68 m^2 in usable floor area for one-bedroom, two-bedroom, three-bedroom and four-bedroom apartments respectively; and those of rooms, in terms of usable floor area, were: 6 m^2 for a single bedroom, 10 m^2 for a double bedroom, 12 m^2 for a double bedroom with living room function; and 12 m^2 for a living room. The Residential Buildings Design Standards enacted in 2011 (MOHURD Public Notice 1093) adopts the same minimum unit size standards but a smaller room standard, with minimum usable floor areas of rooms being 5 m^2 for a single bedroom, 9 m^2 for a double bedroom, 12 m^2 for a double bedroom with living room function and 10 m^2 for a living room.

Residential sizes of commodity housing are measured by Gross Floor Area (GFA) in China, as stipulated by Housing Measurement Standards (GB/T 17986-2000) and endorsed by Administrative Measures on Sale of Commodity Housing 2001 (MoC, 2001). This is different from the convention of usable area widely used before 1990s and at odds with the 1998 *Ordinance of Urban Real Estate Development and Management* (State Council, 1998b). Article 28 of the 1998 Ordinance requires that the sales contract of housing should specify GFA and usable area. Furthermore, it is the norm for planning conditions to specify GFA, rather than usable areas, for sites. It is very rare that the number of units to be delivered are specified in planning conditions. Use of GFA as the only measurement of the size of commodity housing units caused concern among some home buyers because the trouble, or transaction costs, to find out the usable area rested with the homebuyers.

7.3.2 Larger, less efficient designs

In the 1980s and 1990s, the trend was for residential units to become larger and more luxurious to meet the rising aspirations of welfare housing renters, owner-occupiers and investment demand. Costs of renting and buying welfare housing were low. Rents were very low compared to commodity housing. For tenants who participated in the housing reforms, discounts were large for buying housing. There were no housing taxes to pay for occupation and ownership. Buyers of commodity housing at the time were normally higher-middle-income or high-income households, who could only be persuaded to part with their money for spacious housing units.

The measurement based on GFA favours developers who could include non-usable areas, such as common areas and structural area, into GFA of housing units for home buyers to pay for. Developers thus can maximise profits by building bigger apartments to reduce design, construction and installation, and sales costs. As residential buildings became higher due to higher floor area ratio specified in planning conditions, usable floor area ratio (UFAR) becomes less because structural area and common areas, such as bigger lobby, stairs, lifts and other areas, become larger. For high-rise tower blocks, UFAR is normally lower than 80 per cent of GFA, and sometimes as low as 70 per cent in a very inefficient design. Then a two-bedroom apartment of 45 m^2 usable floor area needs to be at least 64.3 m^2 in GFA for 70 per cent UFAR; a three-bedroom apartment of 56 m^2 usable floor area needs to be at least 80 m^2 in GFA for the same UFAR.

In the twenty-first century, commodity housing became the main source of urban housing. Competition between developers normally resulted in new housing being designed and built of higher, more luxurious, standards. Common areas and structural areas increased and UFAR fell. Meanwhile, housing designs became less efficient because developers tried to save on design and construction costs. Less efficient design standards coupled with the trend for luxury meant bigger unit sizes for the same housing function. It was difficult to find three-bedroom units smaller than 90 m^2 in GFA. In fact, many two bedroom units are bigger than 90 m^2. When housing prices, measured by GFA, increased, prices for homes delivering the same function rapidly increased. Home buyers' satisfaction, on the other hand, declined. One common complaint was low UFAR.

7.3.3 New size standard imposed by macro control in 2006

Macro control from 2005 started to change the costs of home ownership in relation to the size of homes by imposing a size standard. Article 5 of the 2005 Measures stipulated that ordinary housing that qualified for preferential tax treatment should be less than 120 m^2 in GFA per unit; in residential districts with Floor Area Ratio higher than 1.0; and lower than 120 per cent of the average housing prices in comparable locations. Local governments could set their own size standards within a 20 per cent range of the recommended size. From the above analysis this size standard is generous because a four-bedroom apartment with minimum usable floor area of 68 m^2 as required in the Design Standards only needs to be 97.2 m^2 GFA for 70 per cent Net Usable Floor Area (NUFA). With 70 per cent NUFA, the recommended 120 m^2 is 23.5 per cent larger than the minimum size required.

However, local governments under pressure from the housing industry almost unanimously chose the higher end of the allowable range as their local size standards for ordinary housing. For example, Beijing and Shanghai set the size standard as 140 m^2 and Guangzhou chose 144 m^2. If Shanghai, under the most severe pressure of land shortage and population density, had chosen 105 m^2 as local size standard for ordinary housing rather than 140 m^2 and adopted this size standard in planning conditions, per unit housing price would have fallen at least 33 per cent for new housing, with significant implications for affordability.

Various data sources indicated the unit sizes of new housing being supplied is too big. For example, the average unit size of urban housing throughout the country in 2003 and 2005 was 77.4 m^2 and 83.2 m^2, respectively (MoC, 2004; 2006b). The 5.8 m^2 increase indicated that newly completed housing in 2004 and 2005 were much larger than 83.2 m^2. In addition, the *Real Estate Blue Book 2006* produced by the Institute for Urban and Environmental Studies of the Chinese Academy of Social Sciences indicated that, by the end of 2005, over 50 per cent of new housing for sale in 23 out of 40 key cities had unit sizes bigger than 120 m^2. Furthermore, less than 10 per cent of new housing for sale in 16 major cities had unit sizes lower than 80 m^2 (Niu *et al.*, 2006).

The new size directive imposed by the 2006 Measures on 24 May 2006 represented the central government's determination to improve the structure of housing supply. Clause 2 of the 2006 Measures stipulated that the housing industry should focus on building ordinary housing; specifically from 1 June 2006, all new housing development projects should allocate 70 per cent of approved floor area to build housing units less than 90 m^2 GFA. Yet this directive, popularly referred to as the 70/90 policy, was equivalent to imposing an across-the-board change in planning conditions for all housing development projects. No consultation was attempted with the housing industry. Even the Ministry of Construction, the national real estate market regulator, was not well briefed. Amid confusion and dissatisfaction by the housing industry and local governments, the Ministry enacted an explanatory document verifying the directive by requiring all new housing projects follow the directive. For projects with existing planning conditions but that had not yet begun work, developers were urged to enter into negotiations with local governments to change the planning conditions to comply with the directive (MoC, 2006c).

Implementation of the 2006 Measures achieved a modest success in altering the housing supply structure. The National Development and Reform Commission indicated housing units with sizes smaller than 90 m^2 accounted for 55.8 per cent in new starts from 2006 to 2007 in 36 major cities (NDRC, 2007). However, the above data should be viewed with caution because many developers covertly resisted the 70/90 policy for high-end housing projects. The most popular tactic was to build twin flats of 75 m^2 each of which could be easily merged by buyers. A less popular tactic was to build *de facto* duplexes, i.e. apartments of less than 90 m^2 with extraordinary ceiling heights to allow easy conversion to provide another floor within those apartments.

7.3.4 Easing of compliance requirements

The slump in housing sales and falling prices in 2008 and 2009 led to the relaxation of the 70/90 policy by local governments. For example, in January 2009, the Beijing Municipal Government formally revised the 70/90 policy in its local real estate policy document, *Implementation Opinions to Promotion the Healthy Development of Real Estate Market in Our City* (BMG, 2009). Clause 11 of the document states: '[F]or commodity housing projects, unit sizes and proportions should be determined according to market demand in the context of the urban master plan, actual housing requirements, efficiency in land use and overall balance [of unit sizes] with the city.' In many cities, the requirements for small- to medium-sized apartments under 90 m^2 per unit were no longer 70 per cent, but more flexible according to the circumstances of sites and local housing requirements.

Macro control measures tabled after 2008 ceased to enforce the 70/90 policy, leaving local governments to make decisions according to local market conditions. However, 90 m^2 has been widely regarded as a unit size standard for more affordable commodity housing, and frequently used in land sales conditions and planning conditions. For example, rapidly rising prices in 2013 in Nanjing resulted in buyers scrambling for units between 80 to 90 m^2, the supply of which was quickly depleted. In response, Nanjing resumed the use of 90 m^2 as the affordable commodity housing size standard in 2014 after abandoning the 70/90 policy in 2008. New residential sites in the city were sold on condition that 50 per cent of permitted floor area be allocated to housing units smaller than 90 m^2, a policy referred to as 50/90 (Xinhua, 2014b). It can be concluded that the 70/90 policy has played its role in increasing the supply of small and affordable commodity housing, but was rendered less effective due to lack of consultation and resistance by developers and local governments.

7.4 Purchase restrictions and price control targets

Purchase restrictions were first imposed in 2006 to restrict housing purchase by foreigners (see Section 7.2) to dampen demand for housing. The 2010 Measures first made top leaders in provincial level governments accountable for housing price inflation, which prompted local governments to find ways within the local governments' power to reduce housing demand. Housing purchase restriction first emerged as a macro control measure when the Beijing Municipal Government promulgated the *Circular on Implementing State Council's Circular on Steadfastly Preventing Rapid Housing Price Inflation in Some Cities* on 30 April 2010 (BMG, 2010), 11 days after the enactment of the 2010 measures.

Purchase restrictions in Beijing were stipulated in two stages. Stage one restrictions, released on 30 April 2010, are as follows. Households with registered household status could buy up to two housing units. Households not registered but with one year's taxes payments or social security insurance could only buy one unit of housing. The number of housing units owned by households is based on Beijing's housing transaction database. Banks were required not to lend to unqualified applicants. Stage two restrictions, released on 26 February 2011, were further tightened as follows. Households with registered resident status could only buy one more housing unit if they already owned one. Households not registered but with five years' payments of taxes or social security insurance could buy one housing unit if they did not own any housing units.

Purchase restriction was quickly imitated by other cities where housing price inflation was likely, because local leaders were concerned about their responsibility to control housing price inflation. Local branches of banks were required not to lend to unqualified applicants. This was effective in mitigating housing price inflation in 2010 and 2011.

However, purchase restrictions drastically reduced housing market activities and triggered mild falls in housing prices in some cities, causing pressure on local governments' land sales. Some local governments were highly dependent on land sales and taxes from housing development and market transactions. In order not to lose momentum in economic growth, they tried to circumvent the restrictions. For example, Foshan Municipal Government tabled a new local policy on 11 October 2011 allowing some relaxation of the purchase restrictions to be effective on the next day. However, it had to table another announcement in the evening on 11 October to postpone the implementation of the new policy indefinitely (*People*, 2011a). There are other cases of local government attempts to relax the purchase restriction policy, which were all without success due to tight provincial oversight.

Like the 70/90 policy, purchase restrictions were made less effective by loopholes in public administration and perverse market behaviour. Abundant anecdotal evidence suggests that housing salespersons advised clients to divorce to qualify as purchasers and to save purchase costs. As a result, those couples who wanted or needed to buy actually divorced each other to qualify for home purchase. After completing purchase they remarried. False registered household status, bogus divorce certificates and fake ID cards were used in home purchases. This was beyond the expectation of decision-makers at the collective choice level, and is the upshot of a society in which dishonest behaviour and corruption were rampant (see Section 2.1). By the end of 2012, housing prices in the cities with purchase restrictions started to rise, signalling that the market had found ways to circumvent the restrictions and that purchase restrictions were starting to be less effective as a macro control tool.

The 2011 Measures required local governments to make housing price inflation targets for 2011 and publicise them in the first quarter of the year. All of the 657 cities complied and set local housing inflation targets. However, the targets published were normally above 8 per cent and many were even higher, which in many cases exceeded the long-term average housing price inflation of 10.26 per cent as measured by Figure 4.3. Thus such targets revealed local governments' preference, i.e. fast housing price inflation (Cai, 2013). Table 7.4 selects the targets from ten cities and lists actual housing price inflation of new housing as measured by the 70 cities price index.

The momentum of setting housing price targets dissipated because very low housing price inflation in 2011 was achieved due to the impact of the 2011 Measures, in particular, purchase restrictions. By 2012, MOHURD no longer required cities to do so and only a few cities declared their targets. This is because of pessimism within the housing industry on housing price inflation in 2012 and the ability of MOHURD to monitor housing prices in only 70, rather than 657, cities. Furthermore, the targets set by local governments in 2011 are meaningless as they are not related to affordability, for example housing price to income ratio, but only local GDP and disposable income growth targets. Nevertheless, MOHURD required local governments to meet price control targets again in March 2013.

TABLE 7.4 Housing price control targets and housing price inflation

Tiers	*Cities*	*2011 Housing price control targets*	*2011 Economic or disposable income growth targets (%)*	*Actual housing price inflation (%)*		
				2011	*2012*	*2013*
1	Beijing	Stable and slight fall	8	1.00	1.60	16.00
1	Shanghai	Lower than GDP and disposable income growth target	8	1.80	0.00	18.20
1	Guangzhou	Lower than GDP and disposable income growth target	11	3.10	2.30	20.10
1	Shenzhen	Lower than GDP and disposable income growth target	10	3.10	0.80	19.90
1.5	Chengdu	Lower than disposable income growth	12	1.30	0.40	9.60
1.5	Hangzhou	Lower than GDP and disposable income growth target	11	1.00	-7.30	11.00
2	Zhengzhou	Lower than GDP and disposable income growth target	13 & 10	3.20	0.80	11.70
2	Jinan	Lower than disposable income growth	12	0.80	0.00	9.40
3	Fuzhou	Lower than disposable income growth	12	2.70	0.80	13.10
3	Kunming	Lower than GDP and disposable income growth target	13 & 10	2.30	1.30	5.80

Sources: local governments' announcements; NBSC, various years for 70 cities price index.

7.5 Discussion: China's housing bubbles?

Are there bubbles in China's housing market? Will the fall of housing prices cause a hard landing for China's economy? Recently, China's rising housing prices (Figure 4.3) have attracted worldwide attention and worries, which led some pessimistic forecasts of market collapse by influential economists,[1] investors and institutions since 2010. This section reports and analyses data and debates on China's housing prices and empty homes.

Housing prices in Chinese cities have started to fall. In June 2014, the 70 cities index issued by the National Bureau of Statistics recorded that housing prices in 55 out of 70 major Chinese cities had fallen. The number of cities recording housing price falls has increased significantly since March, when housing prices in four out of the 70 cities fell. The falls in June triggered more talk of the bubble bursting on China's housing market.

House prices to income ratios (HPIRs) provide some indication of whether housing in China are overpriced. Figure 7.3 indicates that HPIRs were highest in 2007 and 2009, but have been falling since 2009. In particular, HPIRs in 2014 could fall to the range between 6 and 7.[2] This is in line with growth in wages. From 1995 to 2012, the average wage of employees in cities and towns grew 12.8 per cent year-on-year (NBSC, 2013). However, the national average HPIR contains wide local variations. Figure 7.4 shows HPIRs change from 2011 to 2013 for 35 major cities in China, with 15 cities having rising HPIRs and 20 cities falling HPIRs. This is in line with regional and temporal divergence of housing prices and economic growth. In 2013, only 15 out of 35 cities had HPIR lower than 8, indicating housing prices in those cities are expensive. Nevertheless, the expected fall in housing prices coupled with rising incomes will quickly pull down HPIR in more cities. On the other hand, housing prices in many tier 3 and tier 4 cities are lower. For example, Hengyang, a tier 4 city in Hunan Province, had a HPIR of 6.0 in 2013. A small fall in housing prices will make housing more affordable there.

Conventional economic theory holds that a fall in price of a normal good causes an increase in demand. However, housing has been an abnormal good in China, with a fall in prices causing buyers to watch and wait in the expectation of further falls. Such expectation is in the making at the time of writing. Then a fall in prices will initially reduce demand, which in turn will result in the contraction of housing development activities and a reduction in land sale revenues. The diminished fiscal capacity of local governments will impact on the completion of many infrastructure projects. The slowdown of housing development and infrastructure projects will spill over to many industries that provide supplies to housing and infrastructure projects. Hence economic growth will slow down significantly, with consequences in employment and social stability.

However, it should be borne in mind that this round of housing price correction was engineered by the central government's tight money policy, which led to the banks' reluctance to lend. So far, the central government has ruled out rescue packages similar to those in 2008 and 2009. However, the central government has been

FIGURE 7.3 Housing price to income ratio for urban China
Sources: E-house China, (2012; 2014a).
Note: *the figure 2014 a forecast.

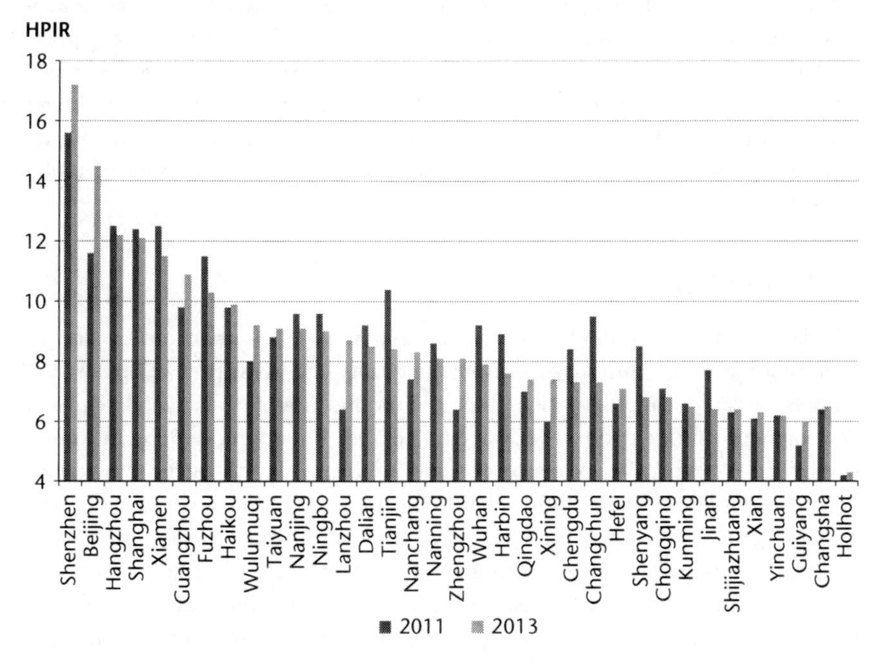

FIGURE 7.4 Change of housing price to income ratio from 2011 to 2013 for 35 major cities in China
Sources: E-house China, (2012; 2014a).

active in mini-stimulus policies to promote the development of other industries and support infrastructure projects. This is to achieve a change of structure in the economy, which was too reliant on the housing market. Furthermore, the central government tacitly tolerates cities' relaxation or even removal of purchase restrictions and allows local governments to borrow in the capital markets. The central bank has urged commercial banks to support home purchase. The logic here is to promote a reasonable speed of economic growth. With continuous growth, the demand of the tens of millions of households who were priced out of the housing market can be released if housing prices fall modestly. Thus, it is likely that the economy will not be seriously affected by a correction in housing prices but instead will achieve a certain degree of restructuring to a more healthy structure. More discussion on the future of the real estate market will be provided in Chapter 14.

A recent article in *The Economist* neatly captures the situation:

> China's building boom has left some parts of the country with too much floor space and other parts with too little. Nearly half of all migrant workers still live in dormitories or on worksites. Where housing is oversupplied, prices will have to fall, inflicting losses on homeowners. But where housing needs remain unmet, scope remains for further construction to fill the gap.
>
> *(The Economist, 2014)*

What the above article does not cover is that every year, 7 million university graduates from all over the country join the workforces in the cities. High housing prices have resulted in tens of millions of them renting and unable to get married.[3] Furthermore, there are several million urban youngsters joining the workforce every year without going to university. These people also need homes of their own. The key is that the constitutional level should act to make housing more affordable to them.

7.6 Conclusion

The housing market administration is one important task carried out by the real estate administration authority to ensure the quality of real estate development and services in real estate transactions, occupation and investment. Good results in development quality control have been achieved, but recent actions to impose market order reveal the existence of many problems in real estate transactions. Parallel to market administration were the efforts by the central government to rein in housing price inflation from 2005 to 2013. Two rounds of macro control were imposed and have brought down housing prices. However, contextual factors such as the international financial crisis and the change of government have resulted in discontinuity of many of the rules imposed by the macro control. Imposition of size standards by the constitutional level, a new institutional arrangement, was done without consultation and met strong resistance from both the collective choice and operational levels. Without the support of local governments, developers were not penalised when building twin flats and *de facto* duplexes to avoid the 70/90 rule. On the other hand, the rule to make local leaders responsible for housing prices encouraged innovation at the collective choice level, with purchase restrictions invented by Beijing Municipal Government and adopted by 45 other cities with rapid housing price inflation. Nevertheless, profit motives resulted in perverse market behaviours such as false divorces and fake identity documents to circumvent purchase restrictions. An evaluation and critique of the macro control measures are presented in Chapter 14.

By June 2014, the long-expected fall in housing prices had materialised throughout the country. However, the pessimistic forecasts of the coming collapse of the housing market and the resultant abrupt economic slowdown are not based on a thorough scrutiny of the fundamentals of the Chinese economy and the realities in the housing market. Rigid demand is still active, due to urbanisation and population growth in the present urban areas. As the engineer to cause the housing market correction, the Chinese government can take the opportunity to implement the long-awaited restructuring of the economy by supporting many sunshine industries to maintain economic growth and thus income growth. Strong pent-up demand will be released for housing when housing prices stagnate or fall if the economy continues to grow at medium speed.

Notes

1 For example, Nouriel Roubini in 2011 and Paul Robin Krugman in 2012 expressed pessimism about China's investment-led economic growth. However, both made their comments without a careful scrutiny of the fundamentals.
2 In Figure 7.3 the house price to income ratio for 2014 is estimated as follows: average house price in 2014 is expected to fall by 4 per cent in nominal terms; and average income to rise by 6.5 per cent in nominal terms. The context for such income growth, which is a conservative estimate, is an estimated 7.5 per cent GDP growth and 2 per cent inflation for China in 2014.
3 In China, home ownership is normally required for a couple to get married.

8

PROPERTY FINANCE AND TAXATION

This chapter surveys property financing on both the supply and demand side, and explores taxation on property development, transactions and holding. It first examines development finance, analysing the sources of development finance and revealing the dominant position of domestic bank lending. The chapter then investigates mortgage finance, both market-based and policy-based, and reviews the use of mortgage deposit and interest rates by macro control to influence housing prices. Taxation on property is then investigated in the chapter. Taxation on property development involves many taxes and a myriad of fees; taxation on housing transactions was first reduced but then substantially increased to control housing price inflation. The chapter ends with a discussion on the delayed real estate tax and the direction of property tax reform.

8.1 Development finance

Property development has been playing a dominant role in the property industry in China since the re-emergence of the property industry in the 1980s. The change of development finance has shaped the direction of property development.

8.1.1 Lending to property development in 1980s

In the beginning, market-oriented property development was in the state plan and was financed by local governments or by loans from state banks arranged by local governments. The state was able to control development activities through orders to local governments and state-owned banks. Market-oriented property development activities reached their peak in Guangdong in the second half of 1980s. The economy was expanding rapidly but the productive capacity of the country was unable to supply that expansion. Inflation rocketed and shops were emptied by shoppers for fear of further price hikes. To control inflation, the central government in 1988 stopped banks from supplying new funds to development projects and required developers to return the borrowed funds in an attempt to cool down the overheated economy and bring down the double digit inflation (see Figure 2.4 for inflation figures in the 1980s). The boom and bust of the small property market in Guangdong constituted the first property cycle in the 1980s.

8.1.2 The rise of mortgage loans

Bank lending in the form of mortgage loans to real estate developers was first practised in Guangzhou. In 1989, the largest office project in Guangzhou at the time, the Guangzhou World Trade Centre, was short of funds because of central government's control of credit supply. The state-owned developer, Zhujiang Industries, was delegated the authority by the Guangzhou Municipal Government to draft the *Guangzhou Administrative Measures on Real Estate Mortgage*, which was completed and promulgated in June 1990 after 11 drafts. This is the first regulatory document on property financing and paved the way for a Singapore consortium and the China Industrial and Commercial Bank to lend US$10 million and US$20 million, respectively, to Zhujiang Industries in July and November of the same year. The loans were crucial to allow the project to be completed in 1992.

8.1.3 First macro control and credit tightening

Tight credit conditions from 1988 cooled down the economy and mitigated inflation. However, the economy was hit hard and GDP growth was 4.1 per cent in 1990, the lowest since 1976 (see Figure 2.1). Development financing was restored to normal when the State Council decided to provide financial support from the state banking sector to tertiary industries, including the property industry, in its move to develop the tertiary sector of the economy in 1992 (State Council, 1992). However, state banks had only limited funds allocated for property projects at the time and could not satisfy the borrowing needs of hundreds of development projects approved by local governments. An active underground lending market developed among banks, firms allocated with borrowing quotas and developers. Banks diverted money from projects on which they had been instructed to lend to property development, and firms receiving quotas of credits for investment projects lent their money to property developers. Interest rates for such underground lending were at 200 per cent or higher than the regulated rates set by the central bank. Thus banks normally lending at the regulated rate could lend at much higher rates and made good profits. Likewise, firms borrowed loans at the regulated rate and could lend the loans at much higher rates, again making good profits. As a result, some local state banks and many state-owned firms diverted the funds earmarked for other purposes to lend to developers for profits, exposing them to high risks and causing a shortage of funds for some key state projects. In some places, the diversion of funds earmarked for the purchase of farm produce caused local governments to issue IOUs to farmers, which led to dissatisfaction and anger among the farmers.

The unregulated bank lending in 1992 and early 1993 to property development was the cause of the 1993 macro control (see Table 7.2), the first of its kind by the government to use both fiscal and monetary policies to target property development. This macro control had major implications for future development financing. In June 1993, the central government forced billions of dollars of loans lent through the underground lending market to be repaid in weeks, resulting in the drying up of funds for hundreds of property development projects throughout Hainan Provinces, Beihai City in Guanxi Province and Huizhou City in Guangdong Province (see Section 3.3 for more details). After 1993, bank lending to the property development sector was under strict control and total lending stagnated until 1998.

The tight control on bank lending to property development was lifted from 1998 when the property sector was designated the new growth engine of the economy. In April 1998, the People's Bank of China (PBoC) promulgated the *Circular on Expanding Credit Supply to Support Housing Construction and Consumption*, which encouraged banks to support development projects and home purchases (PBoC, 1998). At that time, a lot of the bank lending went to state-owned enterprises, many of which were making losses. Many of the loans became non-performing and had to be written off by the central government. Lending to developers and home buyers with land and housing as security significantly lowered the risk of lending, an important factor leading to a significant increase in bank lending to property developers and home buyers. Figure 8.1 shows total funding in property development and year-on-year growth.

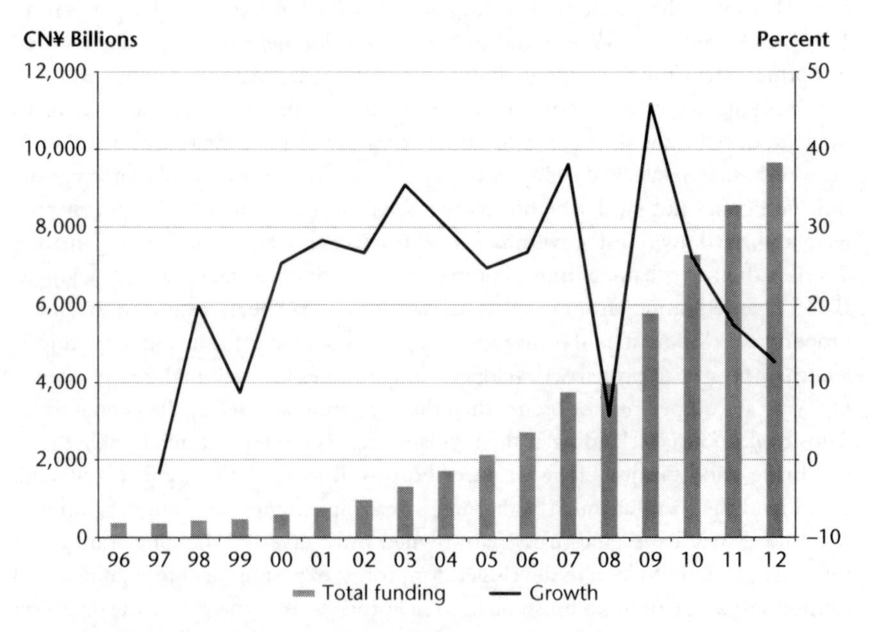

FIGURE 8.1 Financing for property development, 1996–2012
Sources: NBSC, various years.

8.1.4 Requirements for self-owned funds on development projects

To reduce banks' exposure to property risks, in 2003, the central bank first required a minimum amount of self-owned funds for property development projects, limiting bank lending to no more than 70 per cent of the total cost of development (PBoC, 2003). Although unpopular with the development sector, this requirement was upheld by the central government in its pro-property policy document in 2003 (State Council, 2003).

As part of the coordinated effort in macro control, in 2006, a tougher requirement was stipulated by the 2006 Measures to raise the proportion of self-owned funds to 35 per cent of total development costs, limiting bank lending to no more than 65 per cent of the total cost of development. However, the effect of this tougher measure was not significant because development firms had diversified their sources of development funding, with bank lending now having a minor role.

In May 2009, the State Council lowered the proportion of self-owned funds to 20 per cent of total development costs for real estate projects, including affordable and social housing and ordinary commodity housing to increase the supply of such housing (State Council, 2009a). This move provided a large amount of credit to the development industry to expand its scale of operation in housing and commercial property. The impact on accelerated commercial property development is discussed in Chapter 10.

8.1.5 Sources of development funding

Tight bank lending under the macro control since 1993 has shaped a unique funding structure for market-oriented property development in China. Figure 8.2 illustrates the sources of finance for property development from 1996 to 2012, which is divided into four main types, i.e. foreign funding, domestic bank lending, self-raised funds and other sources. Foreign funding consists of mainly foreign direct investment and other foreign funding, which was important as a source of property development finance in the 1990s but has become insignificant since then. Bank lending includes all loans lent by banks and non-bank financial institutions normally secured on land and other property assets owned by the developers. Although from 1998 more domestic loans were made available to developers, the share of direct domestic lending has actually been falling since 1996. In particular, direct domestic lending fell significantly after the central bank issued a warning on the risks of housing price inflation and tightened the amount of lending to less than 70 per cent of total investment in development, requiring developers to contribute at least 30 per cent of own funding (PBoC, 2003). Tight direct lending forced developers to rely mainly on the other two types of funds, i.e. self-raised funds and other sources.

Self-raised funds include developers' equity, retained profits and receipts from the sales of finished property, with the latter taking a bigger share. Other sources consist of mainly deposits and presale receipts from buyers. According to the central bank's 2004 *Real Estate Finance Report* (PBoC, 2005), deposits and presale receipts were

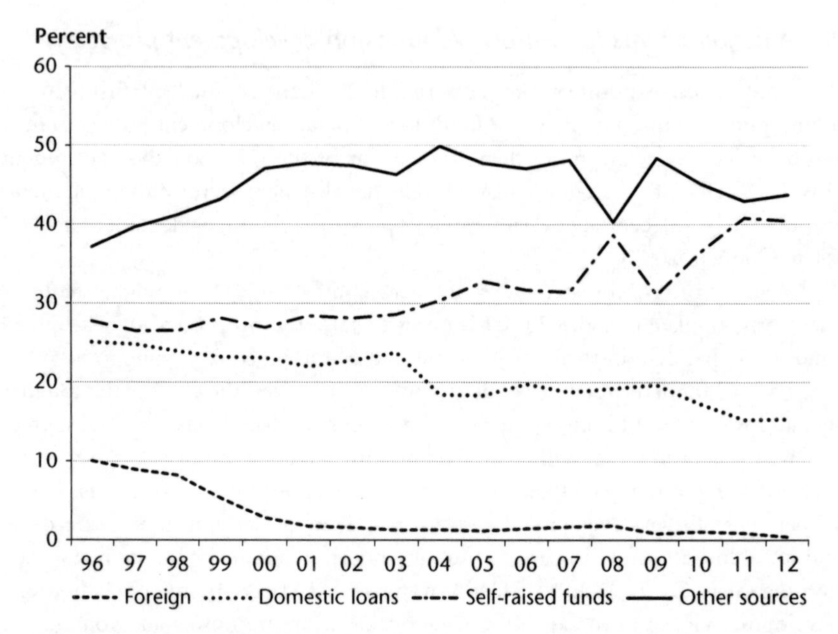

FIGURE 8.2 Composition of the sources of funding for property development, 1996–2012
Sources: NBSC, various years.

CN¥739.53 billion in 2004, accounting for 86.4 per cent of the total of other sources. It is worth noting that about 70 per cent of the receipts from sales are from mortgage loans borrowed by buyers from banks, and 30 per cent of deposits and presale receipts are also from banks loans borrowed by buyers. Thus the real exposure of banks to development finance in 2004 is 55 per cent rather than 18.4 per cent indicated by direct lending (ibid.).

8.1.6 Increase in real estate loans

Figure 8.3 depicts the changes in the amount of total loan balance and real estate loan balance, with the proportion of real estate loans increasing from 13.47 per cent in 2003 to 19.60 per cent in 2011 and to 19.03 per cent in 2013 (PBoC, 2014). The averaged year-on-year growth rate of total loan balance was 15.38 per cent during the period from 2003 to 2013, but for real estate it was 19.06 per cent. A breakdown of the real estate loan balance in 2013 is as follows. Loan balance for property development was CN¥4.6 trillion, with CN¥3.5 trillion for building development and CN¥1.1 trillion for land development. The loan balance for mortgages was CN¥9 trillion.

8.1.7 Presale

The practice of presale, imported from Hong Kong in the early 1990s, involves the sale of contracts of property units under development with a future completion

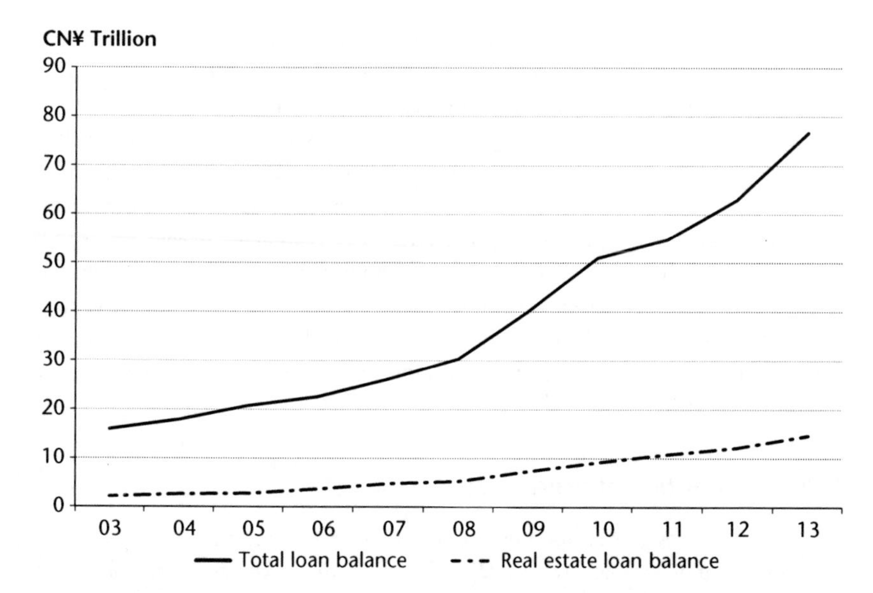

FIGURE 8.3 Total loan balance and real estate loan balance, 2003–13
Source: PBoC (2014).

date. Buyers in fact purchase the rights to occupy the property in the future in exchange for a small discount on the current prices. They can choose to pay a lump sum for bigger discounts, or pay a deposit, with the remainder paid before or after completion and occupation. Presale receipts are held by developers as part of the funding to complete the developments. Through presale, developers pass on part of the risks of property development to buyers and reduce their dependence on bank lending, allowing developers to carry out more projects than allowed by direct borrowing and own sources of funds. However, buyers incurred heavy losses when developers could not complete the projects on time. For example, macro control in 1993 left many projects suspended for more than a decade in Hainan Province and Beihai City in Guangxi Province. Even in Guangzhou and Shanghai, the property market downturn from 1997 resulted in many failed projects, many of which involved pre-sold units (see Box 4.1 for an example).

The regulatory framework for presale receipts, first installed in 1994, consisted of relevant provisions in Urban Real Estate Administration Law (UREAL) and *Administrative Measures on Presale of Urban Commodity Housing*. Presale permits, introduced in 1994 and issued by local governments, require developers to complete 25 per cent of the budget on construction, pay land purchase costs and obtain all necessary statutory approvals such as planning consent (see Section 6.3 for details on planning consents). Local regulations require more concrete indicators of completed investment. For example, in Chongqing, projects for presale must meet the conditions that the main structure of the building is completed in respect of buildings that are eight storeys or lower, and half of total floor areas completed for buildings that are taller than eight storeys. Guangzhou and Shenzhen require completion of the main structure (to the top floor) for buildings that are seven floors or less and completion of two-thirds of all floors for buildings that are taller than seven floors.

Figure 8.4 indicates that the widely adopted presale of real estate in development has resulted in an expanding gap of completion and sale, heightening the risks of non-completion to buyers and to the property industry as a whole. In 2013, the total floor space for all types of buildings under construction was a staggering 6.66 trillion m², which requires enormous financing to complete. In fact, the central bank highlighted the risks of presale and recommended the revocation of the practice in 2005 (PBoC, 2005). However, this proposal was refused by the Ministry of Construction (now MOHURD) which argued that the practice was enshrined in UREAL and the cancellation would have a serious negative impact on the development industry (*People*, 2005).

8.1.8 Cash position of developers

The unique sources of development financing have resulted in the highly volatile cash positions of developers, which depend on the impact of macro control policies, credit supply, interest rates, housing prices, presale and completion of sold property. Figure 8.5 depicts the cash position composite indicator of development firms created by E-House China R&D Institute in China (E-House China, 2014b), which is calculated from variables drawn from the following four areas: sources of financing, utilisation of funds, sales receipts and the credit environment. The movement of the indicator matches the housing price movement. In 2002 and 2003, policy support and loose credit ushered in accelerated price inflation. In 2004, credit was tight but the fundamentals were good. Developers held on to prices, which continued to

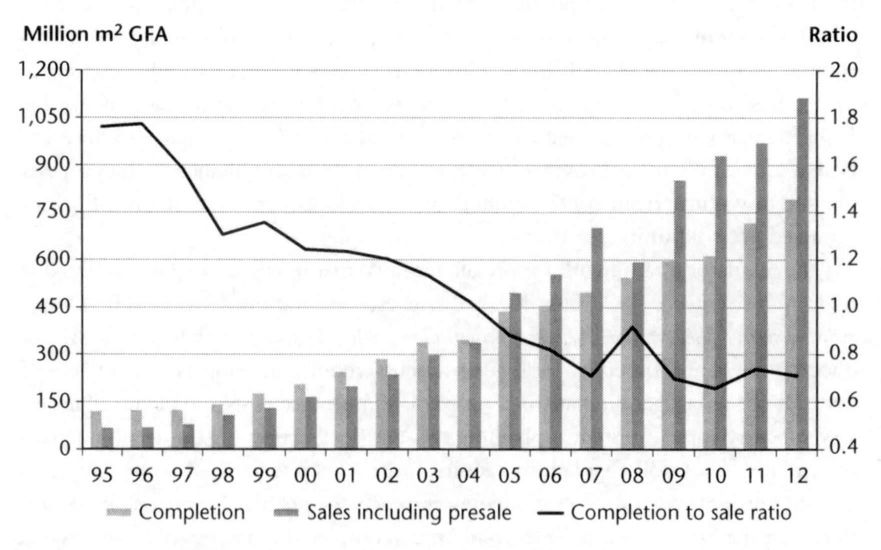

FIGURE 8.4 Completion and sales of commodity housing and completion to sale ratio, 1995–2012

Sources: NBSC (2013, 2014b).

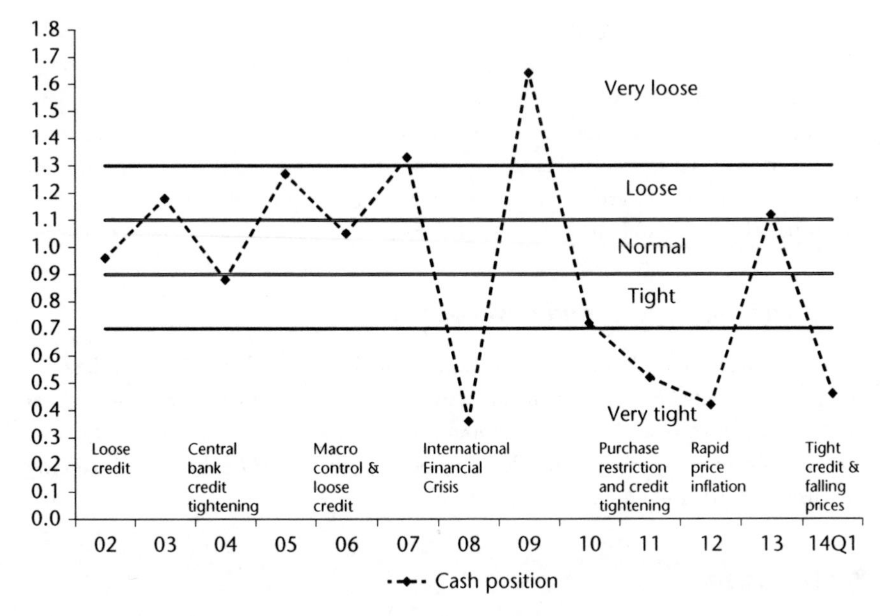

FIGURE 8.5 Cash positions of developers according to E-House indicator, 2002–14 (Q1)
Source: E-House China (2014b).

rise. The first round of macro control from 2005 to 2007 was conducted when developers were in loose, normal and very loose cash positions. With plenty of cash, developers had no incentive to lower prices, generating a situation of rising prices during macro control. However, further tightening of credit, in particular, the new deposit requirement imposed by the central bank in 2007, was effective in reducing sales and thus aggravating developers' cash positions. Housing prices decelerated in November 2007 and started to fall in August 2008. The international financial crisis further tightened developers' cash condition, who responded to lower prices by trying to recoup cash to avoid further deterioration of their cash position.

The release of a rescue package at the end of 2008 created exceptionally loose cash positions for development firms in 2009 due to the ease of borrowing and good sales. With a strengthened cash position and strong demand, developers raised prices sharply and expanded housing development activities. Purchase restrictions from 2010 and the resultant stagnant sales generated very tight cash positions for developers, who cut prices to attract buyers in order to maintain their cash positions. Rapid price inflation and expanded sales improved the cash position of the developers in 2013, who bid up land prices in expectation of further housing price inflation. Nevertheless, tightened credit, a looming oversupply of housing and a slowing economy have dampened the expectation of further housing price inflation, leading to sharply falling sales. In response to a worsening cash position, developers started to lower the prices of new housing, leading to across-the-board housing price deflation from mid-2014.

Further research is needed to scrutinise the relationship of developers' cash position and their decision to lower prices. But from the above analysis, the cash position has a positive correlation to housing price inflation. The failure of macro control in 2005 to 2007 was due to the lack of measures to impact on developers' cash positions. The lessons indicate that mini-control targeting the cash positions of developers is more effective than macro control if the government wants to slow down housing price inflation or cause deflation.

8.1.9 The rise of trusts and private equity

The restriction in bank lending and the unreliability of self-raised funds, deposits and presale receipts gave rise to the growth of alternative financing provided by real estate trusts and private equity. Financing by trusts and private equity has grown substantially and played a major role in property financing. The details of financing by trusts and private equity are provided in Chapter 11.

8.2 Mortgages

Mortgages offered by commercial banks became available in the 1990s on a small scale. In 1997, the central bank encouraged commercial banks to offer mortgage loans in cities which carried out housing projects for low- to middle-income families. Home mortgages were offered to applicants buying housing built by loans from Housing Provident Funds (see Section 8.3 for details). The terms of home mortgages were limited to 20-year maximum length; deposit requirement was 30 per cent; and the interest rate was the central bank's standard fixed asset lending rates (PBoC, 1997).

To support housing purchase by families, in April 1998, the People's Bank of China, the central bank, enacted the Circular on Expanding Housing Credit Supply to Support Housing Construction and Consumption, which deregulated mortgage lending to increase credit supply and delegated all decision-making on mortgage loans to commercial banks (PBoC, 1998). Home mortgage loans became available for home buyers who purchased all types of commodity housing.

To address the violation of regulations by some developers working with banks to offer zero deposit loans to promote property sale, the *Circular on Standardising the Housing Finance Sector*, was enacted in 2001 by the central bank to regulate mortgage loans (PBoC, 2001). The circular set the minimum deposit requirement as 20 per cent. If the mortgage loans were to purchase presale housing, the main structure of the building must be completed for low-rise housing blocks or two-thirds of the floors should be completed for high-rise housing blocks. Mortgage loans to buy commercial property were limited to ten years in length and for completed buildings only, with a minimum deposit of 40 per cent. In 2003, the central bank advised that the minimum deposit for second home purchase should be further increased to dampen housing demand (PBoC, 2003).

To be part of the concerted effort to control housing price inflation, the central bank tightened the money supply to home purchase by raising interest rates (see Figure 4.10) from 2004. In the 2005 Measures (see Table 7.2) commercial banks were required to adjust their lending structure to lower risks and conduct due diligence investigations in lending. In particular, they were urged to conduct an investigation before processing loan applications, an examination at loan approval and checks after the loan approval. Such requirements were in response to the role the banking sector played in financing Shanghai's housing market speculation in 2004. It was very easy to borrow money from banks to buy housing in Shanghai. There is a lack of due diligence in some lending cases. For example, Pudong Development Bank, a local state-owned bank in Shanghai, was involved in a series of fraudulent loans. The crime began to emerge when the auditor discovered one mortgage loan was lent without following the normal lending procedures. On subsequent investigation it was discovered that not just one, but in fact 32 fraudulent loans had been made to the same client over a two-year period since 2004. This was despite not only dubious valuations and incorrect procedures, but he was also using *someone else's* identity card. Ultimately, however, it was the amount involved that rendered investigators speechless: CN¥126 million (US$15.4 million) (Nanfang, 2006).

After 2006, home mortgage loans became an important target for macro control. First, the 2006 Measures, in which the central bank was a signatory, required a 30 per cent deposit for all mortgage loans for homes larger than 90 m^2 from 1 June 2006. A 20 per cent deposit was needed for homes smaller than 90 m^2. Second, when the 2006 Measures failed to stop housing price inflation, the central bank imposed tougher requirements on the deposit (PBoC, 2007) as follows:

- Applicants to supply information on how many housing units he or she has purchased.
- Minimum deposit of 20 per cent for first-time buyer buying units smaller than 90 m^2.
- Minimum deposit of 30 per cent for first-time buyer buying units larger than 90 m^2.
- Minimum deposit of 40 per cent for buyer buying second or more housing units; larger deposit for more housing units purchase, and an interest rate at 1.1 times of standard rates.
- Monthly repayment to be less than 50 per cent of applicant's income.

The measures imposed in 2007 by the central bank were considered instrumental in bringing down the housing prices before the full impact of the Global Financial Crisis hit China.

In the second half of 2008, the operations by the banks rescued the housing market. The restrictions on deposits were gradually lifted. The first rate cut in September 2008 since October 2004 was accompanied by an 85 per cent discount on standard mortgage interest rates, according to the lending banks' discretion. In January 2009, the largest state-owned banks, i.e. the Industrial and Commercial Bank of China, the Bank of China, the China Construction Bank and the

Agricultural Bank of China, provided 70 per cent discount to standard interest rates on home mortgage loans. Such operations contributed to the rebound of housing prices in 2009.

Adjustments to mortgage lending to curb housing prices resumed in the 2009 Measures which raised the minimum deposit to 40 per cent for buyers buying their second or more housing units at interest rates reflecting the risks of such purchases. The 2010 Measures went further, to ban mortgage loans for third homes any families bought, and the 2011 Measures raised the deposit requirement for second home purchases to 60 per cent and the interest rate at 1.1 times the standard loan rate. Coupled with interest rate hikes, housing prices fell in 2011 and the first half of 2012. However, a new round of price inflation started in the second half of 2012 and the tight lending conditions failed to stop the rapid housing price inflation in 2013.

Nevertheless, the credit supply was tightened when the new government, led by Premier Li Keqiang, took office in March 2013. A mini credit crunch occurred in the middle of the year when the central bank reduced the money supply. Secondary banks had to raise cash from the inter-bank lending market, and the Shanghai Interbank lending rate reached new highs. Although the central bank later relaxed the credit supply, all commercial banks started to tighten lending and eventually stopped issuing mortgages with interest rate discounts. By the end of 2013 and the start of 2014, some commercial banks had stopped lending home mortgages, which contributed to slower price inflation in early 2014 and falling prices in mid-2014.

8.3 Policy-related home mortgages

The Housing Provident Fund (HPF), a compulsory saving scheme mainly for help with housing purchase, was first set up in Shanghai following Singapore's experience in 1991. It required most employees and their employers to contribute monthly to the fund managed by the Shanghai Municipal Government. Participants included government agencies, state-owned and collectively owned firms, private and foreign firms, and non-profit-oriented organisations. In 1992, the HPF was adopted by Beijing, Tianjin, Wuhan and Nanjing. The central government at the constitutional level recognised this practice and made the HPF a national policy in 1994 for other local governments to adopt (State Council, 1994a). In 1999, the first national regulation, the *Housing Provident Fund Administration Ordinance*, was enacted to provide regulatory guidance on the operation of the scheme in China.

The HPF became an alternative funding source for home purchase to mortgage loans from commercial banks, offering an average mortgage interest rate which was 2.08 per cent lower than that offered by commercial banks (Figure 8.6). By the end of 2010, the HPF covered 86.06 million employees out of a total of 118.23 million qualified employees throughout the country (Jiang, 2012). Other measures of the scale of the HPF were provided in a briefing by MOHURD, the national administrator of HPF. Total savings held by HPFs nationally was CN¥2.07 trillion by the end of 2008. In that year, 1,313,000 loans with a total value of CN¥203.59 billion

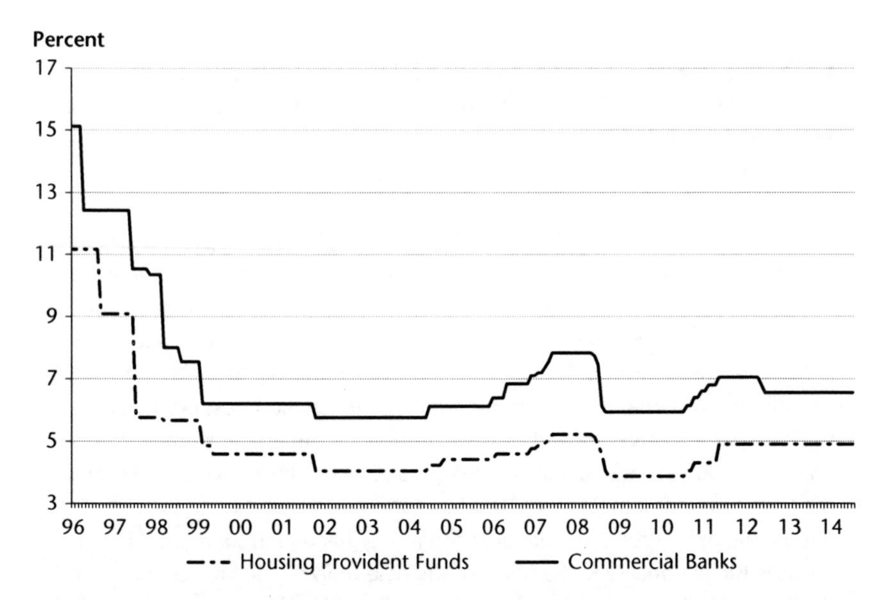

FIGURE 8.6 Housing Provident Fund lending rates compared to commercial headline lending rates, 1996–2014

Source: People's Bank of China.

were lent to home buyers from HPFs. The total HPF home mortgage loan balance rose 20.1 per cent from 2007 to CN¥609.42 billion, and the total home mortgage loan balance from commercial banks was CN¥2.98 trillion in the same year (PBoC, 2008). The total lending of HPF reached CN¥1.06 trillion by 2008 and the total number of loans was 9,611,700 (MOHURD, 2009a).

Local governments at the collective choice level sometimes made use of their power to bend national rules on macro control. For example, housing market inactivity due to the purchase restrictions imposed by the 2010 Measures led some local governments to relax HPF lending to encourage home purchase by qualified purchasers with weak financial standing. For example, on 23 October 2011, Ningbo Municipal Government allowed home mortgage loans from commercial banks to be converted to HPF loans up to the lending limit of CN¥600,000 which would save a significant amount of interest payment due to the difference in interest rates between commercial and HPF loans (*People*, 2011b). Nanjing Municipal Government, on the other hand, raised the borrowing limit of the HPF from CN¥200,000 per person and CN¥400,000 per household to CN¥300,000 per person and CN¥600,000 per household on 24 October 2011 (Sina, 2011).

Nevertheless, demand for HPF loans outstripped the limited pool size of HPF funds in many cities. From the second half of 2013, many applicants in those cities were denied loans due to the funds being exhausted. However, in other cities there were still idle funds in HPF accounts. MOHURD, the HPF policy-maker, is considering proposals from academics to unify the HPF management throughout the country into a housing bank for affordable housing.

BOX 8.1 REAL ESTATE MARKET DEVELOPMENT IN SHANGHAI'S PURSUIT OF AN INTERNATIONAL FINANCIAL CENTRE

Xiaobo Dai

Since becoming a Treaty Port in 1841, Shanghai has developed rapidly into an international metropolis. Since 1949, Shanghai has been a major industrial base in China. The city entered a new era in 1979 when the reforms and the opening-up campaign were initiated, in particular, in 1992 when Pudong New District was established. In the mid-1990s, Shanghai set itself the target of becoming an international economic, financial, trade and transportation centre by 2020, which was endorsed by the central government. Since then, the city has carried out a series of financial reforms and institutional changes to nurture the development of the financial industry, attracting many domestic and international banks to set up regional and national headquarters there. Shanghai became the national securities market, the gold market, the futures market and the Renminbi liquidation centre. In 2013, the central government approved the China (Shanghai) Pilot Free Trade Zone, promoting financial reforms and business expansion such as Renminbi internationalisation, cross-border financing, capital market development and the free movement of foreign currency under capital accounts. These reforms and business expansion mark the swift but steady progress in building Shanghai into an international financial centre.

The impact on real estate of Shanghai's march towards being an international financial centre falls mainly on real estate financing, which is still at an early stage of development. The city already has a bigger real estate financing sector than other cities in China, with 40 per cent of total lending to real estate compared to the national average of 17 per cent. A multiple channel financing system has been established, with shares, trusts, real estate funds and bonds all enjoying rapid growth, especially in the past five years. Due to the restrictive national policy, initial public offering (IPO) on the Shanghai Stock Exchange by real estate companies is still difficult, with most IPOs by real estate firms being carried out in Hong Kong. However, there is great potential in this area when IPO policies are gradually relaxed in China.

After 30 years of rapid development, sales of commodity housing declined in 2014. According to studies conducted by the Shanghai Academy of Social Sciences, China's population will peak at 1.4 billion in 2026 and decline afterwards. Following the increase in existing housing stock, Shanghai and the whole country will enter a new era of the housing market, with existing housing stock dominating housing transactions in the market. In the next ten years, the focus of housing transactions in the country will be the transition from new housing to existing housing. As a result, the focus for real estate financing will change from development financing to mortgage financing. As the monetary policies of the country relax, Shanghai, an international financial centre, will become more important in currency exchange, interest rate setting, inter-banking lending, bills exchange and asset mortgage decisions. Following the relaxation of the country's capital market policies, Shanghai will become more important in IPOs, equity financing, free exchange under capital account, asset securitisation and mortgage securitisation.

The urban area of Shanghai has already seen housing transactions being focused on existing housing. With very limited new supply, average housing prices in the urban area rose to US$8,000 per m^2. On the periphery of the urban areas, average housing prices were US$5,000 m^2. In the outer suburbs, they are US$2,000 per m^2 even though there are limited amenities there. As more migrants arrive, and financing becomes more easily available in Shanghai, the financial market will play a more important role in real estate transactions. Shanghai's real estate market will become a barometer for the financial market.

8.4 Taxation in the property market

Property is a good target for taxation. In China, there is a myriad of taxes at the different stages of property development, transaction, occupation and investment, which together make a significant contribution to the government's fiscal revenues, particularly at the local level. Such taxes contribute to the fiscal reliance of local governments on the property market.

8.4.1 Taxation and fees on property development

In the early days of the property market development, commodity housing was for high-income individuals and cash-rich state-owned enterprises to buy, and could thus be sold at relatively high prices. Many government agencies used their power to impose fees or taxes on property development that was considered lucrative. Government agencies that have a stake in real estate include those responsible for taxation, finance, land, construction, planning, municipal administration, archaeology, transactions, fire, telecommunications, environmental protection, public hygiene and quality control. All public utilities are also involved. For example, the number of taxes and fees were so many in the 1990s that the Guangzhou Real Estate Society (GRES) carried out a study to clarify the situation of fees and taxes. The study reported that there were 62 taxes and fees in the following stages of property development: land purchase, project approval, planning and building regulation consents, construction, completion and sales (GRES, 1994). GRES complained that some of the taxes and fees were not justified at all. All those taxes and fees accounted for 25–30 per cent of development costs. The time and effort spent paying those taxes and fees were quite a burden on developers, constituting high transaction costs in property development.

The taxes and fees later became the battleground between collective choice and constitutional level agents. The central government tried to axe the taxes and fees but the local governments tried to invent more of them. For example, the National Economic Development Planning Commission and the Ministry of Finance repealed 48 taxes and fees in December 1996. But new ones were invented to take their place. At present there are 11 taxes that development companies need to pay, with Business Tax, Land Appreciation Tax and Corporation Income Tax being the main ones. The number of fees varies. Song (2014) reports that Hengda, one of the biggest house builders, found that fees charged by local governments varied from 37 types to 157 types. Eleven taxes accounted for 13–15 per cent of housing prices, and fees accounted for 8–12 per cent. The total ranged from 20–25 per cent of housing prices. Because taxes and fees can sometimes be waived or renegotiated, developers usually keep an in-house team to sort them out.

The tax burden by the development sector has been widely claimed to be one of the reasons for high housing prices. Such a claim, however, is untenable. First, many of the taxes on property development are rather small in monetary terms and have an exemption or rebate provision if certain conditions are met. Many of the fees are legitimate because they are for services provided. For example, the Planning Service Charge at 0.2 per cent of the construction costs is for planning services provided by the planning agencies in the planning application. The Construction Work Drawings Inspection Fee at CN¥1.3 per m^2 of GFA is for building regulation control. Nonetheless, local governments should combine the many fees to reduce the transaction costs in identifying and paying them. Second, housing prices are determined by demand rather than costs. For example, rapid housing price inflation in the second half of 2009 was not accompanied by a significant rise in taxes and fees. Similarly, housing prices fell in 2008 and 2011 even though taxes and fees remained the same.

Avoidance of taxes and fees thus became a profitable business for developers, and investigation of tax avoidance by developers became an annual exercise for government taxation agencies. For example, government taxation agencies investigated 96,300 firms in the property sector in 2010, with CN¥29.45 billion in recovered tax revenues, which is 49.23 per cent of the total amount of recovered tax revenues for the year (Zhao, 2014).

8.4.2 Taxation on housing transactions

The tax burden on real estate transactions fell from the 1990s until 2005 when macro control increased the tax burden on real estate transactions. A notable example is the Deed Tax, which had been 6 per cent of market value since 1950 and was reduced to 3 per cent in 1997. To support the housing market, the Deed Tax was lowered to 1.5 per cent in 1999 for ordinary housing but remains at 3 per cent for luxury housing and villas. In Shanghai, the municipal government reduced it to 0.75 per cent from 1999 to stimulate the subdued housing market until 2001 after its effect in increasing market transactions was achieved. The low transaction costs on housing, in particular in Shanghai when the housing market was inactive, were important to allow the housing market to take off and become the main provider of housing for the urban population. On 1 November 2008, the Deed Tax was further lowered to 1 per cent for housing with unit sizes smaller than 90 m^2 to boost transactions. At present, the rates of Deed Tax vary across localities. Table 8.1 shows the situation in Xi'an. In addition to Deed Tax, there is also a Stamp Duty which is 0.05 per cent of the value of the transactions, and fees for the production of the title deeds.

Macro control to combat housing price inflation since 2005 has substantially increased the tax burden on most types of housing. The 2005 Measures first imposed a Business Tax, which is a turnover tax on housing transactions. Subsequent macro control policies made frequent use of the Business Tax as a policy tool to adjust housing demand, raising the tax when housing price inflation was too high but lowering the tax when the opposite occurred. Table 8.2 lists the detailed stipulations on the Business Tax in four macro control policy documents.

A tax on capital gains in the form of income tax was imposed in 2007 in a follow-up tax regulation after the 2006 Measures. On 18 July 2006, the State General Bureau of Taxation imposed Income Tax of 20 per cent on the capital gains from a housing sale, in the *Circular on Income Tax Levy on the Sale of Second-Hand Housing Held by Individuals* (SAT, 2006). If a seller sold his or her only home after five years of purchase, then no income tax applied. Local governments, however, made use of the provisions in the circular on using 1–3 per cent of the total sales revenues as the amount of income tax when capital gains were difficult to calculate. They charged 1 per cent of the sales revenues as income tax, thus substantially reducing the burden of the income tax and the effectiveness of the regulation to reduce housing investment. However, the 2013 Measures stopped the practice of charging 1 per cent of sales revenues and enforced the 20 per cent of capital gains for housing units where purchase prices were known and capital gains could be

TABLE 8.1 Deed tax rates in Xi'an

Types and conditions	*Tax rates (%)*
Housing	
Unit size: < 90 m^2	1.0
Unit size: 90–144 m^2	1.5
Unit size: > 144 m^2	3.0
Villas	3.0
Commercial real estate	3.0
Floor area ratio < 2	3.0
Price higher than average price in Xi'an	3.0

Source: Xi'an Municipal Government.

TABLE 8.2 Business Tax incidence by different macro control policies

Policies	*Housing types*	*Amount of business tax*	
		Sold within two years of purchase	Sold after two years of purchase
2005 Measures	Ordinary Housing	Full Business Tax	Zero Business Tax
	Non-ordinary Housing	Full Business Tax	Full Business Tax on capital gains
		Sold within five years of purchase	Sold after five years of purchase
2006 Measures	Ordinary Housing	Full Business Tax	Zero Business Tax
	Non-ordinary Housing	Full Business Tax	Full Business Tax on capital gains
		Sold within two years of purchase	Sold after two years of purchase
2009 Measures	Ordinary Housing	Full Business Tax on capital gains	Zero Business Tax
	Non-ordinary Housing	Full Business Tax	Full Business Tax on capital gains
		Sold within five years of purchase	Sold after two years of purchase
2011 Measures	Ordinary Housing	Full Business Tax	Zero Business Tax
	Non-ordinary Housing	Full Business Tax	Full Business Tax on capital gains

Sources: macro control policy documents.

calculated. This change in Income Tax resulted in perverse market behaviour to avoid the tax, which is detailed in Section 8.5.

8.4.3 Taxation on letting

Taxation on property letting is different for housing and commercial property and it changes regularly according to new tax codes from the Ministry of Finance. The tax rates below are indicative of the taxes likely to be charged by recent tax regulations. For housing let by individuals, taxes incurred include Property Tax (4 per cent of rent), Business Tax (3 per cent of rent), City Maintenance and Construction Tax (1–7 per cent of the Business Tax) and Education Sub-charge (3 per cent of the Business Tax), Income Tax (10 per cent of rent) and Stamp Duty (0.1 per cent of rent). For organisations letting non-residential properties, the taxes incurred comprise Property Tax (12 per cent of rent), Business Tax (5 per cent of rent), City Maintenance and Construction Tax (1–7 per cent of the Business Tax) and Education Sub-charge (3 per cent of the Business Tax), Income Tax (according to prevailing corporation tax rate, currently 20 per cent) and Stamp Duty (0.1 per cent of rent).

In reality, local tax bureaux apply different composite tax rates on housing and commercial property incomes. Table 8.3 shows the composite tax rates for property

TABLE 8.3 Composite income tax rates on rental income of individuals

Property types	*Rental income*	*Before 1 Nov. 2011 (%)*	*After 1 Nov. 2011 (%)*
Housing	<1,000	4.00	4.00
	1,000–2,000	6.00	4.70
	2,000–20,000	8.00	6.70
	>20,000	8.00	8.38
Non-housing	<1,000	6.00	6.00
	1,000–2,000	14.00	6.70
	2,000–2,0000	14.00	8.70
	>20,000	14.00	14.30

Note: currency is CN¥.
Source: XKB (2012).

letting by individuals following the change that raised the taxable rental income for Business Tax from CN¥1,000 to CN¥20,000. The adjustment substantially reduced the tax burden on property letting in the city. Taxes on rental incomes of businesses are set at the corporation tax rate, whose details are beyond the scope of this book.

8.5 Market response to tax changes under macro control

Market responses to the imposition of taxes on housing transactions are as follows. First, there were panic transactions before the effective date of the taxes, which served as the best promotion of sales for developers. Panic buying and selling resulted in long queues outside Real Estate Transaction Centres in 2005, 2006, 2011 and 2013, which had to remain open for extra hours to accommodate the demand for registration of the transactions before the new taxes became effective.

Second, the burden of the tax is normally partially or completely passed on to buyers. In some cases, buyers paid for the seller's tax in order to secure the deal. Other cases revealed that buyers and sellers reached an agreement to share the tax. For example, the author witnessed a transaction in which the seller needed seven more months to qualify for the five-year holding of the housing unit and reached a deal with the buyer to finish the transaction by cash and register the transaction seven months later so that the seller could avoid the 5 per cent Business Tax. Such behaviour was so rampant that the 2011 Measures included requirements to stop the 'Yin and Yang', as they were called, or true and false contracts in housing transaction designed to avoid paying taxes. The problem was that many buyers and sellers did not agree with or tolerate the sudden imposition of taxes, and did deals to avoid them. The local governments, on the other hand, failed to investigate and punish such behaviour.

Third, families even divorced to break purchase restrictions and to avoid paying the higher taxes. Anecdotal evidence abounds on such 'divorce-to-buy'. For example, a couple, who were migrant workers in Beijing, bought two flats a few years ago. One flat was for their family and the other was for the husband's parents. The couple then had a baby and wanted the husband's parents to help to look after the baby. The couple's flat was only 46 m^2 in floor area so they needed a bigger flat. When they bought the two flats, they borrowed mortgage loans from banks. Such borrowing records were used by the local government to impose purchase restrictions. Because they were migrant workers, i.e. not registered households, there would be a 20 per cent income tax on the gains if they sold the 46 m^2 flat they were living in. They finally decided to divorce so that one of the couple did not own a home and had the right to buy a new home, rather than selling to incur the tax. They planned to remarry after they had bought a bigger flat (Zhao, 2013).

Another example involved the purchase of a twin flat for which divorce was adopted as a means to optimise the family property finance. As explained in Chapter 7, a twin flat in this context is in fact a single large flat but is officially registered as two smaller ones. This was sometimes done by the developers in order to meet the imposed quota for the inclusion of small affordable flats in the development, while in practice selling larger and overall more profitable size units. The family, already owning one flat elsewhere, wanted to buy this large twin flat also, which would constitute the couple's second and third property. However, the 2010 and 2013 Measures limited borrowing on a married couple's second home to 60 per cent, and on their third home to zero. This mean that in effect, for the purchase of this twin flat, the couple could only borrow 30 per cent of the overall cost and could not afford to proceed on that basis. Finally they accepted the advice of the sales agent to divorce (YCWB, 2012), with the husband fully owning their original small flat. They could then each buy half of the new twin flat, the ex-husband enjoying 60 per cent borrowing on his half – as it was his second home, and the ex-wife enjoying 70 per cent borrowing on her half – as it was her first home. The net result was that they were able to borrow the equivalent of 65 per cent of the value of the twin flat, as opposed to 30 per cent, and were then able to proceed.

8.6 Discussion: real estate tax

One reason for China's housing price inflation is excessive investment-oriented purchases, which led to a significant amount of homes being left vacant by investors due to the very low holding costs. Observation by the author is in line with a large amount of anecdotal evidence on vacant homes in urban housing estates (see discussion of the numbers of vacant homes in Section 9.5). The holding costs of these vacant homes are property management fees, which are normally CN¥2 to CN¥5 per m^2, and interest on mortgage loans. Cash buyers have no interest to pay and the property management fees are normally below 0.5 per cent of the capital values of homes. There is no tax on owners and occupiers of housing in China. The lack of a real estate tax is considered by many to facilitate housing price inflation.

The introduction of a real estate tax has been on the agenda for over a decade. The levy of this tax, a local tax, would have alleviated the fiscal difficulties of local governments after the tax-sharing deal with the central government in 1994 that allocated the bigger share of the country's tax revenues to central government (see Section 2.1). Local governments in China are responsible for economic growth and carry out many social functions, such as provision of affordable and social housing (see Chapter 13). The construction of the infrastructure, for example, is very expensive. With a stable income from a real estate tax, local governments could reduce their reliance on land sale revenues and the taxes charged on property development and transactions, which in turn depend on the housing market activities.

Nonetheless, the real estate tax has been controversial since it was first proposed. There are two types of opposition to real estate tax. The first type is on legal and logical grounds. One popular argument against the real estate tax is that it has 'shaky' legal standing. Some legal professionals strongly believe that real estate tax should target land owners. They claim that all urban private housing ownership in China is on LUR and building ownership. LUR is not an ownership, but just a leasing arrangement. The other popular argument against the tax is that there are already many taxes levied on housing development and transaction stages. If real estate tax is levied, those taxes on development and transactions should be rearranged to decrease the tax burden. The third popular argument against the tax is that China has yet to build a tax regime on wealth. Housing tax should be linked to taxes on other forms of wealth (He, 2014).

The other type of opposition is on pragmatic and emotional grounds. The pragmatic view opposing the tax is the increased burden on home ownership. Many home buyers borrow mortgage loans and have to be disciplined to arrange their daily expenditure to ensure the monthly repayments are met. They are opposed to any additional expenditure on home ownership. The emotional view is that the government induced people to give up the comfort of welfare housing and buy private housing, thus exhausting people's money. It is simply wrong to impose a real estate tax on housing to require people to pay again.

Because of the opposition to the real estate tax, the central government has been cautious in introducing it. Two cities, Shanghai and Chongqing, were picked from six cities that have been dry running[1] a real estate tax on housing for years, to experiment with the real estate tax. To avoid annoying homeowners who are now the majority of urban residents, the real estate tax trial schemes in Shanghai and Chongqing only target a small proportion of housing. In Shanghai, an annual tax rate of 0.6 per cent is applied on the 70 per cent of the transaction price of newly acquired housing by non-registered households and by registered households who already have one unit of housing. In Chongqing, an annual tax rate of 0.5 per cent to 1.2 per cent is applied to the transaction price of detached houses and luxury apartments, the prices of which are two to four times the average housing prices, and second or additional homes purchased by people who have no household registration, jobs, and no other investments in the city. The two trials have had minimal impact on housing prices and fiscal revenues, though some benefits of resource allocation optimisation were claimed (*People*, 2013).

The reform tasks specified in Article 18 of the Decisions by the Third Plenary Session of the Eighteenth Party Congress (ibid.) include the speeding up of the legislation of the real estate tax. This policy document provides a new direction on this controversial tax. First, the imposition of the real estate tax will bring about a tax reform on all the taxes on real estate.. Such a reform can address taxes and fees on real estate development and transactions. Second, the introduction of the real estate tax is by legislation, rather than by a decision of the central government. Legislation will involve much more thorough consultation and debate by society, which will proceed only after solving the controversy surrounding the real estate tax.

8.7 Conclusion

Development finance in China is classified into foreign, domestic banks, self-raised funds and other sources. Foreign funding was important in the 1990s but has declined since. Bank lending, the third largest source, has been declining. Self-raised funds, the second largest source, are becoming more important. Other sources have been the largest source of funding since data became available in 1996. They include deposits and presale receipts, which are mainly financed by banks. Thus bank lending is still the main source of development finance. Housing price changes are closely related to the cash positions of developers. On the demand side, home mortgages are the key to home purchases and have been the tool used to adjust housing demand in macro control. However, domestic banks have started to provide fewer mortgages to home buyers since late 2013 due to the tightening of the credit supply. The Housing Provident Fund, a compulsory saving scheme on urban employees, provides a source of cheap home mortgages, which are important in allowing many families to afford commodity housing.

Taxation on property in China is currently mainly done at the development stage, with 11 taxes and dozens or even hundreds of fees to be paid by developers. The myriad of fees impose high transaction costs for property development in China. Deed Tax was halved to encouraged housing transactions since 1997 and now varies according to the size of homes. Business Tax was first imposed on housing sales in 2005 and remains an important tool for the government to fight house price inflation, with periodic adjustments to discourage investment demand. The tax on holding housing, however, has been controversial and so far trial attempts in Shanghai and Chongqing have been partial and inconclusive. Legislation to develop a property holding tax will cover all property types and result in a reform of the tax system in property in China.

Note

1 Dry running means running the taxation system without actually collecting the tax.

PART IV

Real estate development and investment

9

HOUSING

Starting the focus on real estate investment, this chapter covers investment in housing land, housing development, housing, and housing management. It starts by examining the regulatory constraints on land banking, the crackdown on land hoarding, developers' approaches to building land banks and local government land reserves. The chapter then investigates housing development through the development process: land bidding, housing design and marketing, and housing management. Case studies on innovation are provided. Investment is then studied, with a case study on the rise and eventual failure of speculative housing investment by a city. The chapter finally joins the debate on vacant housing to provide insight on the issue.

9.1 Land for residential use

Given the scarcity of developable land in major cities, land for residential use is a good investment. However, there are legal barriers facing private investors wishing to invest in land for residential uses. On the other hand, land reserves have become a common vehicle for local governments to store land to supply the primary land market.

9.1.1 Legal restrictions on holding land without development

In China only Land Use Rights (LURs) on state-owned land can be bought and sold legally in the open market (see Section 3.1). Those LURs are sold with a specified use and terms laid out in the contract. Land users are required to use the land without delay, i.e. not leave the land vacant. Article 26 of UREAL (amended in 2007[1]) stipulates that those who bought LURs for real estate development must carry out the agreed development within one year; otherwise a vacant land fee up to 20 per cent of the land sale premium for land use may be collected. If not developed after two years, land may be taken back by the government without compensation unless the delays are caused by *force majeure,* the acts of the government or its departments, or the necessary preparatory work for starting the development.

Vacant Land Administration Measures, enacted in April 1999 and amended in June 2012 by the MLR (MLR, 1999), list the following mitigating circumstances by land users holding land undeveloped for longer than one or two years:

- Land is not provided according to the contract, rendering the project impossible to proceed.
- Changes have been made to the urban plan, rendering the project impossible to proceed according to the arrangement, use, planning conditions and construction conditions as outlined in the contract.
- The arrangement, use, planning conditions and construction conditions need to be changed to conform to new government policies.
- The project is delayed in order to deal with residents' complaints to the government.
- The project cannot proceed due to military or archaeological issues.
- The project cannot proceed due to other government actions.
- The project cannot proceed due to a *force majeure.*

Land users can supply evidence to explain the causes of delay according to the provisions in Article 8 of the Measures. If their explanations are accepted by the local land administration agencies, land users can choose the following options under negotiation with the local land administration agencies:

- Extend the deadlines for construction to start, with a maximum extension of one year.
- Change the use and planning conditions with an additional payment or a refund for excessive payment for land.
- Arrange temporary uses by the local government, with a maximum length of temporary use for two years; then the original land user restarts the development.
- Give back the LURs to the local government with compensation negotiated.
- Exchange for a suitable site of the same value.
- Accept other ways of disposal by the local land administration agencies.

Apart from giving back the LURs, land users can renegotiate the deadlines for work to start. If local land administration agencies decide to impose a vacant land charge or take back the LURs, land users have the right to request a public hearing, according to the *Provision for Hearings on Land and Resources* promulgated by the MLR.

Therefore, it is difficult in theory for developers to build a land bank to arrange developments at the right time or simply leave land vacant to capture the benefit of land value appreciation. However, most developers were able to avoid confiscation by resorting to Article 8 of the *Vacant Land Administration Measures* when the local government cracked down on vacant land. As a result, many developers obtained large profits from investing in land.

9.1.2 Crackdown on leaving land vacant

The penalty imposed in Article 26 of UREAL for delay in land use is designed to ensure efficient land use because vacant land had been a major issue since 1993 when failed real estate development projects and the oversupply of industrial land resulted in widespread land dereliction (see Section 3.3). Furthermore, there was widespread unauthorised land occupation by state-owned firms and some private firms, and lower levels of local governments without authority from the land administration. Unauthorised occupation often led to dereliction because development permissions were not received. Large-scale dereliction triggered the first round of national vacant land administration in 1994, resulting in UREAL authorising a Vacant Land Charge and the seizure of LURs by local land administration agencies.

Unauthorised land occupation and leaving land vacant and derelict, however, were not stopped by the first round of dereliction land administration in 1994. Land value appreciation in more and more cities encouraged more occupation. More industrial zones were set up. In Shanghai alone, 207 development zones were founded in the 1990s and only 11 were approved legally (Yin, 1999). Much of the occupied land, including sites in the countryside and in the built up area, remained unused because the finance was not in place. To assist the new round of administration on land vacancy, the MLR enacted the *Idle Land Administration Measures* in 1999 to administer unused and derelict land. The clean-up afterwards resulted in many of the zones being revoked and vacant land being allocated to new uses or returned to agriculture.

For example, Guangzhou had large tracts of unused land in the urban areas after 1994 due to developers lacking funds, failing to achieve sufficient presales, and suspending construction due to lack of demand. Between 1997 and March 2002, to eradicate vacant land in the urban areas that had become eyesores and rubbish disposal sites, the Guangzhou Municipal Government took back the LURs of sites that had not been developed more than two years after the land sale, according to Article 26 of UREAL, confiscating 760 sites (1,118 hectares), mostly without compensation (Hong and Pan, 2002).

Although unauthorised land occupation reduced in the 2000s due to the crackdown by the land administration agencies, land vacancy and dereliction still existed. A survey by the MLR in the planned urban areas of cities and towns throughout the country found that by the end of 2004, there were 71,953 hectares of land lying vacant, 56,160 hectares lying derelict, and 135,630 hectares approved to be converted to urban land, but not yet converted. These three types of land amounted to 7.8 per cent of the total urban areas in cities and towns at that time (Yang, 2005). In particular, 43 per cent of land in development zones (mostly out-of-town) was standing idle or derelict, and over 46,700 hectares of land were left vacant by developers waiting for an appreciation in value. In Beijing, developers were believed to have left unused 14,410 hectares of land with planning permission, and in Shanghai the figure was 3,600 hectares (ibid.).

The enactment of the 2005 Principles and the 2005 Measures provided a new urgency to reduce land vacancy. In particular, the 2005 Measures re-emphasised the penalties on leaving land idle in Article 26 of UREAL. As a result, a new round of administration on unused land took place. In Guangzhou, the Guangzhou Municipal Bureau of Land and Buildings Administration took action on 29 June 2006 to serve 47 confiscation notices to confiscate land from development firms, local governments at lower levels (the Haizhu District Government), local government agencies (the Education Bureau of the Tianhe District Government), and state-owned banks (the Industrial and Commercial Bank of China Guangzhou Branch). Similar examples were found in other cities. For instance, in Beijing, 45 sites with a total area of 67 hectares and permitted construction floor area of 3.09 million m^2, were confiscated in August and October 2005 because of a delay in construction and failure to make full payment of the LUR premium (Qu, 2006). In Chongqing, the municipal government took back 37 sites with a total area of 105.27 hectares in November 2006 in the first round of confiscations of LURs to punish delay of development for more than two years (Zhou, 2006). Nevertheless, Shanghai has been keeping a low profile in land confiscation, allowing developers to negotiate with the government, which usually provided assistance in finding new investors, and so reducing the possibility of loss due to site confiscation (*Shanghai Security News*, 2006). Some sites, delayed up to nine years for development, were aided by the municipal government introducing new investors so that development was possible.

9.1.3 Developers' land reserves

The act of building land reserves, crucial for developers, is unique in China. There are two approaches to build land reserves. One way is to acquire land in the tertiary land market. Figure 9.1 shows the transaction volume in the secondary market where developers sold developed land to other land users. As land prices increase, transferring land is increasingly a profitable business. The other approach is to buy land from the government. The old practice for developers was to negotiate with lower-level local governments without land sale authority, and firms that owned land, to obtain the development right for land, and then apply to and negotiate with local governments with land sale authority to obtain LURs. With the crackdown on illegal land occupation and vacant land, developers have to acquire land in the open market in cities where competition for land is intense.

The strategies by developers to build up their land reserves under strict government oversight are as follows. First, developers normally divide up the development into different phases because the sales contracts of land normally do not stipulate the maximum time for developers to complete development. Phased development has many advantages. Early phases satisfy the government's requirement for development within a specified time limit and serve as to test the market on the estate layout, house types, design and sales strategies. Once the response from the market is satisfactory, developers can raise the prices and even increase development density

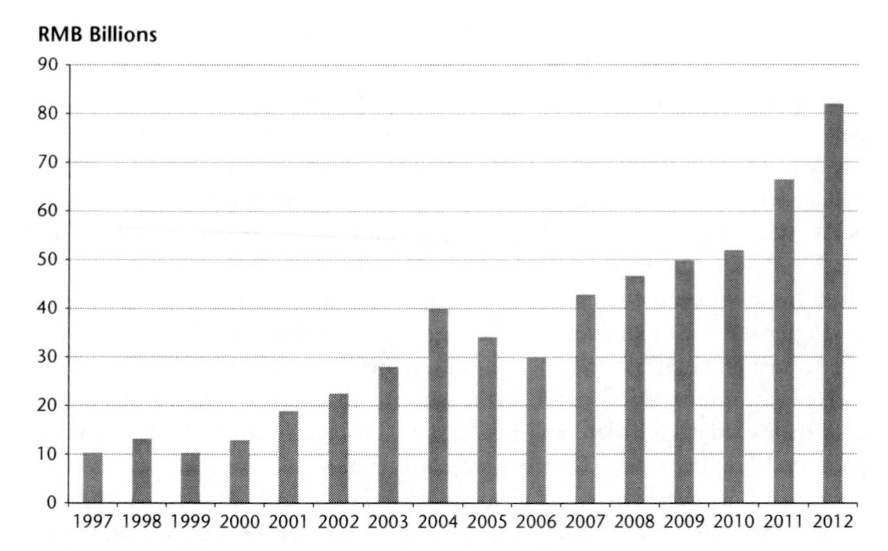

FIGURE 9.1 Land transfer receipts in the secondary market
Source: NBSC (2013).

if possible. If the market response is not ideal, developers can add market-proven attractions while removing unwanted features. Going through all the phases of development takes a number of years and as housing prices in general rise quickly, land appreciation occurs during the development period of the site. Profits are normally quite substantial.

Second, developers try to buy large sites to allow longer development periods and more discretion on planning, design and adjustment. Winning a housing development project covering a big site achieves economy of scale and catches the gains in land price appreciation. There is a financial constraint on buying big sites but as developers increase their operations, they are often able to adjust their cash flows from several, dozens or hundreds of projects to raise the funds to buy new sites on a regular basis.

Third, developers seek to choose cities in the early stages of housing market development to buy land at low prices with less intense or even no competition. In the last decade most large developers operated regionally and nationally, buying sites in many tier 2 and tier 3 cities, or even tier 4 cities. Land and labour costs were low in those cities and local governments there often provided preferential treatment to inward investors. The following example indicates the scale of such land purchase and the time it will take to complete all the permitted housing space. All the data on the projects was checked individually by the author in June 2014.

Qingyuan, a tier 4 city in the north of Guangzhou, is still relatively undeveloped and land prices there are low. However, its southern border is only 60 km from Guangzhou's geographical centre and has been targeted since 2007 by four of the top ten developers in the country. Country Garden, a developer from the same province, purchased four sites with a total permitted floor space of 7.55 million m^2. The first site is 4.85 km^2 in land area and has 2.06 million m^2 permitted low-density holiday housing. Started in 2008, it had obtained presale permits for about 4,000 units by 2014, accounting for less than half of the total permitted developable space. The second site is 0.627 km^2 in land area and has 1.19 million m^2 permitted low–middle density holiday housing. Starting in 2009, it had developed only 20 per cent of the housing space permitted by 2014. The third site is 5.33 km^2 in land area and has 2 million m^2 permitted low-density holiday housing. Started in 2011, its first phase has been on presale since 2013. The fourth site was acquired from a local developer and is 1.17 km^2 with 2.3 million m^2 permitted low–middle density housing. Its first phase, about 250,000 m^2, was on presale from 2012.

Hengda, another developer from the same province, came to Qingyuan slightly later than Country Garden, and now has two projects with 5.48 million m^2 of permitted housing space. The first project occupies 0.67 km^2 land area and has 3 million m^2 permitted low- and high-density housing space. It has currently developed 10 per cent of the permitted housing space on that site. The second project occupies 1.16 km^2 land area and has 2.48 million m^2 permitted low- and high-density housing space. The first phase started presale in 2011 and has now had 15 per cent of the total permitted housing space sold or pre-sold.

Wanke, also from the same province, started work in Qingyuan in 2008, and has currently started work on two projects with 3.05 million m^2 of permitted housing space. The first project occupies a land area of 1.2 km^2 and has 2.5 m^2 permitted low–middle density housing space. It has put about one-third of the permitted housing space on sale and presale. The second project occupies a land area of 0.21 km^2 and has 0.55 m^2 permitted high–middle density housing space. It has put about one-third of the permitted housing space on sale and presale. In March 2013, Wanke purchased a third site, which cost CN¥230 million for permitted development of 493,100 m^2. The unit price was CN¥466 per m^2 of permitted development space. Housing space nearby is sold at over CN¥5,000 per m^2 (Nanfang, 2014).

Poly Real Estate, again from the same province, entered Qingyuan's housing development market in 2011, and has currently one project with a local developer. The project occupies 0.3 km^2 land area and has 1 million m^2 of permitted medium-density housing space. The developer bought five sites within the urban proper area of Qingyuan in 2013, with total land area of 0.5 km^2 and 1.3 million m^2 permitted medium-density housing space.

The above four developers have only committed to a maximum of a third of the permitted development space and their further input depends on the sale of the phases currently on presale. If response from the market is not satisfactory, development work will slow down and the developers will adopt a wait-and-see attitude. Although the housing market slowed down in the first half of 2014, market expectation is that in the medium term, housing prices in Qingyuan will rise significantly because more and more firms are being relocated to Qingyuan to take advantage of low land prices and low wages. The low land prices paid by those developers will prove to be a key factor in the profitability of all these projects.

9.1.4 Local authority land reserves

Legal Land Reserves are set up by local governments with central government authorisation to acquire, service and store land for the purpose of maintaining a sufficient supply of construction land. They are expected to regulate land prices, promote efficiency in land use and increase land supply when needed. The earliest Land Reserve was set up in Shanghai in August 1996 to help state-owned enterprises dispose of their surplus land through market transactions. In August 1997, Hangzhou set up its Land Reserve supported by fiscal allocation and bank loans to acquire land in bankrupt and poorly run state-owned enterprises in an attempt to reduce idle land and increase land use efficiency. Later many other cities set up their own Land Reserves following the Hangzhou model, with compulsory purchase being used by the Land Reserve Centres to accumulate land through the requisition of rural land, along with taking back previously sold LURs due to land vacancy and the acquisition of inefficiently used state-owned land.

The practice of building and operating Land Reserves by Shanghai and Hangzhou were initiatives at the collective choice level. The constitutional level supported the Hangzhou model that made better use of government power. In May 2001, the State Council issued the *Circular on Strengthening the Management of State-Owned Land Assets* (State Council, 2001), which put the emphasis on concentrated and unified management of land to ensure a unified supply of construction land. This circular was widely regarded as an endorsement of the local government Land Reserves. To order the rapid development of Land Reserves, on 19 September 2007, the MLR, in conjunction with the Ministry of Finance and the People's Bank, tabled *Land Reserves Management Measures* to provide policy guidance and rules for Land Reserves (MLR, 2007). The Measures stipulate that all local governments cannot use tax revenues to guarantee borrowing for Land Reserves. The priority of Land Reserves is to acquire vacant, derelict and inefficiently used state-owned construction land. It further stipulates that the requisition of rural collectively owned land should be on land with requisition procedures already completed. To prevent local governments from expanding Land Reserves unnecessarily, the Measures required local governments to make land reserve plans and register such plans with the land administration agencies at higher-level governments.

There are a number of problems emerging from the system of Land Reserves. First, a Land Reserve, once operational, monopolises the supply of new LURs on state-owned land in a city or county. All new supplies of LURs have to be through the local government's Land Reserve Centres, raising concerns about the efficiency of those Centres. The efficiency of the Land Reserves affects the responsiveness of the primary land market to market demand on land. For example, the Beijing Land Reserve and Administration Centre was accused of not being able to supply sufficient land to meet the market needs in 2006, with only 56 per cent of the planned 6,500 hectares of land being supplied by the end of that year (NEWSCHINA, 2007).

Second, Land Reserves present financial risks to local governments because over 80 per cent of the funds used by Land Reserves come from interest-bearing bank loans. Without fiscal support, the land is used as collateral to borrow large amounts of money to finance further land acquisition and servicing, incurring a heavy burden to repay the interest. Some land parcels used as collateral had not completed the necessary legal procedures of land acquisition (Liao, 2013), violating the requirements in the 2007 Measures. Consequently, funding problems have an impact on the Land Reserves' ability to acquire and service land. For instance, the Beijing Land Reserve and Administration Centre raised over CN¥100 billion every year to finance its Land Reserve building from 2019 to 2011, but had to downscale its Land Reserve building in 2012 due to lack of market demand and repayment pressure (Qi, 2013). In Wuhan, there was pressure for the municipal Land Reserves to sell land in large quantities to repay loans, leading to lower prices in an oversupplied primary land market. As a result, developers in Wuhan were able to purchase land at lower prices to add to their Land Reserves and thereby obtain the flexibility to hold land for longer for capital gains (Liu *et al.*, 2010).

Third, Land Reserves in some cities deviated from the not-for-profit principle and became local governments' cash machines. In fact, land value appreciation during land acquisition, servicing and storage by Land Reserves has brought profits to the local government, which generates suspicion that Land Reserves are only the profit centres of local governments to finance their massive infrastructure investment and image projects. Under China's regulations at the time, local governments were not allowed to borrow directly from banks or issued bonds, i.e. a rule restricting local governments from using the financial market to finance local expansion. With Land Reserves, however, local governments could obtain land at low prices through the use of their compulsory purchasing power and sell it at higher prices to finance their infrastructure projects. This led to infringements of farmers' interests in land acquisition. The effectiveness of the Measures tabled in 2007 by the MLR is not clear.

Despite public suspicion and several flaws, Land Reserves, with their ability to carry out government land policies and change the land supply, have become an integral part of the institutional arrangements in China's land market.

9.2 Housing development

Housing development in China has been a profitable business with a modest level of risks. Demand for housing has been strong since the reform and the opening-up campaign started in 1979. It has come from two sources: (1) the desire to improve housing conditions by existing urban residents; and (2) the housing needs of migrants. During the last 35 years, average urban housing conditions have improved substantially. However, housing demand from existing urban residents is still strong due to the demand from a significant proportion of urban families to improve their housing conditions, and the housing needs arising from new families of the younger generation (see discussion below). Demand from migrants, whose total number is approximately 500 million,[2] will be gradually released in the coming two to three decades. Even though there is temporary oversupply in many cities in 2014, in the longer term, much more housing space is needed.

In the late 1990s and early 2000s, profitability in housing development attracted many large firms whose main businesses were in other sectors. Intense competition has been the norm in land bidding, product design, promotion and sales strategy, and after-sale property management. There have been many innovations in estate layout, floor plans, the provision of services, and offering services that create new and different lifestyles for home buyers. To be different from other products, developers in China have studied and in many cases copied housing designs and estate layouts from many countries in Asia, North America and Europe. For example, an English-style town was reproduced in East China as a housing project to provide a different lifestyle and experience to home buyers.

Anecdotal evidence abounds that some developers simply copied foreign practices. The author observed a villa project with a Californian interpretation of Spanish design in the Nansha District of Guangzhou. All the villas were built with basements according to the original design brief from the USA. Because planning conditions did not include a provision for basements, the developer sealed off the basements when the houses were completed and sold, leaving the buyers to discover after occupation that there were large basements under the ground floors of their houses.

9.2.1 The main players

Over the years some local developers have grown to become national giants. Table 9.1 lists the largest developers in 2013 in terms of sales volume and floor area sold (or pre-sold), with housing dominating most of the sales. To understand the scale of development by those developers, a comparison is made between UK house building and Chinese developers. Completion of permanent dwellings in the United Kingdom was 143,680 units in 2012.[3] Assuming an average floor area of 100 m^2 per unit, housing completion in the UK was equivalent to 14.37 million m^2 of floor space in that year. Each of the four largest developers in China built more than the UK equivalent in 2012 (Table 9.1).

There is an increasing concentration of development work being carried out by the largest developers. Table 9.2 shows the market concentration of the largest 10, 20 and 50 developers. From 2011 to 2013, the total floor area sold by the top 50 developers increased approximately 50 per cent. In 2013, the top 50 developers accounted for a quarter of the total sales volume of new property throughout China.

Factors leading to the prevalence of large developers are explained below. They indicate the required competitive edge for a developer to succeed in China's housing development market.

9.2.2 Land bidding

Since 2002, the open market disposal mechanism has resulted in intense bidding for land in good locations. Thus, developers have to develop strategies to buy land in the open market according to their financial standing, their ability to design and deliver attractive products, and their skills and resources in promotion and sales.

TABLE 9.1 The biggest developers in China in terms of sales of new property

Sales volume (CN¥ billion)			*Floor area sold (million m²)*		
Rank	*Firms*	*Volume*	*Rank*	*Firms*	*Area*
1	Wanke	174.06	1	Greenland	16.60
2	Greenland	162.53	2	Country Garden	16.52
3	Wanda	130.11	3	Hengda Real Estate	16.05
4	Poly Real Estate	125.10	4	Wanke	15.17
5	China Overseas Property	117.00	5	Poly Real Estate	10.81
6	Country Garden	109.73	6	Wanda	10.59
7	Hengda Real Estate	108.25	7	China Overseas Property	9.92
8	China Resources Land	68.10	8	China Resources Land	5.89
9	Shimao Property	67.07	9	Shimao Property	5.58
10	Greentown China	55.38	10	Century Golden Resources	5.43
11	SUNAC China	50.83	11	RiseSun	4.72
12	Longfor	49.25	12	Longfor	4.39
13	Gemdale	45.60	13	Honglicheng	3.80
14	CITIC Real Estate	44.51	14	China Futune Land	3.79
15	China Merchants Property	42.80	15	Gemdale	3.63
16	Guangzhou R&F	42.23	16	China Railway	3.57
17	China Futune Land	37.60	17	CITIC Real Estate	3.47
18	Agile Properties	37.40	18	Guangzhou R&F	3.46
19	SPG Land	36.84	19	Agile Properties	3.24
20	China Railway	32.90	20	China Railway Construction	2.99

Source: CRIC and China Real Estate Evaluation Centre (2014).

Prime sites or sites with strong potential are the targets of bidding among developers throughout the country, especially the top 50. Some developers with a lower standing and reputation are keen to gain fame by successfully bidding for famous sites. Table 9.3 lists the 'land kings', or record-breaking land sales, in the first quarter of 2014. Apart from reflecting favourable market expectation in future housing

TABLE 9.2 Market concentration of developers in China in terms of sale of new property

Concentration of sales volume (%)				*Concentration of floor area sold (%)*			
Firms	*2013*	*2012*	*2011*	*Firms*	*2013*	*2012*	*2011*
Top 10	13.27	12.76	10.67	Top 10	8.37	7.72	5.75
Top 20	18.25	17.62	14.87	Top 20	11.12	10.38	7.63
Top 50	25.35	24.56	20.75	Top 50	15.42	14.19	10.51

Source: CRIC and China Real Estate Evaluation Centre (2014).

TABLE 9.3 Land Kings from January to March, 2014

Cities	*Date*	*Price*	*Area*	*Records broken*
Hefei, Anhui Province	9 January	0.42	11.9	Unit price, Hefei
Fuzhou, Fujian Province	10 January	7.70	259.6	Total price, Fujian Province
Zhuhai, Guangdong Province	17 January	1.85	32.0	Unit price, Gongbei District, Zhuhai
Hangzhou, Zhejing Province	22 January	3.38	45.6	Unit price, Hangzhou
Shenzhen	23 January	13.40	50.0	Unit and total price, Shenzhen
Shanghai	28 January	10.10	96.4	Total price, Shanghai
Guangzhou	28 January	0.80		Unit price, Panyu District, Guangzhou
Guangzhou	28 January	1.67	49.4	Unit price, Huangpu District, Guangzhou
Zhengzhou, Henan Province	19 February	0.80	34.0	Unit price, Zhengzhou
Urumqi, Xinjiang Uyghur Autonomous Region	21 February	0.77	180.0	Total price, Urumqi
Guangzhou	21 February	2.99	280.0	Unit price, Liwan District, Guangzhou

Note: price in CN¥ billions; area in thousand m^2 land area.
Source: various reports on land transactions.

prices, the appearance of land kings indicates the financial positions of the developers, the cash requirement of local governments, the scarcity of sites in good locations, and the anticipated demand for high-end housing. In particular, local governments can influence the results of land transactions by promotion of the sites, for example the release of new development plans, infrastructure commitments, and amenities provisions in the surrounding area of the sites. As a result, large amounts of capital have become necessary to bid successfully for sites at good locations. Small developers are being driven out of the open land market for good sites. Thus intense competition in the primary land market increases the market concentration in housing development.

The other way to acquire land is to buy sites in prosaic locations but work on product design, promotion and marketing. Large developers are in a better position to exploit this strategy to buy large sites in normal or even poor locations to add value through product design, promotion and marketing. Such competitive advantages have further pushed small developers out of cities that attract large regional and national developers. Projects developed by four of the top ten developers in Qingyuan City, described in Section 9.1, are examples of such a strategy, which is further explained below.

FIGURE 9.2 Medium- to high-rise residential buildings in Beijing
Source: photo by Albert Cao (2006).

9.2.3 Product design to create lifestyles

Different from very low cost welfare housing, commodity housing has been built to attract home buyers to pay large sums of money for stylish homes, services and lifestyles. There are presently three types of housing demand: rigid demand, improvement demand, and investment demand. Rigid demand derives from new families wanting a roof over their head. Families trying to meet their rigid demand have a budget constraint. Improvement demand derives from families whose rigid demand has been met. Such families look for homes that provide luxury, amenities and lifestyles. Investment demand comes from investors who seek capital gains and rental incomes, and from parents who want to give homes to their children. Housing product design needs to clarify which demand to meet for the particular housing estate, given the constraints from the land sale conditions and, more importantly, the urban planning conditions.

Housing estates to meet rigid demand should provide small to medium-sized units at relatively affordable prices to first-time buyers. They need to contain all the basic facilities, including green open space that is normally in short supply due to lack of public open space, and are normally of high density. Figure 9.2 illustrates residential buildings to meet rigid demand in south-east urban Beijing just outside the third ring road. Although the values of those housing units increased due to

FIGURE 9.3 Features in communal open space in a housing estate in Zhaoqing
Source: photo by Albert Cao (2007).

the good location, they are housing for rigid demand due to the lack of luxury features.

Housing estates to meet improvement demand, on the other hand, need to provide as many luxuries and amenities as possible at acceptable prices to attract existing homeowners to buy. They need to be a good size, have well-designed landscaping, plenty of parking spaces, sports and cultural facilities, a residents club and other selling points that appeal to improvement home buyers. Selling points can be natural endowments such as lakes, waterfronts, slopes, hills and forests, or artificial creations such as fountains, large ponds, waterfalls, large lawns, a golf course, a good variety of vegetation, squares, arches and a combination of natural and man-made features. Figure 9.3 shows an improvement housing estate in a tier 3 city in Guangdong Province. Figure 9.4 illustrates the meticulous attention to small details of design in a very successful housing estate in the capital city of Henan Province. Terraced, semi-detached or detached houses or villas are often found in such housing estates. If situated in new areas and large enough, such housing estates should provide nurseries and primary or even secondary schools. Reputable nurseries, primary and secondary schools (either public or private) are often invited to open branches in large suburban housing estates through financial incentives. International schools are often found in luxury housing estates. For suburban estates, accessibility

FIGURE 9.4 Attention to detail: features in a housing estate in Zhengzhou
Source: photo by Albert Cao (2008).

to motorways and fast public transport such as an underground railway or light railway is desirable.

Many housing estates have been developed to cater for both rigid demand and improvement demand, and offer a variety of unit sizes. They are particularly appealing to the majority of home buyers. Figure 9.5 is such a housing estate when its earlier phases were offered to the market in 2004. Because of its unique golf course, it is beyond the affordability of first-time buyers now and no longer suitable for rigid demand home buyers.

Investment demand mainly targets housing estates to meet improvement demand. In particular, investment home buyers favour housing estates in new development areas with capital growth potential. Figure 9.6 shows a complete housing estate and parts of several others in the background with different architectural styles and estate layouts in the Singapore-Suzhou Industrial Park jointly developed by Suzhou Municipal Government and an investment arm of Singapore. Investment buyers flocked there to take advantage of low prices in the expectation of the eventual maturity of the whole area. When the picture was taken, the area was a 'ghost town' due to the lack of occupiers. However, most of the housing units were already sold so such a high level of empty homes was not a particular issue for either the local government or the developers. Now the area has a much higher occupancy rate and is a normal residential area.

FIGURE 9.5 A high-end housing estate in the suburb of Guangzhou
Source: photo by Albert Cao (2005).

FIGURE 9.6 New housing estates in Suzhou's Sino-Singapore Suzhou Industrial Park
Source: photo by Albert Cao (2005).

9.2.4 Promotion to induce demand

Competition to create lifestyles in housing estates to attract buyers has given birth to a new professional qualification in China: real estate *Cehuashi* or real estate planner (REP). The translation, officially adopted by the China Real Estate Association, however, misses part of the work a REP does. The ambit of the REP covers the whole process of real estate development: market research, design, public relations, financing, project management, marketing, sales, and after-sales property management. Among the critically important issues for an REP, is the creation of a differentiating lifestyle in a housing estate as the key selling point.

With dozens of housing estates under development and for sale at any time in a large city in China, the planning, marketing and management team of a development firm needs to find a niche for the housing estate to appeal to one or several groups of buyers. A large amount of market research needs to be done on the products being provided by rivals. The author has the experience of being suspected of being an REP from rival firms when undertaking fieldwork, and on several occasions was the subject of very unflattering attention from security guards who had not been informed of his visit. Once a niche is found or created, the design, marketing, public relations, project management, and sales teams work together to produce an integrated development proposal. The possibility of presale provides a unique opportunity to test the proposal. During the process, the public relations team, often with other teams, needs to collect market responses and feed back to all the teams involved to fully inform the final positioning of the project.

A number of marketing slogans have proved to be very successful. These slogans become the focus of the promotion, design and sales, which is a creative process. It will be a disaster if the market thinks the slogan does not fit the reality. A number of successful examples are:

- Give you a five-star home.
- Don't let your children lose at the starting line.
- Home holiday home!
- My home, my empire.
- Limitless romance!
- Green ecological bay, world wisdom city.
- See the city's future from your balcony.
- Come to see the standard of a home.

Promotion to induce demand becomes more important as many housing projects are built to attract home buyers to upgrade. Ways to promote include: advertisements in public media such as newspapers and magazines and on popular television shows; activities such as real estate expositions; opening ceremonies and sponsorship of popular cultural activities; and internet marketing and information services. The expansion of internet coverage has been important in promotion, and most housing projects from all over the country are accessible via the internet.

9.2.5 Sales strategy to lead and react

Sales strategy is an integral part of the whole process of housing development. It starts with the presales in phase one of the project and ends when all the units are sold.

It is common practice that the whole sales process follows a 'Start low, end high' pattern of price changes. This is particularly true for projects that are either new in the area or in a brand new area. A number of good units will be on sale in phase one to attract attention. With presale being the norm, there is asymmetry of information. Home buyers are strongly influenced by the information provided by the sales representatives of the developers. The recruitment and training of sales personnel are important to the success of the project. With strong incentives the sales personnel work hard, and there are occasions when the top sales representatives have become millionaires.

Successful promotion often attracts large crowds to the opening sales, with the most successful promotion generating overnight queues and selling hundreds or more units in a day, fetching hundreds of millions of yuan. The sales team should be prepared for unexpected demand, especially when the market is good. If demand is high and there are not enough units available, prepared sales teams will add units to expand sales and use the media to further promote the project.

When demand is not strong, developers will try their best to promote sales without cutting headline prices. One way is by internal sales with discounts, which is claimed to favour buyers with a special relationship to the developers. In reality, many potential buyers are invited to attend internal sales through promotion. The other way is group sales, with some buyers organising group purchases to negotiate with the developer for discounts. However, the organisers could be commissioned by the developer. The sales records by internal sales and group sales will be highly publicised to create the impression that the project is in high demand.

Other promotion measures, such as zero deposit offers and false buyers, have been used but are banned by government administrative agencies. A zero deposit offer contains a loan provided by the developer, which is equivalent to the deposit required by lending banks to a home buyer. A home buyer accepting the offer of zero deposit thereby needs to repay two loans and faces high pressure to make the monthly payments. The central bank has in the past banned banks from collaborating with developers to provide zero deposit offers. Another promotion measure is to use false buyers employed by developers to create a competitive market for a project to influence real buyers.

9.2.6 After-sales property management

Property management after home buyers occupy their residential units is important assurance that the homes they buy can actually deliver the comfort they expect and are well maintained. Thus sound property management is another factor that influences buyers' decisions. With trusted property management, housing projects can sell for higher prices. More details on property management are provided in Section 9.4.

It is worth noting that property management, with charges from CN¥1.0 per m^2 to 5 per m^2 according to gross floor area (GFA), is often not a profitable business. Many developers choose to subsidise property management or simply set up their own property management companies to provide services that cost more than the fees paid. This cost, however, is often paid by home buyers at slightly higher housing prices and is thus a part of the overall development strategy.

BOX 9.1 CREATIVITY IN COMMODITY HOUSING DEVELOPMENT: TWO CASE STUDIES

Since the late 1990s, competition in the housing market has greatly upgraded the quality of new housing. The following case studies illustrate creativity in commodity housing development to bring unique physical and cultural features to housing estates.

The first case is Lingnan Garden in the north suburb of Guangzhou. Situated in Baiyun District, Lingnan Garden is a lower–middle density housing estate occupying a 6.88 hectare site. The permitted development area includes 95,900 m^2 housing, 11,300 m^2 commercial and other buildings, and 17,000 m^2 underground car parking, providing a parking space for every three housing units. A total of 1,195 households live in 51 buildings of six or seven storeys in eight residential clusters. There is an educational cluster with a nursery and a primary school.

A passer-by can tell that this housing estate is different from others: it has traditional Cantonese, or Lingnan, style buildings and native vegetation (Figure 9.7). A short investigative tour will disclose the key feature: the estate provides a traditional Cantonese lifestyle, rare in the hundreds of housing estates in Canton (Guangzhou). The estate features enclosed yards, providing a sense of security and belonging, and all the buildings have partly vacated ground floors providing enhanced ventilation and a roofed open space for shading from the sun and shelter from the torrential rain. The estate has layered vegetation consisting of 96 local species covering 31.55 per cent of the site. The volume of vegetation is controlled to produce enough oxygenated air quality for the entire population of the estate. The commercial street is characterised by the traditional Cantonese verandah, similar to a sotto portico, a Guangdong building style providing shading and protection from heavy rain to shoppers, and a landscaped stream for leisure. The estate includes a small Cantonese temple, a purpose-built space for Cantonese opera, and physical exercise facilities for children and retirees. The walls are made with new insulation materials to increase energy efficiency, and an air-conditioning system uses the cool underground water in summer to save energy. As a result,

FIGURE 9.7 Local vegetation and Cantonese architectural style in Lingnan Garden
Source: photo by Albert Cao (2004).

Lingnan Garden was awarded a Green Housing Estate accreditation (see Section 14.5 for more details) in May 2003 by the Guangdong Provincial Government.

The second case is Hibiscus Ancient Town, a villa project in a western suburb of Chengdu featuring courtyard housing with traditional west Sichuan architectural style and incorporating traditional Jiangsu and Yunnan housing styles. Hibiscus Ancient Town was developed by a local developer in Chengdu, the capital city of Sichuan Province, and was one of the earliest villa projects built in traditional Chinese architectural style offering the same amenities as Western-style villas. The estate adopts a dynamic plan to lay out the semi-detached and terraced (four-house) villas ranging from 150–350 m^2 in a system of traditional style roads, gardens and green belts (Figure 9.8). The houses are normally three storeys in a north to south orientation, with large windows and a large balcony, supplemented by a multifunctional basement and front, rear and side gardens (100–300 m^2 per house). With a neighbouring commercial area built in the same traditional architectural style and with tourist activities that resemble ancient street markets, the estate affords its residents a rich and unique lifestyle rarely found elsewhere. The success of Hibiscus Ancient Town has informed other villa developers and resulted in more Chinese architectural style villas appearing in other parts of China.

FIGURE 9.8 Terraced villas and landscaping in Hibiscus Ancient Town
Source: photo by Albert Cao (2004).

9.3 Housing investment

Before the imposition of purchase restrictions for foreigners in 2006 and for non-registered households in 2010 (see Chapter 7 for details), any individuals, households and organisations could purchase residential property anywhere in the country, which was a means to boost the development of the real estate industry and local economy. Purchase restrictions for non-registered households (see Section 7.4) in 46 cities have since had a significant impact on the housing investment market in China.

9.3.1 Types of housing investment and investors

Individual investors dominate housing investments, with funding from capital gains from selling privatised public housing, non-property income, mortgages and funds from others sources. These investors are partly responsible for the continued housing price inflation. The lack of housing holding tax and widespread avoidance of income tax on rental income and capital gains provide strong incentives for housing investment. The large capital gains in cities without a strong rental market encourages investors to hold on to vacant housing.

Housing investment in China can be classified into two types: buy-to-let and buy-to-hold. The former is self-explanatory and the latter means the holding of housing invested without letting. Buy-to-hold investment in housing arises because of the low holding cost and high capital gains. There is a property management charge for holding housing, which is normally low, but no housing tax and any other taxes on holding housing vacant. Once housing is held for more than five years, the impact of Business Tax imposed from 2005 on selling housing becomes minimal.

Housing investors are normally individuals and families because of the small lot size of housing. In downtown locations there are some high-end housing blocks held by company and institutional investors. Many of those blocks are serviced apartments or holiday homes. The presence of more and more city centre high-end apartments and out-of-town villas provides opportunities for institutional investment in housing. A significant kind of housing investor is the association, or syndicate, set up to bargain housing purchases collectively and in some cases purchase housing collectively. The most famous of such syndicates is from Wenzhou (see Box 9.2 for more details).

BOX 9.2 HOUSING INVESTMENT AND PRICES IN WENZHOU

Wenzhou, a coastal city in southern Zhejiang Province, has a population of 8.07 million (registered households only) and per capita GDP of CN¥49,817 (US$8,044) in 2013. Unlike cities in the Yangtze River Delta, Wenzhou has never been a favourite choice of big businesses, because of its marginal location and lack of modern industry. Since the 1980s, however, Wenzhou has acquired a national reputation for mixed reasons: first, for its manufacturing and nationwide promotion of the small goods it produced; then for investments by its residents in housing, coal, cotton, oil and electricity throughout the country and in some international cities like Paris and New York; and finally for its financial crisis from housing investment by its residents.

Wenzhou's housing market took off in 1997 when the city's private businesses accumulated funds from their successful manufacturing businesses. Average housing prices rose from less than RMB2,000 per m^2 in 1997 to more than RMB7,000 per m^2 by the end of 2004. In 1997, the Wenzhou government worried about the large numbers of unsold and unlet housing, but by 2005 local investment buyers were queuing up whenever there was new supply in the market, even though the price of the high-end housing was over RMB10,000 per m^2.

From 1999, some Wenzhou investors were also attracted to Hangzhou, the capital city of Zhejiang Province, where housing prices rose rapidly; and Shanghai, where housing prices were stagnating. That was the start of outward housing investment. In August 2001, Wenzhou investors arrived in Hangzhou in groups with the title of the Wenzhou Housing Purchase Syndicate (WHPS), and housing prices in the city rose 25 per cent in several months. Since then, the

WHPS became famous and caused rapid housing price inflation in many cities wherever they went. For example, in Spring 2004, WHPC went to Chongqing and prompted housing prices there to rise over a third in the same year. The influence of WHPS is also reflected in the quarterly Wenzhou Housing Show, in which housing projects from many big cities in the country could be found. The inflationary impact of WHPC surprised many people, who attempted to make sense of the ability of Wenzhou investors to cause city-wide housing prices inflation in so many big cities. The answer may lie in the scale: a widely accepted figure is that there were 100,000 Wenzhouers engaging in housing investment throughout the country. Those investors mobilised about RMB100 billion cash during 2008 to 2010, with 60 per cent from borrowed funds. Cash is raised from WHPS members' relatives and friends through a very efficient system of fundraising and underground loans, which included money diverted from bank loans allocated for totally different purposes. The city-wide housing speculation caused the funds to dry up in other industries, causing the city's economy to lag behind other cities in the province.

The national housing market slump in 2008 to 2009 trapped many Wenzhou housing investors. When housing prices fell, many Wenzhou investors sold to reduce their loss. Over several months housing prices rebounded, and many Wenzhou investors borrowed large sums to buy housing; and when prices soared, they borrowed yet more money. Credit was so easily available in 2009 with banks eager to lend the pile of money from the government's rescue package. Funding was also available from high interest rate underground loans. However, the 2010 Measures cooled down the housing market country-wide, forcing many Wenzhou investors to sell to reduce their losses. Many returned home with their money to invest in their hometown housing market because there were no purchase restrictions on them there. Another factor that resulted in the return of investors was the change of state policy on investments in coal, cotton, oil and electricity. Wenzhou investors in those industries in other provinces were forced to return home. With plenty of cash and credit, housing prices in Wenzhou rose to catch up those in BSGS, with average prices rising from CN¥8,405 per m^2 in 2006 to CN¥33,389 in September 2011.

However, 2011 proved to be the tipping point for the city of housing speculation. Higher returns on property investment had undermined the labour-intensive manufacturing industry by leaving it with low technology arising from insufficient investment to upgrade in the past few years. The city had been unable to compete with other cities in the Yangtze River Delta for talent and more competitive industrial projects. To catch up with the city's competitors and to cool down housing prices, the municipal government increased the land supply from less than 40 hectares per year before 2010 to 70 hectares in 2011, and over 200 hectares in 2013 to find the money to upgrade the infrastructure; dicing with the risk of oversupply. When the Chinese Premier, Wen Jiabao, visited the city in October 2011 to sort out underground financial activities, the smarter investors started to sell. The combined result was that housing prices fell every month from October 2011 to August 2014. From 2012, bankruptcies, runaway debtors, auctions of housing due to foreclosure, and large unsold stocks of housing joined to aggravate the selling pressure in the market. By January 2014, average housing prices had plummeted to CN¥19,374 per m^2: a massive 42 per cent reduction on three years earlier. Wenzhou is the only city to have consecutive monthly housing price deflation in the 70 cities index since October 2011 (Figure 9.9). The index measures all housing transactions within Wenzhou City, so the fall in prices is not as large as the figures above that measure housing prices in the urban areas of the city. The message, however, is the same. With unsold stock at a historical high (1.10 million m^2 at the end of 2013 compared to 0.10 million m^2 at the end of 2010), Wenzhou's housing market still needs plenty of time to get rid of the unsold stock. The WHPS disappeared because people started to save in the banks.

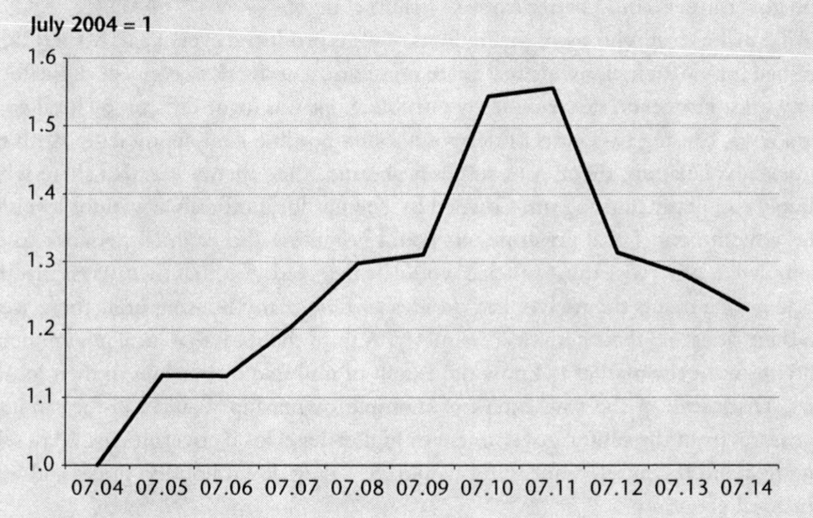

FIGURE 9.9 Housing price index showing falling prices in Wenzhou
Sources: NBSC, various years.

9.3.2 Gains from housing investment

In the new century, housing has become a very profitable yet low-risk investment vehicle due to the phenomenal housing price inflation (see Figure 4.8 for housing price inflation in Beijing and Shanghai). In particular, those who have multiple home ownerships in BSGS are millionaires in dollar terms. A study, *China Family Panel Studies 2014* (Social Science Survey Centre of Peking University, 2014) indicates that the median proportion of housing in total family wealth was 80 per cent in urban areas due to rapid housing price inflation.

9.3.3 Housing registration

There have been two rounds of attempts to require housing investors to use their real names. Housing has been used extensively by people to hide their wealth gained from corruption and other unusual means such as incomes from tax avoidance, and money laundering, and for foreigners to invest in Chinese housing without declaring their identity. It was also found that some developers sometimes use false names as purchasers to push up prices and retain units for sale for higher prices in the future. Thus, the 2005 Measures proposed the compulsory use of real names in housing purchases, and the Ministry of Construction has been working on the issue. Cities like Nanjing and Hangzhou pioneered the system of real names with improved computer software to detect the number of units any individuals purchased. The 2010 Measures again emphasised real names for housing purchase so the government could better exercise macro control.

To make clear who owns the millions of units produced every year, MOHURD pushed forward a nationwide real estate registration network to connect all the local networks. However, this technically possible work was made difficult by local governments, causing two years of delay without a possible completion date. With no official explanation, theories or rumours abound. One theory was that there were simply too many housing units owned by corrupt local officials at various levels of the government. Local governments could not allow the national network to be completed, otherwise those officials would be detected. Another theory was that the local governments themselves had no idea of how many housing units there were in their cities and their networks are messy. A third theory is that local governments did not want the market to know the extent of multiple ownerships in their localities. Disclosure of the true extent of multiple ownership would call for punitive measures from the central government or higher-level local governments. Mass selling to avoid the punitive measures would then drag down housing prices and hurt the local economy.

Nevertheless, the central government is determined to push ahead with a national register for real estate. The task of building the national registration database and the network connecting all local databases has now been transferred to the MLR, which has produced a timetable to complete the work in 2016.

9.4 Property management

The last selling point in contemporary property development is good property management in normally gated communities. In 1994, the Ministry of Construction issued the *Management Measures on New Urban Housing Estates* to introduce professional property management into Chinese cities, after successful attempts of Hong Kong-style property management in Guangzhou from 1992. In 1997,

BOX 9.3 SHANGHAI'S HOUSING MARKET DEVELOPMENT UNDER TIGHT LAND SUPPLY

Xiaobo Dai

Tight land supply in Shanghai is a problem brought about by rapid urbanisation. Shanghai is one of the most favoured cities for immigration from the countryside, due to good employment opportunities and high income levels. However, large-scale immigration causes a shortage of land within the administrative boundary of Shanghai and high land prices, which has a strong impact on real estate development.

Shanghai has a land area of 6,340.5 km^2, with the built-up area close to 3,000 km^2. One-third of the built-up area is in industrial use and two-thirds are in housing, commercial and infrastructure use. Since 1992, Shanghai's built-up area has expanded approximately 100 km^2 per year. By 2010, land shortages were raising strong concerns. Because the ratio of built-up area in Shanghai is already 43 per cent, higher than other international metropolises, Shanghai can no longer expand through increasing new construction land. According to the city's master plan for urban development, Shanghai can have 100 km^2 of new construction land up to 2020. As a result, land shortages will persist in Shanghai.

The main reason for land shortages in Shanghai is population growth. In 1990, Shanghai's permanent population was 15 million. The master plan at that time predicted that the level of population in 2020 would be 20 million. With an estimate of 100 m^2 of residential land and 50 m^2 industrial land per resident, 3,000 km^2 of land would be needed by 2020. However, Shanghai's permanent population reached 24.2 million in 2013. The city cannot support this population level and the quality of urban life is affected by overcrowding.

The municipal government has fully appreciated the impact of population growth on land use and a number of measures have been adopted to tackle Shanghai's land shortage. First, outward relocation of manufacturing has started to encourage outbound migration. For example, notebook computer assembly plants have been relocated to Chongqing with 25,000 jobs relocated and the associated vacation of sites. Second, relocation of industrial land has resulted in all industrial and logistics companies moving to 104 industrial parks occupying over 600 km^2. Through this relocation, one-third of the original industrial land was released for residential and other uses. Third, the city encourages the development of the service industry. Over a hundred creative industry parks have been set up using old industrial properties. Fourth, five to six satellite towns have been under development to release the population pressure in the central city.

Shortage of new construction land has major implications for the housing sector in Shanghai. Some new construction land is still available for housing development at present. By 2020, real estate development will mainly be on brownfield sites under urban regeneration schemes. Housing transactions will change from new housing to second-hand housing, which in 2013 accounted for a third of the housing supply in the market. It is expected second-hand housing will account for 90 per cent of all housing transactions in 2020. The rental market will expand rapidly. Real estate development will mainly focus on urban regeneration, redevelopment and refurbishment. The real estate industry in Shanghai will focus on increasing the efficiency of land use and the output of commercial properties, and transform from a development focus to a service focus.

professional property management was introduced into old housing estates. By 2002, 38 per cent of all urban buildings nationally were under professional property management. By 2005, the coverage in Beijing, Shanghai and Guangzhou reached 80 per cent. The scenario of the rapid dilapidation of new residential buildings, due to the inhabitants installing illegal structures on the outside of buildings and occupying the public space, quickly became a thing of the past. Many of the internationally famous property management firms, like DTZ and Savills, are looking after China's increasing high-specification modern commercial property.

The China Property Management Institute (CPMI) compiled the *Standards on Property Management Services Classes for Ordinary Housing* (CPMI, 2004), which divides services into six categories: basic services; building management; facilities management and maintenance; security; cleaning; external work and landscaping. These operate within three classes that define the level of services. The standards are very high. In the basic services category, class 1 service provides a client's reception centre with 24-hour telephone operators, and class 3 services to have 8-hour telephone operators. In the security service category, class 1 service requires 24 hours guarding of main entrances, patrol of important areas every hour and 24-hour CCTV monitoring of key points, and class 3 service requires 24-hour security guards and patrol of important areas and key points every 3 hours. The CPMI's Standards do not apply to non-ordinary housing estates. In fact, many high-end residential estates claim to have the service provision at a five-star hotel level.

However, institutional factors have resulted in chronic problems in professional property management in the ordinary and low-end housing and commercial property, which lead to poor protection of the rights of property owners. Similar to developing the housing market to alleviate the state's burden on housing provision, the government is eager to introduce and expand professional property management to alleviate the state's burden on public services such as policing, cleaning, care and cultural activities. Such services, however, have to be paid for if the government shies away from provision. With the retreat of the state, the burden of payment falls on the owners of new and old private housing and commercial property. In higher value housing estates, owners and occupiers are able to foot the bill. In new housing estates, low maintenance and repair costs will also make property management firms more profitable. For housing estates mainly for low-income households, especially not new ones, owners and occupiers are heavily burdened and sometimes refuse to pay, causing property management companies to quit, leaving acute management problems behind.

In addition, the local governments' hands-off practice in guiding and supervising housing estate development and management has further enhanced the difficulties of professional property management. Local governments neglect the social implications of gated communities and thus have no policy to guide their establishment. Many ordinary housing estates (according to size, facilities and locations) have been developed into gated communities, with the consequences of high costs in maintenance of private roads, private communal gardens, refuse collection and security. The trend of bigger size flats also takes its toll, because property management fees are normally charged on a GFA basis.

The author was asked by reporters in an interview in June 2014 in Shanghai: why does property management find it difficult to satisfy the housing owners? The author indicated that the cost of property management is too high if the standards issued by CPMI are to be followed. Property management firms are mainly privately owned and thus profit-oriented. They provide poor service if fees are not paid in full. They may quit if there is no hope that the owner-occupiers will be prepared to raise the fees when costs rise. The solution is public services, which are now a buzzword for the new generation of leaders. Reforms of local public services should alleviate the burden of property management companies of providing too many services for capped fees.

9.5 Discussion: how many empty homes are there?

The issue of empty homes has been the focus of attention since macro control started in 2005 due to the observation of empty homes in some cities. Many new housing estates were found to have only a few lights on throughout the night, hence the suspicion that most of the flats were empty. Because there is no consensus on the definition of empty homes and no official statistics, various estimates by wild guess and systematic studies have been produced. However, these eye-catching figures are all found to be problematic.

In 2010, a message appearing on the internet became a sensation. It claimed that the national grid conducted a survey in 660 cities in China and found 65,400,000 housing units showed no record of electricity usage for six months; concluding that these units were thus empty (Xinhua, 2010). The national grid later denied that it had conducted such a survey. Academics and property practitioners found the figure beyond belief. However, the National Bureau of Statistics could not provide data on empty homes, and market-based real estate data providers and consultancy firms were also unable to provide such data. The lack of official and industry data encouraged more guess work but also systematic research on the issue.

An example of a systematic study on China's housing conditions, including vacant homes, is the Vacancy Rate of Urban Housing and Trend in the Housing Market by the China Household Finance Survey (CHFS) of South-Western University of Finance and Economics (CHFS, 2014). CHFS claimed that over 1,600 students (undergraduates, master's-level and PhD students) participated in the survey, which covered 1,048 neighbourhoods in 262 counties of 29 provinces.[4] Some 20 to 50 households were selected randomly from each neighbourhood and a total of 24,182 households were interviewed.

Based on the survey, CHFS reported that 48,980,000 housing units (in private ownership) were vacant in 2013 in cities and towns, 8,420,000 more than in 2011. In addition, there were 3,500,000 unsold commodity housing units. CHFS defined vacant housing as: (1) home owned by one-home family with all family members working and living elsewhere; and (2) home owned by multiple-home family with neither a family member nor a non-family member living inside. The former accounted for 5.10 per cent and the latter accounted for 17.3 per cent of all urban homes, making a total vacancy rate of 22.40 per cent. Therefore, CHFS asserted that the vacant stock of housing in China's cities and towns was able to meet all the rigid demand (23 million units) and improvement demand (9.5 million units) from existing urban households. Only 40 per cent of current housing supply was needed to satisfy all new demand from both existing urban households and new immigrants from the countryside.

The results of the CHFS study, however, are problematic. First, some data do not make sense. For one thing, in 2011 and 2012 numbers of commodity housing built were 7,220,000 and 7,460,000 units, respectively (NBSC, 2013). Completion in 2013 is estimated as 8 million units by a real estate market data provider (Yang, 2014). If CHFS's claim that 8,420,000 units became vacant between August 2011 and August 2013 is true, then roughly half of the new completion units during this period became vacant. This is different from what practitioners experienced, as revealed by the author's interviews in August in 2013. For another, the report showed the percentage of families in multiple ownership grew from 18.6 per cent in August 2013 to 21 per cent in March 2014. It means that the 5 million commodity housing units built during that period need to be bought by 5 million families in single-home ownership to achieve this effect, or all new housing were sold to families in single-home ownership.

Second, some data are self-contradictory. CHFS reported that in August 2013, multiple-home ownership was 18.6 per cent. The sixth population census, conducted in 2010, indicated that 21 per cent of residents in cities and towns rented their housing (ibid.). Government statistics indicated that by 2012, 12.5 per cent of urban registered urban households were covered by Affordabe and Social Housing (ASH).[5] A rough estimate is that about 8 per cent of those families were provided by LRH and Public Rental Housing (PRH), and rental subsidies. Then the remaining 13 per cent of the 21 per cent need to rent. If 17.3 per cent out of the 18.6 per cent families in multiple-home ownership left at least one housing unit vacant, in what housing did those 13 per cent families live?

Yang (ibid.) reported that a National Bureau of Statistics survey in Guangzhou in 2011 found one hundred households in Guangzhou owned 119 housing units, with 13 let to tenants. If all the remaining units were left vacant, then the vacancy rate was 5 per cent. However, the author observed that many families in multiple-home ownership do not leave their unlet additional homes vacant. They use them as holiday homes or second homes for easy access to work, for their aging parents to stay in, or for many other uses.

Third, the sample size of 24,182 households in the CHFS study accounted for only 0.01 per cent of the total urban households. However, a National Bureau of Statistics survey in Zhejiang Province in 2007 for the same purpose covered 31,000 households (ibid.). In that survey, 14.3 per cent of urban households were found to be in multiple-home ownership, with 1 per cent owning more than two homes.

Fourth, the report's claim that only 40 per cent of current housing supply was needed suffers from a conceptual problem. Even if vacant, homes in multiple ownership are the private property of their owners and do not constitute a new supply to the housing market. Housing policy could be adapted to encourage homeowners to make better use of their second homes, say, to let them out, and to provide incentives for them to sell. A real estate tax, discussed in Section 8.6, could induce some multiple homeowners to sell the units they do not use often. However, the decision to sell or let rests with the homeowners.

A key issue is the reliability of the sampling approach in CHFS study. Chinese families are reluctant to talk to outsiders about their family wealth, in particular to young strangers from a university that few knew. Unlike surveyors from government departments, those young students had no way of checking the validity and reliability of the data provided by the families they surveyed.

In his fieldwork, the author had seen over a dozen housing estates in several cities with less than half of the units lit up at 7pm to 8pm in the evening. This sort of observation prompted many to argue for the case of high home vacancy rates. Wang reports that the total urban household number in China was slightly under 220 million in 2010 (Wang, 2013). A researcher in a central government ministry confirmed that there was one housing unit for one urban household in fieldwork in 2013. Thus a 1 per cent vacancy is equivalent to 2.2 million units. These vacant units tend to concentrate in newly completed housing estates. Thus observation on how many units are lit up in the evening, suggested by some in the media, to derive a vacancy rate is not a rigorous approach to find out the real vacancy rate.

More and more academics and practitioners have joined the call for the central government to conduct a housing census to provide accurate data on housing conditions in the country. However, housing vacancy should not be equated to available supply. It has no direct link to housing price movement and profitability of housing investment. The housing industry should continue to build new homes for the tens of millions of new migrants and urban young people.

BOX 9.4 TRENDS IN HOUSING DEMAND IN CHINESE CITIES

Lin Liu

From a medium- to long-term perspective, factors that increase housing demand in China's cities and towns include urbanisation, economic growth, income growth and urban renewal. Aging, on the other hand, is a factor that reduces housing demand. China's urbanisation level, indicated by the proportion of urban population to overall population, passed 30 per cent in 1996. After 1996, urbanisation speeded up, with the urban population increasing by more than 20 million every year. Calculated by purchase power parity, per capita GDP in China is estimated to grow from US$8,000 to US$18,000 from 2010 to 2012. This will be in line with urbanisation in the 1960s to the 1980s in major European countries and in the 1930s to the 1960s in the USA as measured by 2000 dollars. At this development stage, improvement housing demand is strong and urban renewal activities increase, leading to strong demand on housing. From the age structure, however, China's old age dependency rate reached a turning point in 2013 and is expected to rise afterwards. Furthermore, population aged between 20 and 45 peaked in 2008 and decreased afterwards. This change in population structure will lead to a reduction in housing demand.

The above factors will lead to housing demand remaining at high levels but with slower growth. At a micro level, China's urban housing demand originated from migration from rural areas, young urbanites getting married, demand from families whose homes were to be demolished, and improvement in housing standards. Analysis of population survey data shows that the overall demand for housing from 2006 to 2010 in cities and towns was 3.85 billion m^2, or 0.77 billion m^2 on an annual basis. Demand from marriages, demolition and migration was 2.82 billion m^2, accounting for 77 per cent of the overall demand. Demand from improvement in housing standards, on the other hand, was 1.01 billion m^2, accounting for 23 per cent. It is estimated that the total demand from 2011 to 2015 is 1.1 billion m^2 per year from 2011 to 2015, which is significantly higher than that from 2006 to 2010. However, the estimate of total housing demand from 2016 to 2020 is 1.1 billion m^2 per year, only slightly more than that from 2011 to 2015.

With the progress in Redeveloped and Improved Housing provision, the number of families living in poor housing conditions will decline significantly in 2015. Families living in poor housing conditions will be migrants from the countryside, whose number is estimated to grow at 5 million per year. Large-scale affordable and social housing construction (Chapter 13) will be phased out in the period from 2016 to 2020. With plenty of housing, housing provision for low-income families will change to rental subsidies.

9.6 Conclusion

Land price inflation in the last 30 years has made urban land in China an attractive investment. Developers in China have found ways to avoid the country's very strict regulation on land use to build land reserves. Local governments, on the other hand, have built Land Reserves to facilitate land supply and to regulate the housing market through land supply. In the past three decades, housing development, a profitable business, has seen increasing competition which has resulted in growing market concentration of housing development in big firms that can bid for land regionally or nationally on the open market. Competition has resulted in very sophisticated housing and estate design, marketing, sales and after-sales management, raising housing standards for China's home buyers. Commodity housing in China provides not just a place to live, but a lifestyle that is different. Thus, innovation in the housing industry is essential for property developers to be competitive. Housing investment, on the other hand, has low risks and high returns due to rapid housing price inflation. However, there is excessive risk taking in housing investment by some investors, who incurred big losses when housing prices fell due to macro control campaigns. Housing management, on the other hand, has been heavily regulated and unrealistic standards are imposed on ordinary housing management, leading to affordability problems among homeowners. The issue of vacant homes, a negative outcome of institutional inadequacy in China, has been controversial. However, both wild guess and academic estimates of home vacancy rates are problematic. Action by the central government is essential to provide answers to the vacancy issue, which is important to the future housing development and investment.

Notes

1 Article 25 in UREAL 1994.
2 This is a rough but conservative estimate by the author. The figure comprises about 100 million urban residents who are former migrants currently living in rented accommodation, and about 400 million rural residents who will migrate to cities and towns for jobs, education, and better healthcare in the coming two to three decades. The figure of new migrants can be greater if taking into account migration of urban residents from small cities and towns to larger cities.
3 See Life Table 241 from Department of Communities and Local Government UK.
4 Xinjiang, Tibet, Hong Kong, Macau and Taiwan were not covered in the survey.
5 This figure is from the 2013 Work of the Government Report, announced by the premier at the March General Meeting of National People's Congress.

10

COMMERCIAL PROPERTY

This chapter investigates the commercial property investment market in China, which covers offices, retail, and industrial including logistics properties. It first looks into the issue of LUR renewal, and analyses the potential and risks for Chinese cities as destinations for property investment. An institutional analysis is conducted to explore the mechanisms that power commercial property development in Chinese cities, with a case study on the decisions to site major office districts in Xi'an. The chapter then surveys the office and retail property markets by analysing historical rental growth, the supply and the supply pipeline, and explores the pending oversupply in office and retail space through a number of case studies. The preference to develop HOPSCA, an office, retail and leisure complex, by local governments is investigated. Later the chapter examines the situation of logistics property and its prospects in Chinese cities. Finally, early symptoms of severe oversupply in retail space in Shenyang are examined and the actions by Hangzhou to mitigate the oversupply situation through collective choice-level action are considered to be a short-term solution to this commercial real estate bubble. The long-term solution lies in reforming the system of local governance.

10.1 Land for commercial uses

As China's economy continues to expand and urbanisation continues, demand for commercial property has been increasing. As a result, land for commercial uses is a good asset for investment.

As explained in Chapter 5, the maximum length of LURs for commercial use is 40 years; that for industrial use is 50 years; and that for mixed uses 50 years. LURs for commercial uses are renewable. LUR owners should apply for renewal at least one year before the LURs expire (see Section 3.1). However, there are no detailed regulations enacted to govern LUR renewal, because the great majority of LURs were granted after 1990 with over 20 years to run.

A special case of allocated LUR renewal in Shenzhen provides a useful observation on the government's approach to LUR renewal. Land was allocated to users with allocated LURs of 20-year terms in Shenzhen Special Economic Zone in the 1980s. For instance, the first high-rise office building in Shenzhen, the International Commerce Building completed in 1983, came with a 20-year allocated LUR. After 2000, the building experienced difficulty in letting because the LUR was about to expire. The Shenzhen Municipal Government enacted an *ad hoc* regulation, the *Stipulations on Renewal of Expired Real Estate*, to renew the expired allocated LURs and convert them to granted LUR (Shenzhen Municipal Government, 2004). The renewal premiums were 35 per cent of the market land prices at the time of renewal in Shenzhen, with the new terms being the difference between the maximum terms allowed by Decree 55 and the original terms of the expired LURs. For example, the expired allocated LUR of a site allocated in 1984 for mixed use was converted to a granted LUR in 2004 with a 30-year term, with payment of 35 per cent of the market land price of the site. The implication arising from this case is that the government should allow for renewal well before LUR expiry to provide flexibility to LURs owners to apply for renewal.

There is a legal risk in holding land without immediate development, which is explained in Section 9.1 along with ways to avoid this risk.

10.2 Commercial properties investment in China: potential and risks

Chinese cities have been the core of the country's spectacular economic growth in the past 35 years. More and more of these cities are ready for commercial property investment. This section discusses their potential and analyses the risks imposed by the regime of governance.

10.2.1 Potential

The rise of Chinese cities has been an integral part of China's economic success. The rapid changes in property market transparency of Chinese cities were captured by JLL's transparency index.[1] In the 2004 and 2006 indexes, Chinese cities were considered in tier 4 with low transparency. In 2008 the index recognised the variations within Chinese cities and divided them into three tiers, with tier 1 cities in semi-transparency and tier 2 and tier 3 cities in low transparency. The 2010 index upgraded tier 2 cities to semi-transparent. The 2012 and 2014 indexes regarded tier 3 cities as semi-transparent and tier 1 cities very close to transparent (Table 10.1).

The increasing attractiveness of Chinese cities for real estate investment is indicated by research dedicated to them. For example, JLL increased the number of

TABLE 10.1 Level of property market transparency of Chinese cities

Years	*Cities*	*Transparency scores*	*Transparency category*	*Transparency description*
2004	Whole of China	3.79	4	Low transparency
2006	Whole of China	3.50	4	Low transparency
2008	Tier 1 cities	3.33	3	Semi-transparency
	Tier 2 cities	3.68	4	Low transparency
	Tier 3 cities	3.97	4	Low transparency
2010	Tier 1 cities	3.14	3	Semi-transparency
	Tier 2 cities	3.38	3	Semi-transparency
	Tier 3 cities	3.73	4	Low transparency
2012	Tier 1 cities	2.83	3	Semi-transparency
	Tier 2 cities	3.04	3	Semi-transparency
	Tier 3 cities	3.31	3	Semi-transparency
2014	Tier 1 cities	2.73	3	Semi-transparency
	Tier 2 cities	3.04	3	Semi-transparency
	Tier 3 cities	3.26	3	Semi-transparency

Sources: JLL (2004; 2008; 2010; 2012b; 2014).

Chinese cities worthy of investment, excluding BSGS, from 30 in 2007, to 40 in 2009 and to 50 in 2012, indicating rising property market maturity in Chinese cities (JLL, 2012a). The 50 cities recommended by JLL are listed in Table 5.1. However, JLL's recommendations are based on economic development and property market activities, but not explicitly the administrative factors that are important in China.

An alternative way to assess Chinese cities is to take into account the administrative status of cities, which indicates the level of decision-making power and resources available to those cities. Table 10.2 provides basic information on 36 cities: 31 being provincial capital cities (including the four provincial level municipalities: Beijing, Chongqing, Shanghai and Tianjin) and five other cities designated in the national plan (Dalian, Ningbo, Xiamen, Qingdao and Shenzhen). A comparison with JLL's recommendations is as follows. These 36 cities include the four tier 1 cities, eight out of the nine tier 1.5 cities, eight out of the ten tier 2 cities, eight out of the 13 tier 3 (emerging) cities and five out of the 13 tier 3 (early adopter) cities. Only three cities in Table 10.2, namely, Yinchang, Xining and Lhasa, are not in the 50 cities named by JLL. These three cities are the capitals of China's three smallest provinces in terms of population: Ningxia; Qinghai and Tibet. On the economy side, 16 out of the 36 cities had per capita GDP higher than US$10,000 in 2012.

When assessing Chinese cities, institutional factors are more important than existing levels of economic development and property market maturity. Cities as provincial capitals have much higher potential than other cities with lower administrative status, because they are more resourceful and have greater powers to

TABLE 10.2 Basic information on provincial capital cities and cities designated in the national plan (2012)

Cities	*Population*	*GDP*	*Per capita GDP*	*Fixed asset investment*	*Total retail sales*	*Students studying at university*
	(millions)	*CN¥ billion*	*US$*	*CN¥ billion*	*CN¥ billion*	*(thousands)*
Beijing	20.18	1,787.9	14,063	646.3	770.3	581.84
Changchun	7.68	445.7	9,211	314.0	174.0	387.66
Changsha	7.04	640.0	14,430	401.2	252.2	523.17
Chengdu	14.05	813.9	9,195	589.0	331.8	685.64
Chongqing	28.84	1,141.0	6,280	938.0	403.4	670.17
Dalian	6.88	700.3	16,156	565.4	222.4	263.69
Fuzhou	7.12	421.8	9,404	326.6	232.0	305.39
Guangzhou	12.70	1,355.1	16,937	375.8	597.7	939.21
Guiyang	4.32	170.0	6,247	248.3	68.3	391.07
Haikou	2.05	81.9	6,340	51.0	43.6	115.50
Hangzhou	8.70	780.2	14,235	372.3	294.5	459.18
Harbin	10.64	455.0	6,788	395.0	239.5	482.21
Hefei	7.08	416.4	9,336	400.1	129.4	417.21
Holhot	2.91	247.6	13,504	130.1	102.2	227.19
Jinan	6.81	480.4	11,197	218.6	242.0	659.87
Kunming	7.26	301.1	6,583	234.6	149.4	361.00
Lanzhou	3.62	156.4	6,857	123.9	74.9	306.64
Lhasa	0.56	26.0	7,372	28.5	12.5	19.58
Nanchang	5.04	300.1	9,450	240.3	99.5	509.24
Nanjing	8.11	720.2	14,095	468.3	310.4	651.95
Nanning	6.66	250.3	5,966	258.5	125.6	318.05
Ningbo	7.64	658.2	13,675	290.1	232.9	145.36
Qingdao	8.90	730.2	13,023	415.4	263.6	296.65
Shanghai	23.47	2,018.2	13,649	525.4	741.2	506.60
Shenyang	8.11	660.3	12,923	564.3	280.2	369.29
Shenzhen	10.55	1,295.0	19,484	231.4	400.9	75.57
Shijiazhuang	10.16	450.0	7,031	372.9	191.6	395.54
Taiyuan	4.20	231.1	8,735	132.1	113.0	358.20
Tianjin	13.55	1,289.4	15,104	887.1	392.1	473.11
Wuhan	10.20	800.4	12,455	503.1	343.2	946.99
Wulumuqi	3.35	200.4	9,496	101.0	83.4	135.95
Xiamen	3.67	281.7	12,184	133.3	88.2	143.96
Xian	8.47	436.6	8,182	424.3	226.4	723.96
Xining	2.23	85.1	6,058	70.0	31.7	61.86
Yinchuan	2.05	115.1	8,911	91.9	31.6	82.47
Zhengzhou	9.10	555.0	9,680	367.0	229.0	698.19
Total	303.90	21,497.9	11,229	12,435.5	8,524.4	14,689.15
% of the country	22.44%	41.64%	185.53%	33.19%	40.53%	61.43%

Source: NBSC (2013).

mobilise resources to achieve their development targets. For example, Kunming was considered a tier 3 city (emerging) by JLL, and Dongguan a tier 2 city (Table 5.1). In terms of economic development and property market activity level in 2012, Kunming was inferior to Dongguan. However, being the capital city of Yunnan Province, Kunming is able to designate and build its CBD and relocate its airport in 2012, releasing a large tract of land ripe for development. It has the state's support as well. Furthermore, Kunming would become an international railway pivot should the Kunming–Bangkok–Singapore high-speed railway[2] and the Kunming–Rangoon high-speed railway go ahead. Changsha was named a tier 2 city by JLL (2012a). Being the capital of Hunan Province, it has a larger economy and much higher development potential.

10.2.2 Risks

Despite the increasing potential of Chinese cities as investment destinations, property investors need to understand the risks imposed by a particular set of institutional arrangements on commercial property investment in China. The institutional arrangements include the lack of national policy on commercial real estate, the bias towards GDP growth as a performance indicator of local leaders, local dependence on land sale revenues and tax receipts from property development, and insufficient checks and balances on local development planning.

First, there is a lack of national policy on commercial real estate development and investment. Chapters 4, 7, 10 and 13 explain that housing has been the focus of the central government at the constitutional level, because housing availability and affordability affect social stability and the success of the country's urbanisation. In contrast, commercial property is mainly left up to local governments at the collective choice level to decide. There is neither national control on the amount of land allocated to commercial real estate development, nor national planning guidance to guide commercial property development. Macro control only applies to residential property. As a result, local government discretion has shaped commercial property development in all cities.

Second, there is a bias towards GDP growth against other growth indicators because local leaders are incentivised to achieve high GDP growth and mobilise all resources to promote economic growth. Development campaigns, initiated by each new local leadership, normally involve dozens or even hundreds of projects being given the go-ahead at the same time, with preferential treatment or even subsidies to encourage state-owned, private and foreign firms to take part. However, local leaders are less concerned whether such growth is ahead of time and whether so many developments will bring risks to developers and property investors. An oversupply will result in falling rental and capital values, high vacancy rates and project failures. Poorly located and inappropriately specified buildings can suffer from long-term vacancy. In the 1990s, when the market was small and financing was limited, there were failed projects in the hundreds in both Shanghai and Guangzhou (see Section 5.4), which were costly to revive (see Box 5.1). The collapse of the local property market and the damage to the local economy in Beihai (Box 4.1) were the consequences of local government failure in adopting a wrong commercial property policy.

Third, local governments have been expanding land sales and pushing forward property development to generate funds for infrastructure construction. Infrastructure is essential for cities to attract inward investment, which is crucial to economic growth. Section 6.2 analyses the dependence of local governments on land sale revenues and taxes from property development and sale. To obtain land sale revenues and tax receipts, local governments often approve plans for large numbers of commercial property development projects to go ahead at the same time, regardless of whether the local property market can absorb the resultant new supply.

Fourth, lack of clear national policy on commercial real estate has resulted in the absence of checks and balances in local development planning. The issues of planning failures are discussed in detail in Section 6.3. In the 1990s, the central government was concerned with the inflationary impact and financial risk brought about by the large number of commercial property projects under development, and imposed some temporary bans and restrictions on the approval of new projects. In the 2000s, the scale of commercial property development was dwarfed by that of housing. The master plan approval system described in Table 6.7 had limited success in preventing some cities with low standing in China's urban hierarchy from implementing their plans to expand their cities and build them into international metropolises. Hence only the housing sector became the target of the macro control from 2005 (see Sections 4.5 and 7.2).

However, the very strict macro control measures imposed in 2010 resulted in local governments and developers shifting the development focus to commercial property. At that time, there was market demand for commercial property development and many Chinese cities could benefit from large-scale commercial property development. Nonetheless, local governments at the collective choice level were not bound by national planning guidance on commercial property development. As a result, they set the market rules to promote commercial real estate development to achieve growth targets. These rules include various preferential policies on development projects in designated CBDs, retail districts and industrial parks, and on the occupation of completed buildings in those districts and parks should market demand be inadequate.

A major problem is that many local governments did not put urban development in the context of local economic development stage and population trends, and approved and promoted commercial property developments with insufficient regard to need and demand. In particular, unrealistic ratios between commercial and residential floor space were imposed in many development projects, which resulted in a glut of commercial floor space. Too many projects were planned and given the go-ahead at the same time. Sometimes insufficient effort was spent planning and promoting commercial property schemes in locations where the market demand existed, leading to geographical imbalances of commercial property.

At the operational level, some developers did not follow due diligence procedures or conduct rigorous market research. The design and layout of commercial properties at the project level was often done in a hurry so that the project could reach the market ahead of its competitors. Developers without any expertise in developing and managing commercial property took advantage of favourable market sentiment at the time to take on development schemes. On the other hand, developers with the expertise and experience of commercial property development were overconfident that their schemes would win in any competition.

With hindsight, an institutional arrangement such as the scrutiny of major urban development projects by a system of planning inspectorates would have stopped many overambitious local development plans for commercial property. The existing rule is approval of master plans by the State Council, which is conducted on an *ad hoc* basis by people who are not particularly qualified to inspect such master plans. Once the master plan is approved, there is no further national guidance or checks and balances on commercial real estate developments approved by local governments, even though those developments alter the master plan. Box 10.1 describes decision-making in Xi'an when developing its CBD.

BOX 10.1 XI'AN'S NON-CENTRAL CBDS

Xi'an, the capital city of Shaanxi Province, had a population of 8.45 million in 2012 and the third highest number of university students of Chinese cities (Table 10.2). It is the most important city in north-west China due to its history, location and industry. In the 1990s, the city recognised its need to build a modern city centre similar to those in Shanghai and Guangzhou and submitted its latest urban master plan (1995–2010) to the State Council for approval. Approval came in May 1999. The approved urban master plan designated a CBD at the southern boundary of the city's built-up area. The city's television transmission tower and the Qujiang International Convention and Exhibition Centre were built there as the anchor infrastructure.

Such a decision was perplexing to the author, who was doing fieldwork in the city at the time. There is already an office cluster at the south of the walled city (South City in Figure 10.1), which was close to the traditional city centre and had good transportation links. Another office cluster had been formed at Xi'an High-Tech Industries Development Zone (High-Tech Zone in Figure 10.1) in the south-west of the walled city. A sensible proposal would be to combine the two clusters, which were less than 2 km apart, to make a large business district at a more central location to the city. In contrast, there was mainly farmland on the CBD site and no infrastructure. Interviews with experts revealed that the decision was to use the new CBD project to develop the area south of the existing CBD to generate economic growth and land sale revenues. However, the designated area (Qujiang in Figure 10.1) has insufficient market appeal compared to South City and High-Tech Zone.

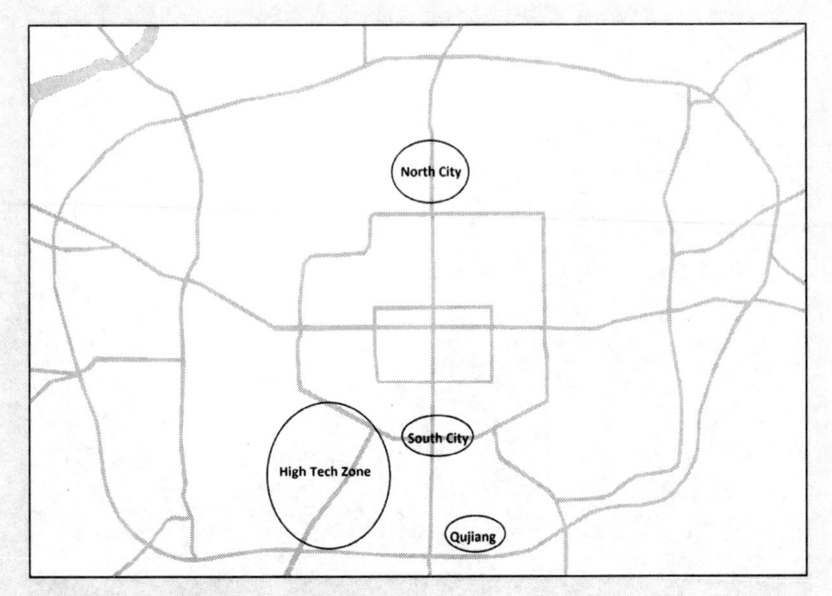

FIGURE 10.1 Office clusters in Xi'an
Source: Albert Cao (2015).

The local government soon designated another zone at the northern boundary of the city as the city's future CDB (North City in Figure 10.1) in its new urban master plan (2004–20). The northern zone was an industrial zone, which was far away from the existing city south and High-Tech Zone office clusters (Figure 10.2). Again the purpose of the new master plan was to use the CBD to justify massive urban development in the north of the city. The relocation of the municipal government to the north resulted in many office projects being completed there. The northern urban expansion was completed. Being far from potential clients, however, the northern office cluster has suffered from high vacancy rates since its completion.

Limited availability of land in the city south office cluster restricted its growth, but there is plenty of land in the High-Tech Zone and more and more new office buildings were built there. By quarter one of 2014, 57 per cent of the city's 1.04 million m^2 Grade A office stock was located in the High-Tech Zone; 19 per cent in South City; 12 per cent in North City; and 12 per cent in Qujiang.[3] The country's heavyweight developers further strengthened the High-Tech Zone as the *de facto* CBD by entering into development projects there. Greenland's 270-metre twin office towers, the tallest in northwest China, will be completed in 2016, providing 170,000 m^2 Grade A office space in the High-Tech Zone. The problem with the High-Tech Zone is that it is not a central location. The metro, with line 1 operational in 2013, will

FIGURE 10.2 Office buildings in Xi'an High-Tech Industries Development Zone
Source: photo by Albert Cao (2013).

provide a solution because lines 3, 5 and 6 will run through the High-Tech Zone, improving accessibility.

In the end, it is the market, not the local leaders' command, that had the ultimate say. The south-western High-Tech Zone is the *de facto* 'central' business district.

10.3 Offices

Large-scale office development has been carried out due to rising demand from occupiers, in particular, domestic occupiers, and the favourable rules set by local governments for developers to build offices in Chinese cities. Purchase restrictions imposed in 2010 on housing forced many national and regional developers to diversify their activities from housing to commercial property development. Local governments, on the other hand, encouraged the expansion of office development in their cities to maintain economic growth, increase the provision of high-quality business space, and promote a modern city image. Since 2010, China has had a commercial real estate development boom.

In order to generate sufficient land sale revenues to fund the large-scale infrastructure development, local governments designated many office development schemes and sold sites for the schemes at the same time. They then invested the revenues to build the infrastructure, and the expected improved infrastructure, the increased economic activities from infrastructure and office development, and investment in offices combined could boost local economic growth. At the same time local governments actively engaged in place promotion and attraction of inward investment. Depending on the balance of supply and demand, developers and future investors bear different levels of risks on future letting and sales in different cities.

For the country as a whole, the vacancy rate of prime office space fluctuated around 15 per cent from 2003 to 2014 (CBRE, 2014a). There is a distinction between tier 1 cities and other cities. The office markets in tier 1 cities perform better in general than their counterparts in tier 2 and tier 3 cities. In Beijing, rental values rose sharply when supply became limited after the glut of the pre-Olympic building boom was absorbed; in Shanghai, rental values also rose significantly due to limited supply (Figure 5.3). In Guangzhou and Shenzhen, rental values started to rise after the slack from the previous cycle in 1990s was exhausted in the early years of the twenty-first century (Figure 5.4). Table 10.3 provides a snapshot of office rents at the best locations in Beijing, Shanghai, Shenzhen, Guangzhou and Chengdu, whose office rents are the highest in the country.

Three case studies on the office markets in tier 2 cities are presented below. The first case is Chengdu's office market (see Table 10.2 for basic information). The city, the capital city of Sichuan Province, has been competing with Chongqing, a provincial-level municipality, to become the hub of south-west China. One way to win the competition is to have faster economic growth and more modern office space on offer. Hence the city made a series of ambitious plans for office development. In the past ten years, the local government designated four major office development districts: the CBD in the city centre, East Avenue, Financial Town and Dayuan Area, in addition to existing and new office development along Renmin South Road. Many of the approved office projects were to be completed in the period from 2013 to 2016: more than a quarter of a million m^2 quality office space was completed in

TABLE 10.3 Office rents in Beijing, Shanghai, Shenzhen, Guangzhou and Chengdu in quarter one, 2014

Cities	*CN¥/sq m/month*	*US$/sq ft/year*	*Location*
Beijing	387	69	CBD
Shanghai	331	59	Jing'an District
Guangzhou	179	32	Pearl River New City
Shenzhen	260	38	Futian CBD
Chengdu	122	20	CBD

Source: Carlby Xie, of Colliers International Shanghai Office, personal communication, 15 July 2014.

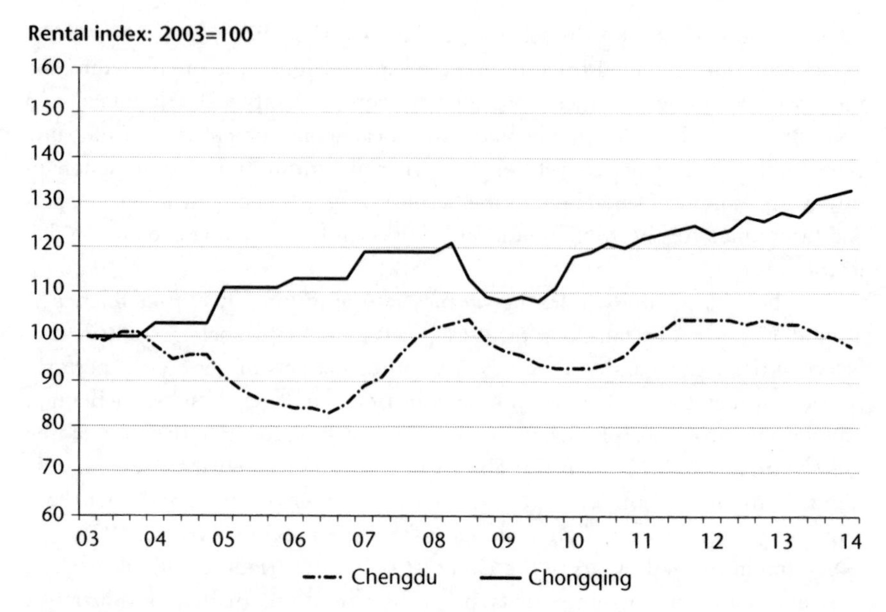

FIGURE 10.3 Grade A office rental index for Chengdu and Chongqing, 2003–14
Source: adapted from CBRE (2014a).

2013; more than 1 million m^2 are being completed in 2014; and close to 1 million m^2 and 2 million m^2 new space will be completed in 2015 and 2016, respectively (Colliers International, 2014a).

However, new completions from 2013 have pushed the office vacancy rate to 45.1 per cent in quarter two of 2014 (CBRE, 2014a). High vacancy rates and falling rents are expected to result in some of the projects being delayed due to a poor leasing market and the financial difficulty of the developers. Yet such delay will not ease the pressure of oversupply in any major way. Figure 10.3 shows the Grade A office rental index in Chengdu and its competitor, Chongqing. It is expected that both cities will experience falling rents and rising vacancy rates in the coming years due to the large numbers of office buildings being completed. Table 5.3 lists the current stock of Grade A office space in both cities. Figure 5.10 traces office vacancy rates in Chongqing from 2005 to 2013.

Severe oversupply also occurs in many other cities. The second case study city is Qingdao, a non-provincial city designated in the national plan (see Table 10.2). The city has a long-established office district at Hong Kong Middle Road (Figure 10.4), which accounts for over half of the total office stock in the city in early 2014. There is also a large office cluster at Haier Road and a couple of small office clusters elsewhere. Yet the municipal government designated a CBD at Shandong Road in 2005, and achieved good results in land sales there in 2007. The large numbers of new supply from all these office clusters are entering the market from 2011, pushing up vacancy rates (Figure 5.10) and exerting downward

FIGURE 10.4 Hong Kong Middle Road office cluster in Qingdao
Source: photo by Albert Cao (2013).

pressure on rents (Figure 10.5). For example, in the second half of 2014 alone, over 200,000 m^2 new office space will be completed (CBRE, 2014a). Being a non-provincial city, the existing office demand has been satisfied and new demand is limited.

Changsha (see Table 10.2 for basic information), the capital city of the economically backward Hunan Province (see Figure 2.9 for per capita GDP), had lagged behind in economic development since the 1990s due to its inland location and slower pace of economic reforms. In 2009, the 995 km Guangzhou to Wuhan high-speed railway (where 60 per cent of the track is set on bridges, on elevated platforms and in tunnels), became operational and greatly eased the accessibility problem of Changsha, which lies in the middle of the railway line. The construction of the Hangzhou to Changsha high-speed railway line and the Changsha to Kunming high-speed railway line provided fast mass transit connections to the east and west of the country, making the city the only high-speed railway pivotal junction in South China.

To take advantage of the improvements in transportation, Changsha has been expanding its urban areas at a fast pace in the past decade (see Table 6.3 for growth in built-up area) and designated its first CBD at Furong District in 2006. Following a period of rising office rental values (Figure 10.5), several new office districts were given the green light by the municipal government, resulting in the development of multiple office clusters in a city without international and national firms as the main

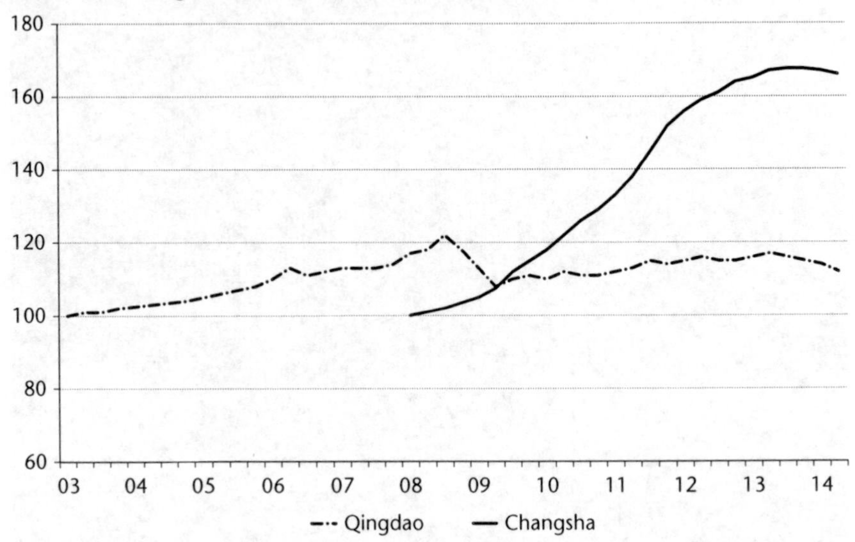

FIGURE 10.5 Grade A office rental index for Qingdao and Changsha, 2003–14
Source: adapted from CBRE (2014a).

office occupiers. These office clusters are competing with each other. For example, the office cluster at Keifu District exceeded the CBD in terms of Grade A office space in 2014. Two landmark projects, the 268-metre North Star Times Square and the 228-metre Pomp International Plaza, were completed in early 2014. In the second quarter of 2014 alone, completion of 140,000 m^2 Grade A office space and 180,000 m^2 secondary space pushed the total Grade A office space to 370,000 m^2 and the total office space to 1,790,000 m^2 (Colliers International, 2014b). The Grade A office vacancy rate in quarter 2 of 2014 was 32.4 per cent. If all the projects in the development pipeline go ahead, by 2017, there will be 3.65 million m^2 Grade A office space and 7.12 million m^2 total office space. There is no sign that the city can attract enough occupiers who can afford the rents. A national picture of office investment indicates that the scale of investment in offices has risen substantially. Figure 5.5 indicates the purchase of office space in 2012 was 22.54 million m^2, a figure 660 per cent of that in 1997. The average price of office space in 2012 was 263 per cent of that in 1997. The national picture of office price inflation is shown in Figure 10.6.

Existing and looming oversupply in office space is widespread in Chinese cities in mid-2014. This demand and supply imbalance will no doubt dampen investor appetite for purchasing completed office buildings in the short term. Prime office yields in China's tier 1 cities are low at the time of writing, with those of Shanghai and Beijing being at 5–6 per cent. Faced with pressure from oversupply, vendors started to lengthen the disposal process to find the buyers they wanted, and there was already evidence of yield decompression in the second quarter of 2014 (CBRE, 2014b).

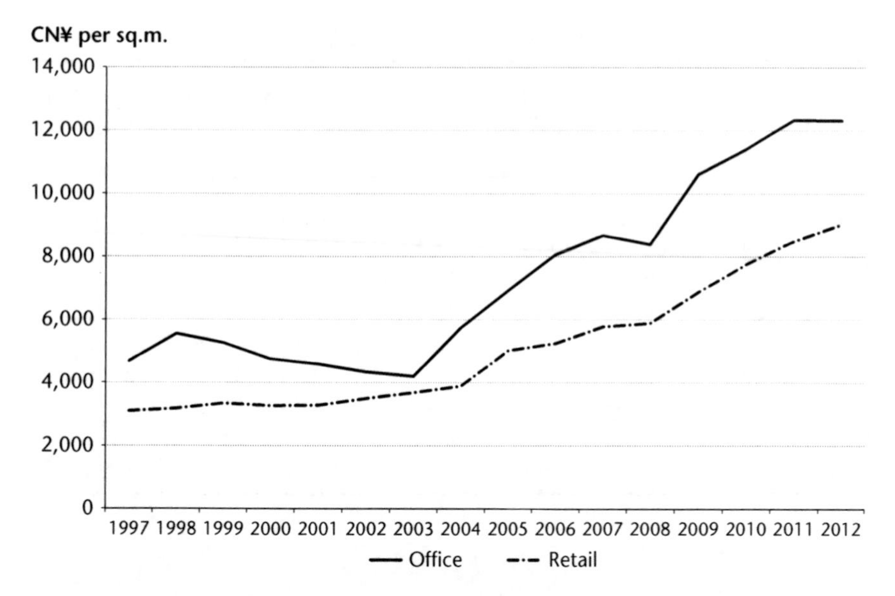

FIGURE 10.6 National average prices of office and retail space
Source: NBSC (2013).

For tier 2 and tier 3 cities, deals are normally limited due to high expectation by vendors and lack of investment interest in the large office projects. The falling rental values and rising vacancy rates as a result of oversupply may further dampen investment interest. Consequently, some developers may be forced to start strata title sales, i.e. selling office units within office buildings. Strata title sales will lead to management difficulties in the future and detrimentally affect capital values. However, investors interested in long-term growth are able to find opportunities in the Chinese office market, especially when vendors become realistic.

10.4 Retail properties

Rising disposable incomes in China have led to a rapid increase in consumption (Figure 5.11), making retail property sought after by retailers. This in turn has led to a rapid increase in development and investment in retail property. Figure 5.5 indicates the purchase of retail space in 2012 was 77.59 million m², a figure 1,223 per cent of that in 1997. Average price of retail space in 2012 was 292 per cent of that in 1997. Table 10.4 provides data on chain stores in the country as a whole and in Beijing and Shanghai. Chain stores grew faster in the country as a whole than in Beijing and Shanghai, a catching up that is still going on. In 2012, the number of stores was 248.55 per cent of that in 2004. This is made possible by the large amount of new quality retail space developed in tier 2 and tier 3 cities during this period and the penetration of many foreign, as well as national brands, into those cities. Sales per store grew 70 per cent from 2004 to 2012, which is consistent with the rapid

TABLE 10.4 Chain stores in China as a whole and in Beijing and Shanghai

	Years	*Number of stores*	*Retail space (million m²)*	*Space per store (m²)*	*Sales (CN¥ billion)*	*Sales per store (CN¥ million)*	*Sales per m² (CN¥)*
National	2004	77,631	72.03	928	839.36	10.81	11,653
	2008	168,502	101.98	605	2,046.65	12.15	20,069
	2012	192,870	147.66	766	3,546.21	18.39	24,016
Beijing	2004	4,726	4.34	918	103.56	21.91	23,862
	2008	6,059	4.96	819	136.01	22.45	27,421
	2012	6,810	6.68	981	241.10	35.40	36,093
Shanghai	2004	10,146	7.28	718	164.56	16.22	22,604
	2008	16,913	7.30	432	238.16	14.08	32,625
	2012	19,310	10.27	532	360.56	18.67	35,108

Sources: NBSC (2006; 2009; 2013).

increase in total retail sales and smaller average store space. The national average of sales per m² was slightly below US$4,000. In 2012, the national average of sales per m² was about two-thirds of those of Shanghai and Beijing

In the country as a whole, the prime retail vacancy rate has been around 10 per cent from 2003 to the first quarter of 2014 (CBRE, 2014a). The retail trade has been dynamic due to: the entry of foreign brands, the decreasing competitiveness of department stores; the growth of fast fashion, luxury, affordable luxury and Food and Beverage (F&B) brands; and the explosive growth of e-tailing. To combat the expansion of e-tailing, retailers and retail property managers are working hard to adjust tenant mix to enhance the shopping experience and make shopping a 'lifestyle consumption'. The key is to attract shoppers to come to the shops and to spend as much time there as possible. One step forward in making shopping a pleasant experience is to focus on the 'consumer experience', which comprises the shopping environment and services offered in the whole purchase process. In addition to expanding store networks, retailers need to embrace new business models and e-commerce.

The proliferation of hyper-shopping malls and, more significantly, HOPSCA by heavyweight domestic developers is instrumental in such a trend. HOPSCA stands for hotels, offices, parks, shops, clubs and apartments. It is a type of integrated development in urban communities, where people live, work, play and shop in proximity. The emergence of HOPSCA is due to four reasons. First, China's economic development had reached a stage when cities became richer and more prosperous. On the business side, the shift to the service industry generated demand for modern offices in good locations and with sufficient services. On the consumer side, the upgrading of consumption demand introduced opportunities to better meet such demand in air-conditioned, landscaped, multi-functional shopping and leisure complexes with good connectivity to roads and metro. Second, many developers in China had accumulated expertise and ability to design, build and operate very large multi-functional buildings in a relatively short time. As discussed in Section 9.2, many developers were capable of meeting the housing development needs of the equivalent of a medium to large European country. Delivering a building complex at the scale of 100,000 to 1,000,000 m^2 in several years or less is within their capacity. Third, poor prospects in the housing market after 2010 prompted many large residential developers to seek opportunities in commercial real estate. Such developers, however, are keen to include housing elements in commercial property projects. Fourth, participating in HOPSCA schemes is politically correct – meeting local governments' needs to maintain economic growth and generate property-related taxes and other incomes when residential development grew at a slower pace.

The rise of HOPSCA was facilitated by macro control on the housing market. The numbers of HOPSCA were 2 in 1995, 7 in 2000, 18 in 2005, but 131 in 2010. In 2015, the number is estimated to rise to 330 (Wang, 2012). By the end of 2011, three cities, Beijing, Shenyang and Chongqing, had over 10 million m^2 floor space in completed HOPSCA. By the end of 2015, 27.57 million m^2 of HOPSCA space will be added to Chengdu, and 15 million m^2 of such space added to Kunming, Xi'an, Shenyang, Hefei and Chongqing (ibid.). The sizes of HOPSCA can be seen from three examples. Figure 10.7 shows a HOPSCA under construction in Kunming. Figure 10.8 presents the atrium of a shopping mall that is part of a

FIGURE 10.7 A HOPSCA under construction on the southern verge of the city centre in Kunming

Source: photo by Albert Cao (2013).

FIGURE 10.8 A shopping mall inside a HOPSCA in Tianhe District, the retail heart of Guangzhou

Source: photo by Albert Cao (2013).

HOPSCA in Guangzhou. Figure 10.9 shows a newly completed HOPSCA in downtown Urumqi.

The retail ratio of HOPSCA floor area is approximately one-third, so new retail space produced by HOPSCA is going to be substantial. Data from international property consultants confirm the rapid increase of shopping malls, many of which form the core of HOPSCA. For example, the number of shopping malls completed in 2012 and 2013 in 11 major cities[4] in China was 52 and 68, respectively, bringing 6.1 and 6.5 million m^2 retail space to the market (Knight Frank, 2014). These shopping malls, in particular within HOPSCAs, incorporate more entertainment and leisure facilities for adults as well as families, for example, cinemas, gyms and other entertainment outlets. There are cases of mall operators providing promotion coupons on their websites for shoppers to spend in malls, encouraging footfall in the malls they operate. The increase of F&B units proves to be effective, and shopping

FIGURE 10.9 Century Jinwa Plaza in Urumqi
Source: photo by Yuanze Gao (2014).

mall operators also introduce various activities, such as art shows and live music events. Nonetheless, there have been cases of mall closure and withdrawal of stores, in particular, department stores.

The government's anti-corruption campaign has led to falling sales of many luxury brands and high-end F&B, which affects the desire to invest in retail property.

However, this is a short-term phenomenon. The demand for luxury goods by Chinese shoppers is strong, as revealed by the purchasing power of Chinese tourists in Europe and North America. Shopping overseas offers different shopping experiences, which includes quality of products and services, and is encouraged by changes in custom duties and exchange rates. If brand operators pay enough attention to the Chinese market and provide similar quality and services in China, shoppers will find shopping at home a better alternative because they can save the travel costs, time costs and travel fatigue. If fact, a few brands have already adjusted their operations. For example, there are products that are specially designed for the Chinese market and shoppers are unable to find those products in Hong Kong and other places such as London.

The four tier 1 cities, BSGS, are the leading cities in total retail sales (Table 10.2). Strong demand for retail space in these four cities is reflected in the high rental growth records. Figure 10.10 indicates that from 2003 to quarter two of 2014, prime ground floor retail rents doubled in Beijing and Shanghai. Similar rental growth occurred in Guangzhou and Shenzhen over the same period (Figure 10.11). The retail rents at the best locations in quarter 2 of 2014 for the four tier 1 cities and Chengdu are shown in Table 10.5.

In tier 2 and tier 3 cities, the retail property market at the time of writing (mid-2014) is suffering from oversupply, a consequence of the property-led growth model. Similar to office property (see Section 10.3), too many retail projects were approved at the same time, with preferential policies to induce developers to commit to large projects in secondary locations. Retail development, many within HOSPCAs, in Chengdu is an example of potential oversupply due

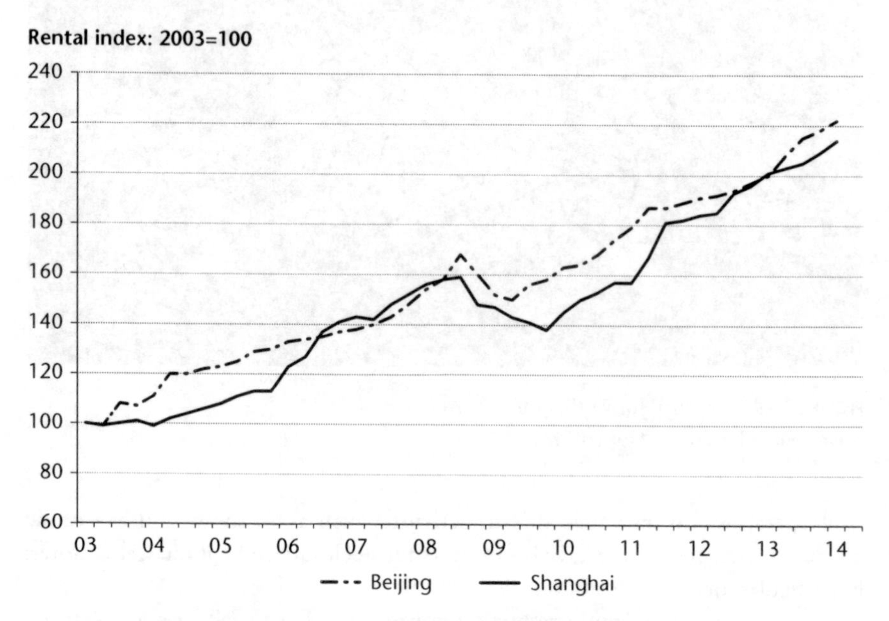

FIGURE 10.10 Prime retail property rental index in Beijing and Shanghai, 2003–14
Source: adapted from CBRE (2014a).

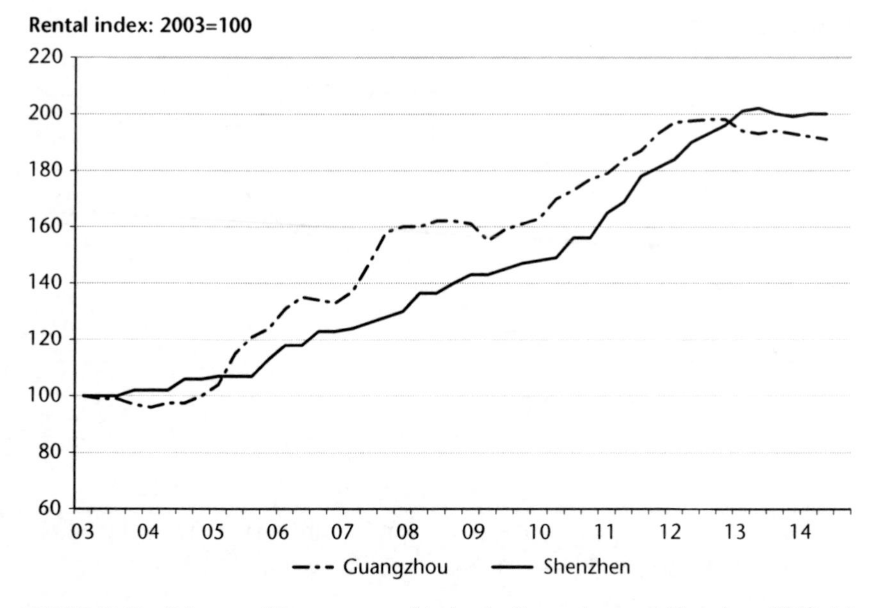

FIGURE 10.11 Prime retail property rental index in Guangzhou and Shenzhen, 2003–14
Source: adapted from CBRE (2014a).

TABLE 10.5 Highest ground floor retail rents in Beijing, Shanghai, Shenzhen, Guangzhou and Chengdu in early 2014

Cities	*CN¥/sq m/month*	*US$/sq ft/year*	*Location*
Beijing	2,600	466	Wangfujing
Shanghai	2,310	414	Jingan
Guangzhou	2,160	387	Tianhe Road
Shenzhen	1,606	288	Luohu
Chengdu	1,782	236	Chunxie Road

Source: Carlby Xie, of Colliers International Shanghai Office, personal communication, 15 July 2014.

to massive retail development engineered by the local government's urban development campaign to become the retail hub of south-west China. Completion of retail property was about half a million m^2 in 2012, and three-quarters of a million m^2 in 2013. In first half of 2014, four major retail schemes were completed, bringing a total of 524,597 m^2 of new retail space to the city. The total retail space in the city reached 3,383,795 m^2 as a result. Nevertheless, it is expected that another half a million m^2 of retail space will be completed in the second half of 2014, and over a million m^2 completed in 2015 (Colliers International, 2014a). It is worth noting that retail rents in Chengdu have been stagnating since 2007

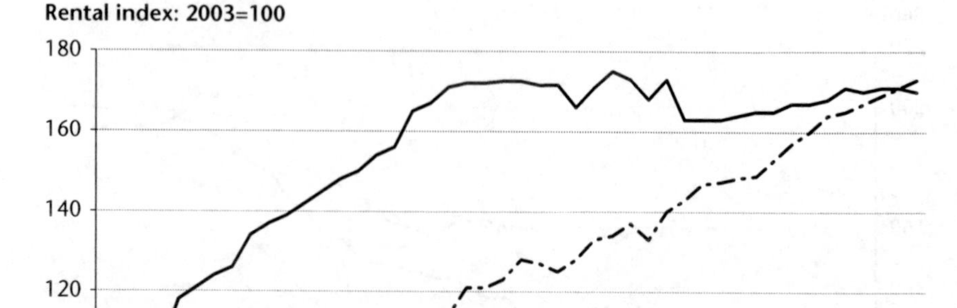

FIGURE 10.12 Prime retail property rental index in Chengdu and Wuhan, 2003–14
Source: adapted from CBRE (2014a).

(Figure 10.12). Large amount of new supply will drive rental values down in the near future.

Another case study of the retail market in tier 2 cities is the retail market in Wuhan, with rental value rising since 2005 (Figure 10.12). In the third quarter of 2013, the city had a total 3,437,460 m^2 of quality space, with new completion in 2013 as a whole predicted at 681,300 m^2. New supply of quality retail space, many within HOPSCAs, from 2014 to 2016 is estimated to be 1,164,700 m^2 (*Hubei Daily* and JLL, 2013). In the first half of 2014, 300,000 m^2 of retail space was completed, and another 110,000 m^2 is expected to be complete in the second half of 2014. The slower pace of new supply of retail space in Wuhan is likely to generate a more healthy development of the market.

Oversupply in tier 2 and tier 3 cities and the decline of demand for luxury goods have resulted in fading investment interests in China's retail properties in the middle of 2014. Taking on retail property in many tier 2 cities incurs a high leasing risk. Rents for prime retail property are believed to have reached their peak, and the investment yield at about 5–6 per cent for Beijing and Shanghai is considered to need adjustment. However, this supply and demand imbalance is likely to decompress investment yields and present opportunities for long-term investment to gain the benefit from the ongoing increase in personal disposable income and urbanisation.

BOX 10.2 MORE HEADWINDS FOR CHINA'S COMMERCIAL REAL ESTATE MARKET

Carlby Xie

What will China's commercial real estate market look like by 2017?

There is a widening split in China's commercial real estate market between tier 1 cities, which are commonly known as Beijing, Shanghai, Guangzhou, Shenzhen, and tier 2 cities, which are typically provincial capitals or economically strong regional hubs, such as Shenyang and Dalian in north China, or Chengdu and Chongqing in south-west China. For example, in China's office real estate market, the average vacancy rate for tier 1 cities was 11.7 per cent by the end of first half of 2014. In tier 2 cities, this was nearly 20 per cent. In the retail property market, these figures were 9.6 per cent for tier 1 cities and 7.2 per cent for tier 2 cities.

In the office market, this difference reflects the fact that major domestic companies and most multinational corporations (MNCs) prefer to be located in tier 1 cities. Consequently while the total office stock in tier 2 cities is low, relative to the tier 1 cities, the vacancy rate is higher. In the retail property market, the low vacancy rate in tier 2 cities reflects a relatively limited amount of suitable stock, and retailers' expansion to these cities, where retail sales growth often outpaces the growth in tier 1 cities as a result of its lower base.

Looking to the future, there will be a spike in the supply in tier 2 cities in both office and retail real estate as these markets catch up to the levels of development already seen in tier 1 cities. However, the pace of supply may outstrip demand in the early years. In the office market, the vacancy rate for tier 2 cities will be in the 30–35 per cent range. For the retail property market, it will be in the 10–15 per cent range.

Assuming no major economic changes, the vacancy rates in tier 1 cities are expected to increase more modestly, to the 18–20 per cent and 10–12 per cent range, respectively, given the schedules of new developments.

Fundamental reasons for oversupply

While oversupply in China's commercial real estate market is a complex problem, it is essentially the result of a combination of market forces and urban planning.

In the office sector, similar to the 'tertiarisation' trend that continues to intensify in China's tier 1 cities, many tier 2 cities have been actively re-engineering their economic growth model towards the service sector and high-value manufacturing – sectors that provide high wages and increased tax contributions.

In the retail sector, retail sales have been growing in double digits for several years (faster than tier 1 cities) as the income of their residents rapidly increases from a low base.

However, urban planners may have been overly optimistic in how much commercial real estate these trends will actually support in the coming five years, and developers may have overestimated demand from companies and retailers. This has been complicated in some cases by certain local governments who required developers to construct mixed-use projects beyond the city centre as part of their city's broader urban planning scheme. In the short term, this has led to excess capacity in many lower-tier markets as urbanisation, infrastructure development and consumer sophistication catch up.

Softening asset performance, stable capital values

There are many consequences of this oversupply, with the main ones being intensified leasing competition, difficulties in attracting desired occupiers and direct cuts in rentals by landlords. However, oversupply is not a major problem in tier 1 cities, where there is a healthy balance between demand and supply, and capital values have remained stable.

This balance will fluctuate in the short term, as tier 1 cities receive new completions, particularly Shanghai and Guangzhou. Competition from this new supply, as well as other trends such as decentralisation and the development of business park markets, may constrain rental growth to some extent. Beijing will be an exception, as demand for office property in the city is stable but supply is limited, potentially supporting further rental growth.

For the tier 2 cities, more subdued rental growth is expected, given the slow growth of demand but the fast pace of new supply. Leasing competition among existing and new office developments will be intensified. In some cases, city governments have begun to offer incentives such as rental rebates to stimulate demand. However, this will not be enough to absorb the large amount of supply in the pipeline, and vacancy rates for the office real estate market in tier 2 cities are expected to grow substantially in the coming two to three years. This will place great pressure on rental growth in these cities.

The retail sector may be even more challenging, with certain cities facing even heavier oversupply in this sector. In some instances, landlords have offered revenue guarantees in addition to fitting-out subsidies in order to attract their desired tenants. Overall, the coming supply in tier 2 cities will lead to a correction in the average rent. New stock will outpace retailers' expansion plans, and the vacancy rate will rise. In some developments, this may lead to operational problems.

Investors move back to safety of tier 1 cities

Market observations and sales transaction volumes suggest that both international and domestic investors remain keen to invest in China's commercial real estate market, given the country's rapid economic development in the past decade or more, and its future prospects. However, they will be increasingly cautious towards tier 2 cities in the coming two to three years, a trend that has already been seen in the relatively limited number of transactions and a slowdown in the number of inquiries. Instead, they are more likely to seek investment opportunities in either traditional commercial markets in tier 1 cities, or in the development sectors of these markets, such as business parks or mid-end, regional retail hubs.

10.5 Industrial and logistics properties

Industrial and logistics sectors play a crucial role in the world's largest manufacturing, export and e-commerce sectors that China boasts. Domestic consumption has become the key drive in the next phase of the country's economic growth. In particular, retailing and e-tailing development in China's inland cities requires the support of a modern logistics sector. Yet China's logistic sector needs further modernisation for domestic consumption in order to become the leading sector of the economy. This section focuses on logistics property that has strong potential and for which market forces are important.

Institutional factors matter in the development of the logistics sector. The lack of national policy until 2011 took its toll on the development of logistics property. Local governments with responsibility to achieve high growth targets were reluctant to allocate the increasingly scarce industrial land for logistics use. Compared with manufacturing and R&D sectors, logistics generates a relatively small amount of jobs and tax revenues. As a result, there is less attention to and support from local governments at the collective choice level for logistics property.

To promote the development of the logistics sector, the central government enacted the *Circular on Adjustment and Invigoration Plan for Logistics Industry* (State Council, 2009b), which is the first national policy on the industry. The plan recognised the importance of the logistics industry that accounted for 6.6 per cent of the GDP in 2008, but pointed out that acute problems remained in the industry. First, the industry was inefficient in general compared to its counterparts in developed countries. The ratio of total expenditures on logistics to GDP fell from 19.4 per cent in 2000 to 18.3 per cent in 2008, but was twice that in developed countries. Second, insufficient demand for logistics services coincided with insufficient supply capacity of specialised logistics provision. Third, there is inadequate infrastructure for logistics. An integrated logistics network needed to be developed. Fourth, local barriers and an industrial monopoly prevent a competitive logistics market from operating efficiently. Fifth, technology, talent and standards in logistics were not satisfactory. The plan called for the expansion of logistics demand and the professionalisation and marketisation of logistics services, with the government

TABLE 10.6 Rental growth of logistics properties in major Chinese cities

Cities	*Rental growth (%)*
National	60
Tier 1 cities	55
Tier 2 cities	65
Beijing	70
Tianjin	0
Dalian	80
Shenyang	220
Qingdao	30
Shanghai	30
Nanjing	75
Hangzhou	70
Ningbo	60
Guangzhou	40
Shenzhen	80
Chengdu	210
Chongqing	90
Wuhan	60

Note: The period is from Q1, 2003 to Q1, 2014 (for Shanghai: Q1, 2007 to Q1, 2014).
Source: adapted from CBRE (2014a).

providing a favourable business environment. To speed up the growth of the logistics sector, the central government enacted another policy in 2011 (State Council, 2011a) to offer tax cuts to the logistics sector, to provide subsidies to key firms and important infrastructural developments, and to increase the land supply to the logistics sector.

Yet land remains the biggest impediment to the growth of the logistics property market in China. As housing, commercial, infrastructure and ecological uses consume large amounts of land in China's continued urban expansion, land for logistics uses surrounding the bigger cities becomes difficult to obtain. In particular, to improve the efficiency of industrial land use, the four tier 1 cities, BSGS, and tier 2 cities such as Qingdao and Hangzhou, have tightened industrial land uses by shortening the terms of LURs to 20 years from the maximum period of 50 years and setting new guidelines on uses (see Section 5.1 for the new guidelines in Shanghai). Such regulations vary and evolve on a local basis. Land prices increase faster than rent in many local markets (see Figures 3.5 and 3.6 for industrial land price inflation nationally and in Beijing), driving developers to look for efficiencies to achieve target returns. For example, the requirement for higher plot ratios will lead to new developments to 'go vertical'.

Local industrial land availability and prices vary (see Figures 3.7 and 3.8), which affect rental growth of logistic properties. Table 10.6 shows the rental

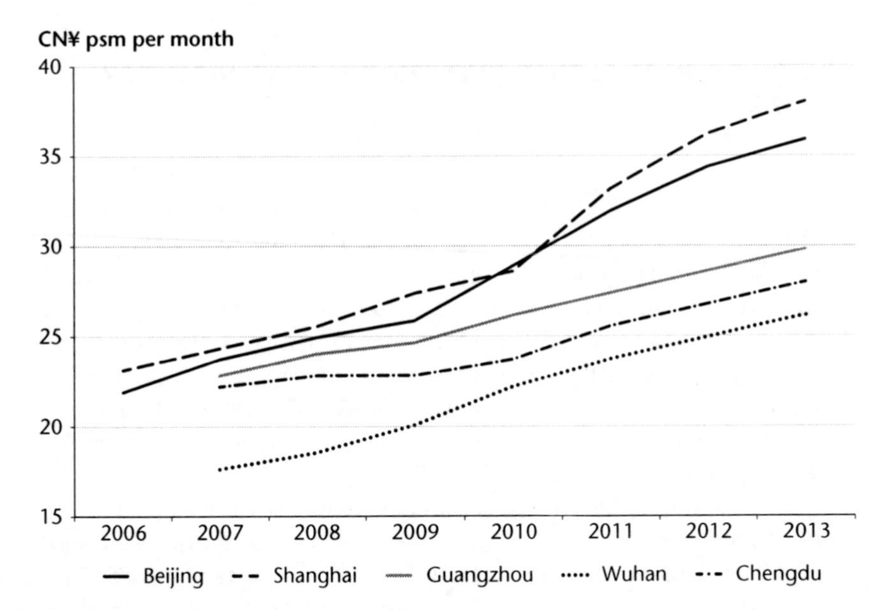

FIGURE 10.13 Logistics property rents in Beijing, Shanghai, Guangzhou, Wuhan and Chengdu

Source: Colliers International data.

growth of logistics properties in selected Chinese cities since 2003 and Figure 10.13 illustrates the movement of logistics rents in Beijing, Shanghai, Guangzhou, Chengdu and Wuhan. It can be seen that the rental index in Shenyang and Chengdu doubled due to tight logistics land supply and strong demand. In Tianjin, rents fell for a few years after 2003 and bottomed in 2010 when Tianjin Binhai New District provided a large amount of land. However, due to near depletion of land for industrial and logistics use, recently rental values have recovered to their 2003 level.

Unlike the housing, office and retail sectors, the logistics sector has the highest concentration of foreign developers, with many of the world's largest 3PL (third-party logistics provider) operators having large-scale operations in China. For example, Schenker has been in China for over 30 years. Domestic property developers such as Ping An and Wanke are joining the business of logistics development. An interesting trend is that the internet retailer Alibaba, the parent company of China's largest retail platform, Taobao, is going to build a national logistics network. Investment interest was high in 2013 in the Chinese logistics market, and this trend is going to be continued in the near future. Currently the logistics market is very active with plenty of development and leasing activities. Investment yields are expected to be between 6 per cent and 7 per cent. However, in the short term, the level of new supply, particularly in tier 2 and tier 3 cities, may lead to oversupply that will discourage short-term investment activities.

BOX 10.3 REAL ESTATE AND THE CREATIVE INDUSTRY IN SHUNDE

Junshi Yuan

Shunde is a district of Foshan City in Guangdong Province with a per capita GDP of US$16,778 and has been ranked number 1 of the 100 top districts in China for the last two years. It is one of the largest manufacturers of electrical appliances in China, and has been famous for air conditioners, refrigerators, water heaters and electric dish sterilisers for over ten years. However, land shortage has hindered the district's further expansion in manufacturing. As a result, Shunde has not had new growth drivers in the past decade and has struggled to upgrade its key economic sectors from its current white goods manufacturing.

To overcome land shortage and upgrade to higher value-added industries, Shunde has developed a new growth initiative to focus on the so-called 2.5 industries, which are between manufacturing (2) and services (3). The 2.5 industries comprise business services and trade, research and development, and modern logistics, with creative industries at the core. The South Wisdom Valley (SWV), a 7.5 km^2 business park, is the key project to deliver Shunde's new initiative. Using One-North, a similar business park in Singapore as a reference point, SWV features many research and development parks and a residential area for up to 60,000 residents. It is adjacent to the capital town of Shunde, which provides further logistic support.

To attract research and development talents from the country and around the world, SWV has signed a strategic collaboration agreement with Tsinghua University, a coalition between Carnegie Mellon University and Sun Yat-Sen University, Xi'an Jiaotong University, Beihang University, the China Academy of Sciences, the China Academy of Engineering and other institutions. It plans to rival the Silicon Valley in the USA in 20 years. Land becomes an essential source of funding for this big project. At the time of writing, five sites for company headquarters have been sold to developers and the money invested in infrastructure. For example, a 40,000 m^2 site at B district has been sold to build five office buildings, each of which will rise up to 150 metres, as headquarters for companies.

The new development initiative in Shunde follows the property-led development model and focuses on the development of commercial property for the creative industries. It attempts to provide state-of-the-art accommodation for R&D activities to achieve economy of scale and thus upgrade Shunde from a manufacturing base to a base of creative industries, pushing forward further economic growth.

10.6 Discussion: leaving the commercial property bubble to pop?

Discussion of bubbles in the housing market has been topical in China. However, commercial property bubbles, the by-product of macro control from 2010 on housing prices, have not been a popular topic. The lack of attention to the problems of commercial property is not enough to put pressure on the central government at the constitutional level to tackle the looming commercial property crisis.

Shenyang, the capital city of Liaoning Province (see Table 10.2 for basic data), provides the first example of the disadvantage of massive oversupply of office and retail space (Figure 10.14). Seeking to consolidate its position to be the largest city in the north-east of the country, Shenyang has been implementing a massive urban redevelopment and expansion plan, requiring developers to build HOPSCA with large shopping malls. From 2000 to 2007, the city had an average annual supply of 111,000 m^2 of mid- to high-end retail space. From 2008 to 2013, however, the annual supply of mid- to high-end retail property grew from around 400,000 m^2 in 2008 and 2009 to around 800,000 m^2 in 2010 and 2011 and to around 1,000,000 m^2 in 2012 and 2013 (Savills, 2014c). There is over half a million m^2 new retail space poised to enter the market in 2014 (JLL, 2014). Rents have been falling since 2011, and the highest average ground floor rent, achieved at the Shenyang Municipal Government Square, was CN¥600 per m^2 per month, or US$107.5 m^2 per square

FIGURE 10.14 Offices towers of two HOPSCAs in construction in Shenyang
Source: photo by Albert Cao (2013).

foot per year (see Table 10.5 to compare like-for-like rents in other cities). The overall average vacancy rate was 19.1 per cent in quarter 2 of 2014 (CBRE, 2014a). In particular, over half of the shopping centres that had been opened for two or three years still experienced difficulty in letting, with vacancy rates over 20 per cent. Malls located in the vicinity of metro stations and pedestrianized shopping streets also experienced letting difficulties. Famous brands, for example Dalian Wanda, fared much better. Consequently, some projects faced adjustment in the tenant mix and business model after a couple of years of operation. One had to end up in an *en bloc* sale. The key problem, however, is that there is insufficient purchasing power to support all these malls and shops.

The local governments at the collective choice level can intervene in the market to alleviate the problem of oversupply. Hangzhou (see Table 10.2 and Figure 10.15) sets an example through its decisions to adjust the development of HOPSCA. In 2008, the city enacted a plan to develop over a hundred HOPSCA. In 2012, these projects started to enter the market. In 2013, 535,700 m² of retail space, mostly in malls, entered the market (Savills, 2014e). The density of these HOPSCAs is reflected in 2013 when two HOPSCAs and a mall were completed in the northwest of the city in an area filled with university campuses: the 400,000 m² IN CITY, a HOPSCA, opened in May 2013; the first phase of the 60,000 m² Xitian City, a shopping mall, opened in June 2013; and the 120,000 m² Shenghua

FIGURE 10.15 A HOPSCA in Hangzhou
Source: photo by Albert Cao (2013).

IN TIME CITY opened in August 2013. The three projects were within 5 km of each other. It is expected that close to half a million m^2 of retail space will open in 2014, and about a million m^2 will open in 2015. Although a vacancy rate of 1.7 per cent, modestly rising in the first half of 2014, was the lowest in major cities in the country, intensive competition has resulted in postponement for adjustment in some projects. The severe loss and closure of WAOW Plaza in April 2014, the newly created retail arm of Hangzhou's Wahaha, China's largest and the world's fifth largest beverage company, heralded the impact of oversupply to retail operators.

In June 2014, Hangzhou Municipal Development and Reform Commission adjusted the city's retail development policy. First, a new HOPSCA development in the northern new town and eastern new town was banned, and strict control was imposed on districts with retail space considered to be saturated. Second, upgrading of retail facilities was allowed in the town centre and Qiantang New City CBD to safeguard their position as the retail hubs of the city. Third, community HOPSCAs were encouraged in other university campus areas without sufficient retail provision.

Collective choice-level decisions in Hangzhou represent a good example of adjustment of local retail property policy. However, there is no sign that the central government at the constitutional level will take action to prevent the oversupply in some cities getting out of control. The new doctrine of the central government after the Third Plenary Session of the Eighteenth Congress of the CCP is to leave market decisions to the market. However, the oversupply of commercial property is partly due to local government intervention and the visible hand of the central government should help and force local governments to act to let some air out of the commercial property bubbles. What Hangzhou did is a good start, but more needs to be done.

10.7 Conclusion

Although there is no explicit provision on LURs' renewal, *ad hoc* practice in Shenzhen shows the government is keen to renew LURs. In the last 35 years, Chinese cities have achieved a relatively high degree of development and are increasingly recognised as offering potential for real estate investment by international property consultants. However, Chinese commercial real estate has a significant systematic risk imposed by local governments to obtain land sale revenues and property development-related taxes in pursuit of infrastructural development and economic growth. This systematic risk has been ignored by the constitutional level decision-maker, i.e. the central government, which has not set policy and rules for local governments to mitigate the local growth-related systematic risk.

As the economic structure shifts to favour the service sector, demand for offices by domestic firms has been growing rapidly, fuelling large-scale office development in major cities in China. However, local governments have manipulated urban plans and provided preferential policies to encourage the development sector to carry out prestige projects and large numbers of office developments.

This was particularly so when the housing market was under tight restrictions in 2010 and 2011. As a result, a wave of oversupply hovers around almost all large Chinese cities. For some cities, long-term vacancy rates at high levels are inevitable and the threat of a large number of failed projects looms. The situation in retail property development is similar, with local governments imposing a ratio of residential and commercial property on development schemes that leads to oversupply of retail space. In particular, the obsession with the HOPSCA type of development by Chinese cities is going to cause the severe oversupply of retail space. Logistics property is currently undersupplied in China and the central government has tabled policies to encourage the provision of logistic properties. However, there is risk of future oversupply because of the enthusiasm in the market and that shown by local governments.

Although in the medium to long term, Chinese cities need the amount of commercial property being developed at the time of writing, local governments and developers have developed the property simply ahead of time. There is a bubble in the Chinese commercial property market and the bubble will burst when oversupply drives rents and prices down. Some local governments have attempted to intervene in the market to mitigate the oversupply situation. Without central government attention and policy, however, many local governments may not act proactively to address the situation in a timely and effective manner.

Notes

1 A composite index is calculated from scores in five aspects with different weights: performance measurement – 25 per cent; market fundamentals – 20 per cent; governance of listed vehicles – 10 per cent; regulatory and legal – 30 per cent; and transaction process – 15 per cent. There are five tiers of transparencies: highly transparent; transparent; semi-transparent; low transparent; and opaque (JLL, 2014).
2 The Kunming–Bangkok railway line has been approved by the Thai government.
3 Data from Savills: Xi'an Office Market Briefing, Q1 (2014g).
4 The 11 cities include the four tier 1 cities (BSGS), and the seven tier 2 cities: Chengdu, Chongqing, Dalian, Harbin, Shenyang, Tianjin and Wuhan.

11

INDIRECT INVESTMENT VEHICLES

Indirect investment vehicles in property in China include shares, trusts and funds. In this chapter China's share markets are first surveyed and property shares are examined to explain why Chinese developers chose to list outside Mainland China. Then the chapter scrutinises the rise in and trouble with trust companies in the 1990s and the 2000s, and explains the threat of falling housing prices and implicit guarantee to the trust industry. Real estate funds are later explored to reveal their structure and investment preferences. The chapter ends with a discussion of the lack of long-term finance for property in China and real estate companies' response to this institutional deficiency.

11.1 Shares

In China, domestic investors (individuals and organisations) and qualified foreign investors can freely purchase shares and other financial products on the Shanghai Stock Exchange (SSE) and the Shenzhen Stock Exchange (SZSE). A plan for Hong Kong investors to purchase designated shares in the SSE and for Mainland Chinese investors to purchase designated shares in the Hong Kong Stock Exchange is being implemented in the second half of 2014, which provides opportunities for foreign investors without qualified foreign investor status to invest in Chinese shares.

The Shanghai and Shenzhen Stock Exchanges started operations on 1 and 19 December 1990, respectively, as the fruits of China's economic reforms in the 1980s. Since then the two stock exchanges have grown rapidly, with about 2,500 firms listed in both stock exchanges' main boards and second boards (Figure 11.1). By the end of July 2014, SSE had 966 companies offering 957 A shares and 53 B shares. A shares are priced in CN¥, while B shares are quoted in US dollars. Initially, A shares were for

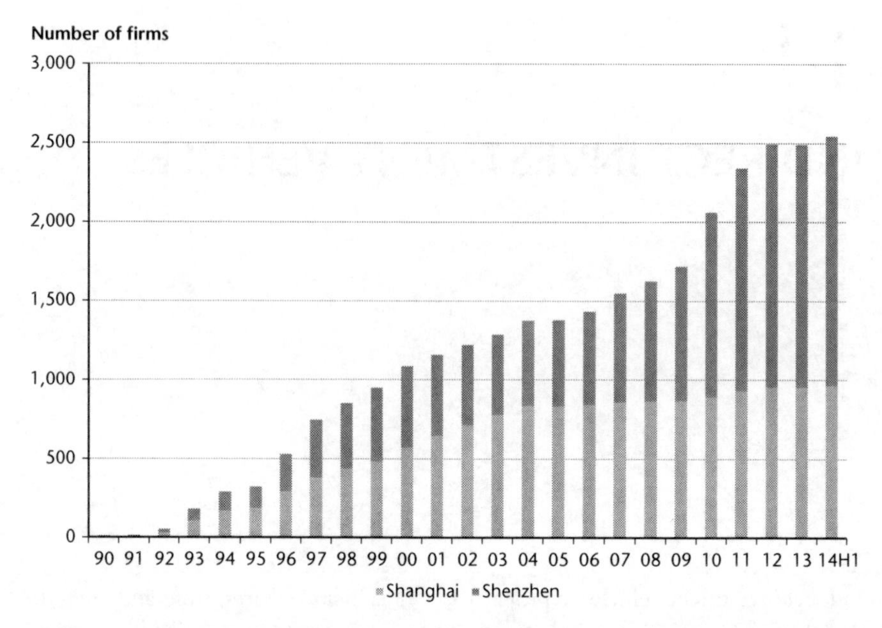

FIGURE 11.1 Number of firms in Shanghai and Shenzhen Stock Exchanges, 1990–2014
Sources: NBSC (2013); official websites of Shenzhen and Shanghai Stock Exchanges.

domestic investors only while B shares were for foreign investors. Since December 2002, foreign investors have been able to trade in A shares under the Qualified Foreign Institutional Investor (QFII) programme launched in 2003. At present, a total of 98 foreign institutional investors have been approved to trade in shares under the QFII programme, with a quota of US$30 billion. The two types of shares will be merged in the future.

SZSE had 480 companies on its main board, 722 companies on its medium and small companies board, which was set up in 2004, and 387 companies on its NASDAQ-style second board, ChiNext, for starter companies.

There are a number of companies incorporated in Mainland China whose shares, designated as H shares, are traded on the Hong Kong Stock Exchange (HKSE). Many companies float their shares simultaneously on the HKSE and one of the two Mainland Chinese stock exchanges.

The Shanghai Stock Exchange issues a SSE Composite Index, which bases its starting value of 100 on stock prices as at 19 December 1990. SSE's total market value was CN¥16.05 trillion on 8 August 2014. Shenzhen Stock Exchanges issues a SZSE Composite Index, which bases its starting value of 100 on stock prices on 3 April 1991. On 8 August 2014, SZSE's total market value was CN¥10.20 trillion. The performance of both indexes in the last two decades is depicted in Figure 11.2. It can be seen that both stock indexes are highly correlated. A more detailed performance of the SSE Composite Index is depicted in Figure 11.3. It is clear that both indexes reached very high peaks in 2007 and have been depressed since then.

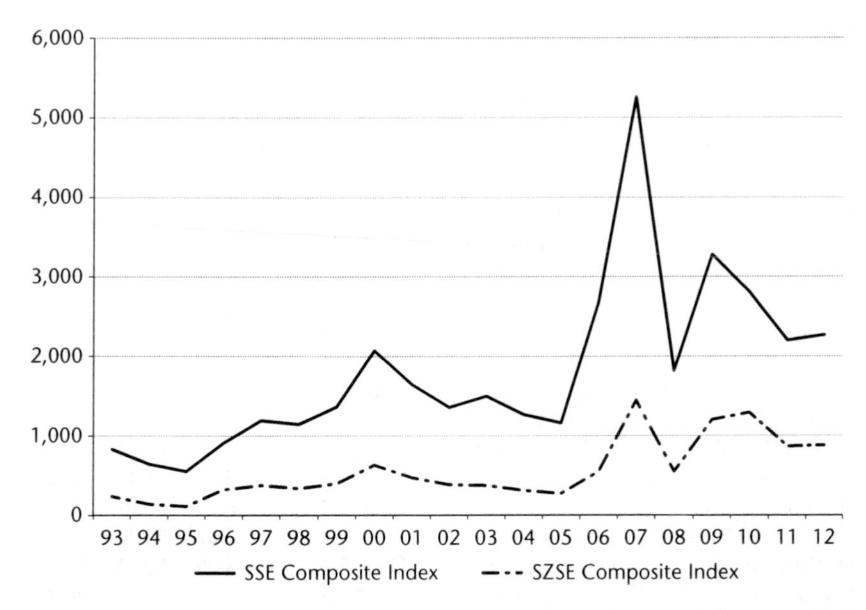

FIGURE 11.2 Annual closes of SSE Composite Index and SZSE Composite Index, 1993–2012
Sources: NBSC (2003; 2013).

There are 71 and 64 property companies listed in SSE, and SZSE's mainboard, respectively. In addition, over 50 Mainland Chinese property development firms are listed in HKSE, and several are lining up for initial public offering (IPO) in 2014. Table 11.1 lists the basic data of the top 20 developers according to sales revenues

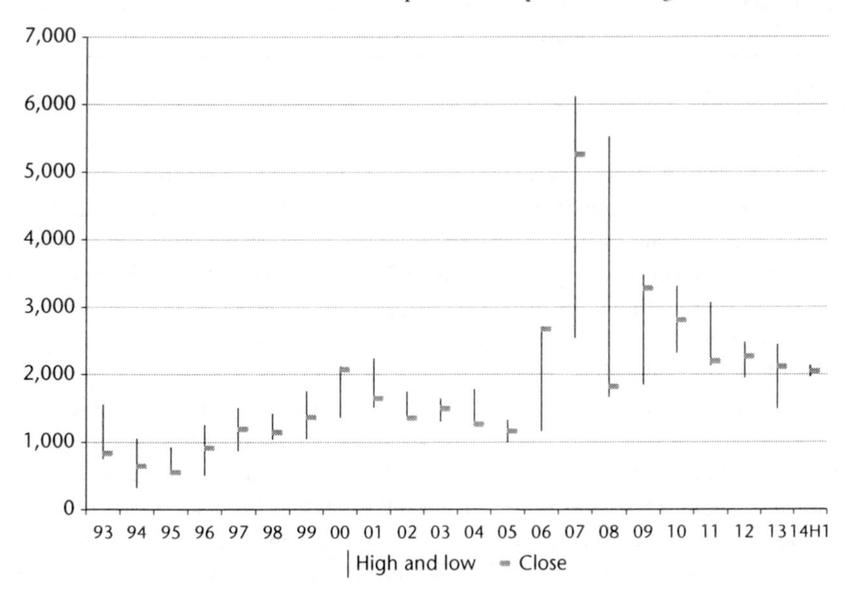

FIGURE 11.3 Performance of SSE Composite Index, 1993–2014
Sources: NBSC (2003; 2013) and Shanghai Stock Exchange (www.sse.com.cn).

TABLE 11.1 Status of listing for top 20 developers according to revenues in 2013

Company name	*Stock market*	*Market value (6 Aug 2014)*	*PE Ratio*	*Sales 2013*	*Sales 2014H1*
Wanke	Shenzhen	CN¥105.5	7.02	174.06	101.8
Greenland	Hong Kong (partially listed)	HK$4.4	148.80	162.53	83.0
Greenland/Jinfeng	Shanghai (backdoor listing)	CN¥4.5	126.24		
Wanda	Hong Kong	HK$11.5	33.46	130.11	50.7
Poly Real Estate	Shanghai	CN¥63.2	5.83	125.10	62.7
China Overseas Property	Hong Kong	HK$187.6	8.14	117.00	57.5
Country Garden	Hong Kong	HK$71.5	6.28	109.73	58.8
Hengda Real Estate	Hong Kong	HK$48.5	3.26	108.25	71.4
China Resources Land	Hong Kong	HK$99.7	6.78	68.10	24.9
Shimao Property	Shanghai	CN¥11.4	6.85	67.07	32.0
Greentown China	Hong Kong	HK$18.5	3.07	55.38	19.8
SUNAC China	Hong Kong	HK$20.7	5.00	50.83	25.8
Longfor	Hong Kong	HK$57.5	5.57	49.25	20.2
Gemdale	Shanghai	CN¥41.6	11.92	45.60	16.5
CITIC Real Estate				44.51	10.6
China Merchants Property	Shenzhen	CN¥31.2	7.52	42.80	17.2
Guangzhou R&F	Hong Kong	HK$10.8	9.22	42.23	25.8
China Futune Land				37.60	20.0
Agile Properties	Hong Kong	HK$22.0	3.53	37.40	19.8
SPG Land	Hong Kong	HK$31.9	5.65	36.84	10.9
China Railway	Hong Kong	HK$18.0	7.58	32.90	11.0
China Railway	Shanghai	CN¥17.1	31.78		

Note: market values and sales in billions.
Sources: SSE and SZSE websites and company websites.

in 2013. Eleven of these twenty firms were listed in HKSE, three in Shanghai, two in Shenzhen, one in both Shanghai and Shenzhen, and one in both Shanghai and Hong Kong. This shows an interesting phenomenon: HKSE has more top Mainland China property companies than SSE and SZSE combined. One of the reasons is that it was more difficult for property companies to be listed in SSE and SZSE in recent years, with a virtual ban on IPO and allotment of shares by listed firms. The 2010 Measures (Table 7.2) required the China Securities Regulatory Committee (CSRC) to suspend IPO approval, allotment of shares and major restructuring applications of property companies that are found to conduct land hoarding and speculation (State Council, 2010b). As a result, there has been no IPO of property companies in both SSE and SZSE since 2011. Backdoor listing, a popular form of getting listed by development firms, was also strictly controlled, being frozen after May 2011. One exception was made in March 2014 to allow Greenland, one of the biggest property firms and a state-owned company, to purchase Jinfeng Investment to complete its backdoor listing. The ban resulted in property companies seeking listing in HKSE, and 20 have been successful so far, including Wanda by way of backdoor listing in 2013.

The second reason that Chinese property companies favoured IPO in Hong Kong is that listing in Hong Kong provides access to vast international capital. In addition, the corporate governance requirements of HKSE provide greater assurance for investors than mainland stock exchanges. The downturn of the stock market in Mainland China since 2011 has reduced its fundraising capacity, which peaked at CN¥960.1 billion in 2010 but fell to CN¥312.8 billion in 2012 (NBSC, 2013). Thus, many Mainland Chinese firms, including property companies, found listing in Hong Kong and New York advantageous in fundraising at and after IPO.

However, listing in Hong Kong has a couple of disadvantages. Investors in Hong Kong, many of whom are of international origin, attach lower values and higher discounts to net asset value to Chinese property companies than their counterparts do from Mainland China. Thus IPO in Hong Kong brings much less to those companies than in Shanghai and Shenzhen. Easier and lower-cost fundraising after the IPO provides some comfort to many of those companies initially dissatisfied at the time with their IPO. Furthermore, the price to earnings (PE) ratio in HKSE is lower than in Shenzhen and Shanghai. As Table 11.1 indicates, shares of three of the top 20 firms have PE ratios lower than 4.

The link between SSE and HKSE is likely to bring benefits to both stock exchanges. The prices of Mainland Chinese property shares listed in Hong Kong are likely to benefit from purchase by Mainland Chinese investors. This may encourage more property companies to seek floatation, improving the finances of those companies with equity capital. Existing companies listed in HKSE will find fundraising more robust due to higher share prices. More equity capital will allow many property companies to reduce strata title sales of commercial property and hold on to good commercial property assets they developed as long-term investments.

11.2 Trusts

Trusts in China are different from trusts in the West because they are allowed to operate in a much wider area of financial services and become part of the shadow banking system. They fill the gap in financing left by banks and are important to the economy. By the end of 2008, the scale of assets managed by the trust industry in China reached CN¥7,470.6 billion, exceeding that of the insurance industry. Thus the trust industry became the second largest financial sector after banks (DYXTW, 2014).

11.2.1 Rise of trusts and bankruptcies in the 1990s

The importance of trusts in China was due to the inability of banks to provide sufficient financing to meet the country's needs. In the 1980s, many provinces and some major cities followed the business model of the China International Trust and Investment Company (CITIC) in setting up International Trusts and Investment

Companies (ITICs) to raise funds for major local projects from domestic and international sources. In Guangdong Province alone, there were three major ITICs, i.e. Guangdong ITIC (GITIC), Guangzhou ITIC (GZITIC) and Shenzhen ITIC.

In the 1990s, ITICs became important sources of finance for the country's rapidly expanding real estate sector, in particular after the 1993 macro control measures (see Chapter 4). In particular, GITIC and GZITIC funded a large number of real estate development projects, many of which went foul after 1997. In 1999, GITIC became insolvent and went through a bankruptcy procedure. The 494 debtors, 96 of which were foreign, were surprised because they thought GITIC, a state-owned company, would be bailed out by the Chinese government. Nonetheless, this implicit guarantee was an illusion. At the end, 12.54 per cent of the US$3.2 billion debts were repaid from GITIC's assets (LZMP, 2003). The bankruptcy of GITIC and several other ITICs led to attempts by the government to separate the remaining ITICs from high-risk investments and irregular operations, reducing the exposure of ITICs to real estate after 1999.

11.2.2 Expansion and regulation in the 2000s

In 2001, the enactment of the Trust Law and other regulations provided the much-needed regulatory framework for the trusts. For example, the *Provisional Measures on Cash Trust Schemes of Trust and Investment Companies* (PBoC, 2002) stipulated that a cash trust scheme based on contracts could have only 200 trust contracts with a minimum value of CN¥50,000 for each contract. However, the trust industry had problems with their position in the rapidly developing financial industry in the early 2000s. The reason for their emergence, i.e. the lack of funds in local governments, eased as local government finances improved by way of land sale revenues and increased tax incomes. They had to face competition from banks in deposit taking and lending; and from insurance funds and stock brokerage firms in fundraising and securities investment.

Regulatory changes in the property industry then provided the opportunity for the trust industry to expand. Tightening of bank credits from June 2003 by the central bank (PBoC, 2003) resulted in increased demand from real estate developers for alternative financing. In 2003, a total of 66 real estate trust products were issued. Among them, 37 were issued in the fourth quarter, raising CN¥3.5 billion. By contrast, the total number of schemes offered between July 2002 and June 2003 was only 20. The rise of real estate trust products dented the effectiveness of the central bank's credit control.

Real estate collective trust products were popular with individual investors, who were attracted by high interest rates. This created a new form of indirect property investment. However, these investors were often not aware of the risks of the underlying assets. To them, the collective trust products were like a high-interest fixed term deposit, offered by reputable state-owned financial institutions, and thus without risks. The reality, however, was different. Trust companies had limited responsibility for the property assets for which they provided finance. According to Clause 8 of the

Provisional Measures on Cash Trust Schemes of Trust and Investment Companies (PBoC, 2002), trust companies were not responsible for losses arising from the property under trust so long as they had complied with the trust contract in managing and using the cash. The loss would be borne by the property, the underlying asset or company in the trust, which means that investors, not the trust, bear all the risks if the underlying asset loses value or the underlying company becomes insolvent.

The problem of poorly perceived risk exposure by investors in real estate trust soon received the attention of the regulator. On 28 August 2005, the China Banking Regulatory Commission (CBRC) issued the *Circular on Strengthening the Risk Remainder on Some Operations of Trust and Investment Companies*. The circular pointed out that some collective trusts' products failed to liquidate in time. Among those products, 61 per cent were real estate products. Trust companies had to divert money from elsewhere, for example newly offered schemes, which could derail other healthy products. The circular imposed a set of conditions for developers to obtain trust finance: 35 per cent of the required investment must be funded by the developer's equity; possession of all required title deeds and consents for development; and class 1 and 2 qualifications for development (see Section 7.1). On 22 July 2006, CBRC issued another document, the *Circular on Further Strengthening of Real Estate Credit Management,* to require trust companies to strictly comply with the 2005 Circular and to disclose relevant information. Regulatory agencies under the CBRC were required to strengthen the monitoring of trust companies. Tightened regulations eventually slowed down the scale of fundraising by real estate collective trust products by 2006 and 2007 (Figure 11.4). Tightening credit contributed to the fall in housing prices in 2008 and 2009 (Figure 4.3) because developers needed liquidity to stay afloat and cut housing prices to speed up sales.

11.2.3 Further expansion in the 2010s

Tight credit for developers in 2008 resulted in the resumption of interest in real estate trusts' products (see Figure 11.5). The rapid housing price inflation resulting from loose credit in 2009 prompted developers to seek funding to start new projects, increasing demand for real estate trusts' products. It is the macro control imposed from 2010 that provided the opportunity for the trust companies to become a major financial sector. The 2010 Measures, referred to as the toughest macro control, restricted both bank lending to and the stock market floatation of development firms. To avoid the purchase restrictions, most national and regional developers shifted their development focus to tier 3 and tier 4 cities where land prices were cheap and construction costs low. To obtain land there, most developers, big and small, made use of funds raised by real estate trusts, both collective and single, to purchase land and carry out development. With rapid price inflation in those tier 3 and tier 4 cities and very low costs, developers could afford to pay 20 per cent to 25 per cent costs annually to real estate trusts, which were sold at 10 per cent to 15 per cent interest rates to institutions (single trusts) and individual investors (collective trusts). Funds raised by trusts in 2010 and 2012 allowed the Chinese housing market

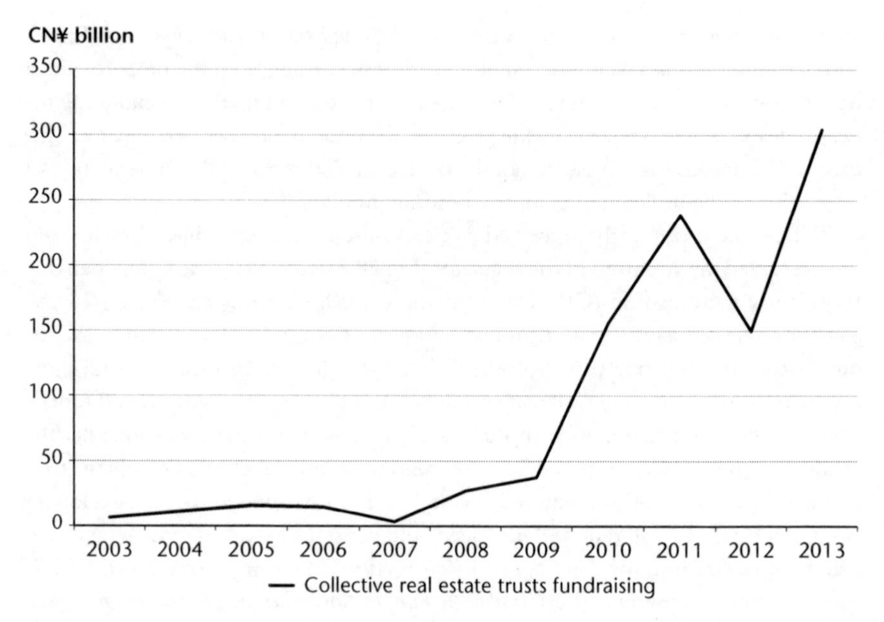

FIGURE 11.4 Funds raised by collective real estate trust products
Source: China Trustee Association.

to expand (Figure 4.1). In 2013, when housing prices rose rapidly in cities with purchase restrictions, demand for trust money surged to a new level (Figure 11.4).

The expansion of collective trusts schemes, however, was made possible by the implicit guarantee offered by trust companies. The CBRC strictly bans any guarantee offered by trust companies because those companies effectively offer high yielding bonds to investors, who are expected to bear the risks. The same as in the 2000s, the risks of trust products are dependent on the underlying operations, for example property development, which is a high-risk activity. Nevertheless, competition among trust companies and the need to attract investors resulted in the provision of implicit guarantees. Interviews with practitioners revealed that such guarantees were hinted at to customers during the sale of trust products. As a result, collective trust products (not just for real estate) became very attractive on the understanding that they were low-risk, high-return investments.

The risks of property development were exposed from 2012 when some completed housing projects in tier 3 cities and tier 4 cities failed to sell as expected. For example, some real estate trust products were for housing development in Ordos in Inner Mongolia. However, there were so many projects built in poor locations that the sale of housing in those projects was very slow. In fact, sales in some projects were so poor that those projects became 'ghost towns'. As a result, these trusts failed to liquidate those products when they expired. Trust companies had to pay the investors of these failed trust products the principal from internal resources with the hope that the distressed assets could be sold in the future to recover loss. Yet such

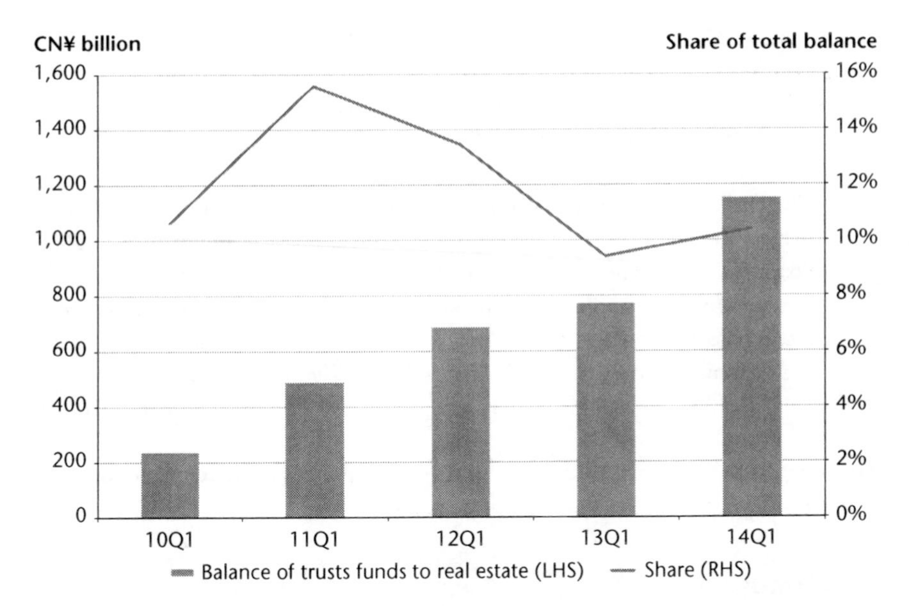

FIGURE 11.5 Balance of trust funds to real estate in recent years
Source: China Trustee Association.

projects are highly illiquid. However, at the time of writing, none of the trust companies has broken the implicit guarantee for two reasons: protection of their reputation; and protection of the industry from state intervention. If a failed trust product leads to social unrest, then the state may have to intervene to set a new policy to order the activities of the trust companies. Such intervention in 2005 and 2006 led to the contraction of the real estate trust businesses (see above).

In quarter 1 of 2014, the total assets of trust companies were CN¥12,727.2 billion. An internal meeting of the CBRC revealed that assets at risk in July 2014 were over CN¥70 billion, which was about 0.55 per cent of total assets (Zhang and Liu, 2014). As housing prices have fallen in the first half of 2014, more real estate projects may be at risk. Total amounts of money in trusts' products for the real estate category maturing in 2014 were CN¥247.9 billion. Because part of the CN¥211 billion maturing trust products for the business operations category is for real estate companies, the total amount of money that needs to be paid in 2014 is more than CN¥247.9 billion. The total balance of real estate trusts products is shown in Figure 11.5, with the balance in quarter 1, 2014 in excess of CN¥1 trillion. The trust industry is vulnerable to a systemic price fall in China's property market, in particular the housing market.

Is there a systemic risk in the trust industry? Interviews in June 2014 indicate that practitioners in the trust industry were not expecting a systemic price fall in the housing market so the risks to trust products are controllable. Housing prices fell across the board in the first half of 2014, but such a fall was milder in tier 1 cities and tier 1.5 cities. With purchase restrictions being lifted by more and more cities and credit increasingly eased, the fall in housing prices is expected to be limited in depth. This is in line with the analysis in this book (Sections 2.8; 4.7; 7.5; and 9.5). Furthermore, collective real estate trust products accounted for less than half of the total real estate products.

For the trust industry, removal of implicit guarantees needs to be high on the agenda. In its *Global Financial Stability Report*, the IMF states that:

> The challenge for policymakers is to manage an orderly transition toward more market discipline in the financial system, including the removal of implicit guarantees. In this process, investors and lenders will have to bear some costs of previous financial excesses, and market prices will need to adjust to more accurately reflect risks. Pace is important. If the adjustment is too fast, it risks creating turmoil; if too slow, it will allow vulnerabilities to continue building.
>
> *(IMF, 2014: x)*

However, at the time of writing, there is no sign of policy and action from the constitutional level to remove the implicit guarantee.

11.3 Funds

Tightening of bank credits from June 2003 by the central bank (PBoC, 2003) resulted in a funding problem in the property industry and prompted a new type of real estate investment and financing channel, real estate funds, to emerge in China. To alleviate the funding challenge, domestic developers used a voluntary industry platform, CURA[1] Investment, formed in 2002, to explore alternative sources of finance. In 2002, the company had 12 shareholders; all of them were developers. From 2003 to 2005, the company explored investment and financing among shareholders. It became an investment fund (Curafund) from 2006 and changed to a fund management company from 2009, with fund management centres in Beijing and Tianjin. The company, pioneered in China to set up private equity funds to invest in the Chinese property market from 2004, set up the first limited partnership investment fund in 2006. Now it has 56 shareholders, all of them developers, and runs a dozen or so investment funds.

China's real estate investment funds sector as a whole took off in 2008, when bank finance was difficult to obtain, and has grown steadily since then (Figure 11.6). It should be noted that due to the lack of official statistics and the opaque nature of some of the funds, data in Figure 11.6 were incomplete. Nevertheless, the data available shows many funds are still short term in their duration. In 2013, a significant number of funds reached maturity and exited the market. The amount of money raised grew from a few CN¥ billions in 2008 to over CN¥100 billion in 2013 (Figure 11.7). Such an amount accounted for over 30 per cent of the private equity market in China. Developer-based general partners (GPs) and independent brand GPs managed roughly three-quarters of the number of funds and assets under management in 2013. Funds based on financial institutions accounted for slightly less than a quarter of the market. Funds set up and managed by foreign GPs accounted for less than 5 per cent of the market (Figure 11.8).

Real estate funds are relatively small in terms of their capital (Figure 11.9), with 70 per cent of the funds under CN¥500 million (US$81.3 million). This is because

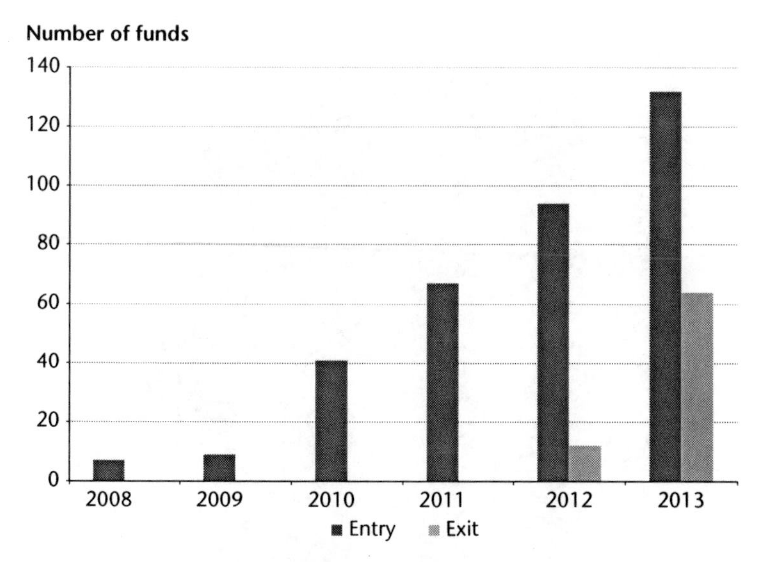

FIGURE 11.6 Number of real estate funds in China
Source: CREA (2014).

many funds were set up to finance a particular project. Housing accounts for the biggest allocation by all real estate funds (Figure 11.10); HOPSCA ranks second. Because HOPSCA has a bigger share of commercial than residential property, it can be deemed that real estate funds have an equal share of residential and commercial.

The rise of real estate funds, in particular the expansion in 2012 and 2013, has brought a change in real estate financing and investment. Decision-makers at the collective choice level, the CBRC and Insurance Regulatory Commission, relaxed market rules for financial institutions under their ambits to carry out asset management. There has been a proliferation of real estate funding products in the market, leading to strong

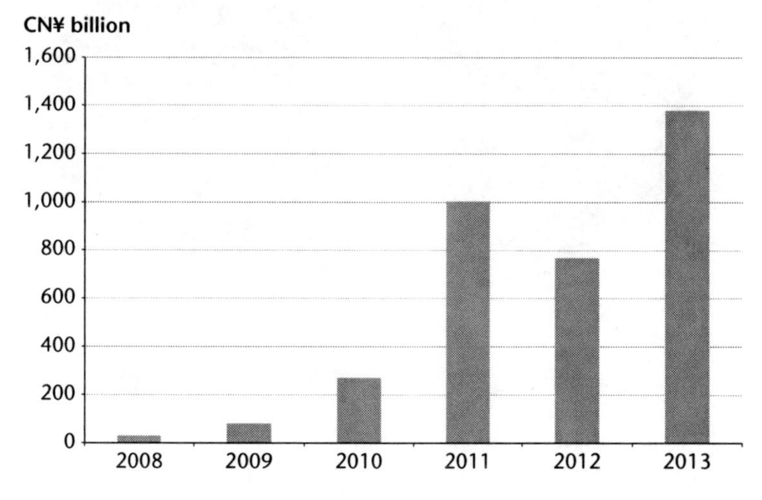

FIGURE 11.7 Amount of money raised by real estate funds
Source: ChinaVenture (2014).

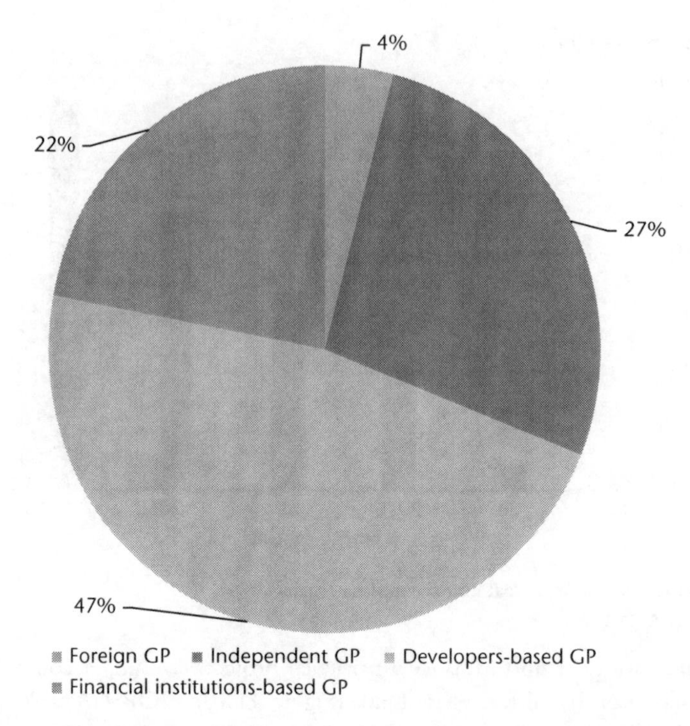

FIGURE 11.8 Classification of real estate funds according to management
Source: ChinaVenture (2014).

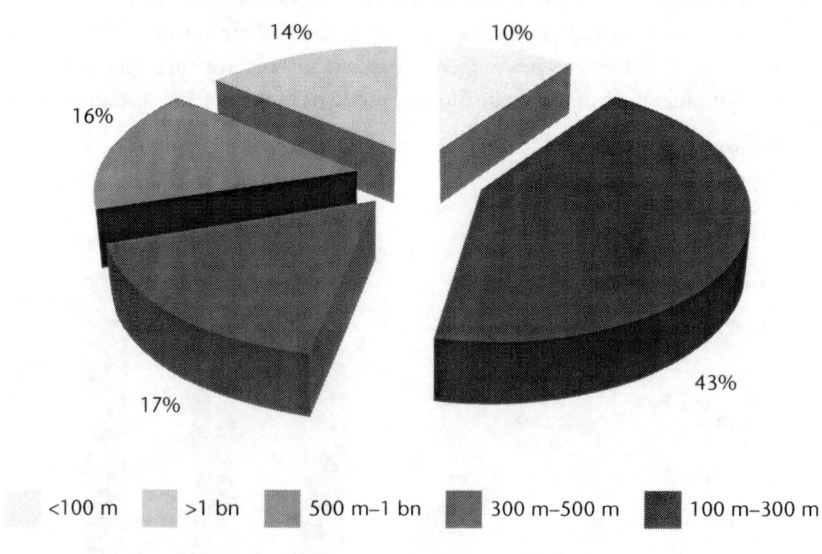

FIGURE 11.9 Sizes of real estate funds in terms of CN¥ millions
Source: ChinaVenture (2014).

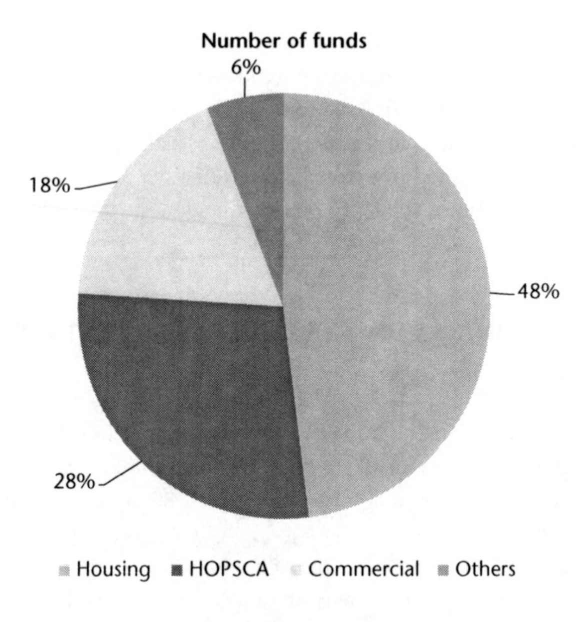

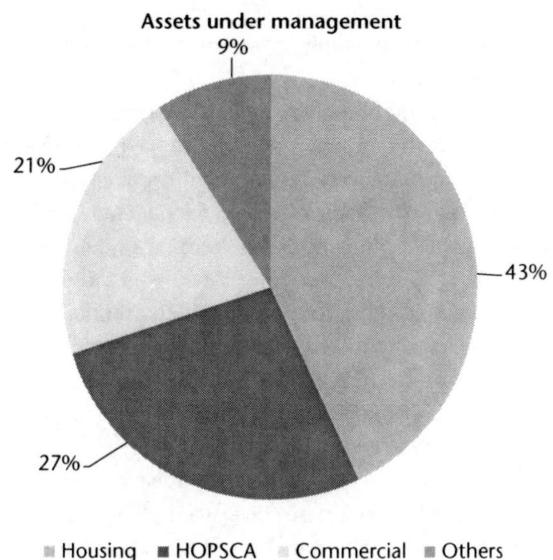

FIGURE 11.10 Investment destinations of real estate funds
Source: ChinaVenture (2014).

competition for trust companies. For example, the top ten developers in China have already set up their own real estate funds. Most of the funds still play the role of financing developments and the choice of clients, i.e. developers, are crucial for their market share and success. The location of projects is important, too. Tier 3 and tier 4 cities are generally found to have an abundant supply, which takes years for the market to absorb. Many of these cities should become no-go areas for funds for the near future.

As market risks, caused by oversupply and a slowing economy, increase, the issue of implicit guarantee becomes more important. Affected by the implicit guarantee in the trust industry, investors in real estate funds also expect that there is implicit guarantee in the funds. Thus the real estate funds industry needs to screen investors, choosing only those qualified investors, and remove the implicit guarantee. It will be wrong to make the funds appear to be risk-free deposit-taking businesses for the general public.

BOX 11.1 CHINA'S PRIVATE EQUITY REAL ESTATE FUNDS

William Dong

In 2009, two years after China's law governing partnership enterprises came into effect, private equity real estate (PERE) funds started to take shape, raised in Chinese currency and managed by local firms. Before that, funds denominated in foreign currencies and managed by offshore managers had been actively investing in China's property market, with varied returns.

Similar to the foreign funds, almost all local ones have adopted opportunistic strategies, investing mostly in development schemes. However, sponsors and fund managers vary considerably. In summary, there are three major types of managers active in the arena.

The first type is new firms backed by major developers with the key persons owning a small stake. Such a model is believed to have the advantage of having a steady pipeline and a proven development capability from the developers, as well as alignment of interest of a professional team, which is critical in terms of mitigating conflict of interest with the sponsor, i.e. the developer.

The second type is so-called boutique firms, which usually have a track record managing foreign capital. They are regarded as more independent and prudent. Having been around for a while, they are expected to know where to look for deals and partners.

The third type is fund managers backed by groups of local businesses. These sponsors can either be an established alliance of developers or a group of private enterprises trying to diversify their business. Obviously the developer-alliance model offers a more diversified pipeline and potential conflict of interest is of less concern if members of the alliance are not allowed by contract to interfere with fund-level issues.

More recently, a few placement agents have tapped into the fund management business, leveraging on their investors network.

One predominant feature of these local funds is that their terms are usually short. Anywhere between one to three years is regarded as the norm. This is mostly due to the fact that the majority of their investors are high-net-worth-individuals (HNWI) who are not willing to bet anything longer. A 2013 survey showed that only 5 per cent interviewed have tolerance for a term over three years.

The risk appetite of these investors took a downturn around 2012 after the first batch of funds raised in 2009 and 2010 began to show performance that was not as rosy as expected. Ironically, when compared with the next wave, which was crowded with structured products skewed more towards debt investment in one single asset, these early funds were more standard *per se*, in the sense that they took equity position in a portfolio diversified by developers and locations. Some reports would put the market share of such bridge financing deals as high as 95 per cent, and the overall cost to the borrower is anywhere between 10–30 per cent per annum.

Institutional investors started to emerge after two supplements to the insurance law were issued in 2010 regarding equity and real property investments. Although insurance funds had been investing in income-producing commercial properties, some are now taking a more proactive approach by becoming a limited partner, albeit not quite silent.

11.4 Discussion: financing long-term property investment

Chinese development firms have grown substantially over the years and accumulated a large amount of equity (Figure 11.11). They expanded at the fastest possible pace they were allowed. For example, the central bank required development firms to invest at least 35 per cent of equity to qualify for bank lending and trust lending. But these firms have been keeping a high-liability asset ratio to take on as many

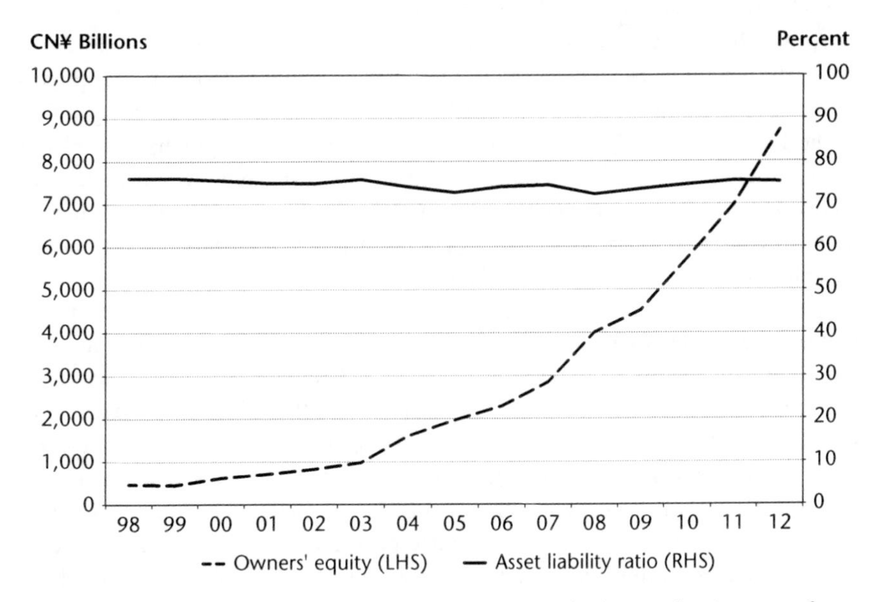

FIGURE 11.11 Owners' equity and liability asset ratio of Chinese development firms, 1993–2012

projects as possible, and have relied on trust companies and real estate funds to provide short-term finance. Now the housing market is cooling down, and developers need long-term finance to keep the commercial properties they develop and invest in other commercial properties. However, there are few long-term financing options available. Banks, trusts and funds all provide short-term finances only.

To solve the issue of long-term finance, the Chinese property industry has been calling on the government to set up US-style real estate investment trusts (REITs) since the early 2000s. The call for the introduction of REITs was greeted with enthusiasm by the Ministry of Finance, the Ministry of Commerce, and the Ministry of Construction in 2005. Concerned with the difficulty of holding many large retail properties, the three ministries recommended to the State Council the introduction of REITs (Gu and Xing, 2005). The recommendations were taken seriously by the State Council, and the PBoC conducted a detailed investigation into the issue. By 2009, the PBoC had designed the framework of REITs in collaboration with other relevant ministries. Yet the regulatory environment is still not ripe for REITs to start in China. The two possible regulatory bodies for REITs, i.e. the CSRC and the CBRC, are still not directly involved in any such discussions. Furthermore, the communiqué of the Third Plenary Session of the Eighteenth Congress did not mention anything related to REITs (CCP, 2013).

Another source of long-term finance is insurance companies, which were banned from real estate investment until 2010. Since then, the proportion of allowed investment allocation to real estate has grown from 10 per cent in 2010 to 20 per cent in 2012 and to 30 per cent in 2013. By 2013, the insurance companies had invested CN¥100 billion in real estate, which is 1.5 per cent of the investment allocation (CE, 2014). This amount is still small compared to other sources of funding.

Although there is insufficient offer from the collective choice and constitutional levels, the market level has been creative in finding a solution. Domestic developers went to Hong Kong or New York for IPO floatation, or attempted backdoor listing in the HKSE, the SZSE and the SSE. The largest developers purchased banks or formed strategic partnerships with banks. So far, 30 developers have such arrangements. For shorter-term financing, trusts and funds have been a major sources of money. Many developers seek financing in international monetary markets to take advantage of lower interest rates available there.

The domestic situation in long-term financing may get worse before it gets better. At the time of writing, the CSRC is conducting a public consultation exercise to amend the 2011 *Measures on Major Asset Restructuring of Listed Companies*, which intends to rule out easy backdoor listing. With IPO difficult, the restrictions on backdoor listing will further squeeze developers' ability to obtain long-term finance inside Mainland China. It is obvious that the lack of long-term finance causes inefficiency in China's property market. For example, strata title sales result in property management problems and reduce property values. The central government at the constitutional level and regulatory bodies at the collective choice level need to embrace institutional innovation to solve the problem of long-term funding to property development and investment.

11.5 Conclusion

Indirect investment vehicles in China's real estate include shares, trusts and funds. Shares of Chinese development companies are listed in the SSE, the SZSE and the HKSE. Both the SSE and the SZSE have over a hundred property companies listed through IPOs and backdoor listing. However, the ban on IPO and backdoor listing imposed by the 2010 Measures prevented many large developers from listing on the SSE and the SZSE, forcing many of these companies to list on the HKSE. It is possible that all backdoor listing by property firms will be banned in the future.

On the other hand, money from trusts has been used by developers since 1990s, with a high-profile bankruptcy case breaking the implicit guarantee offered by trust companies. However, tight financing conditions provided opportunities for trust companies to lend on property development in the 2000s, in particular by collective trust products. Risk control by the central bank resulted in such lending declining substantially in 2007. Yet trust lending recovered and grew to become the second largest source of development finance from 2008 due to tight credit and high profits in property development. However, the recent slowing down of the housing market, in particular the severe oversupply in tier 3 and tier 4 cities, has resulted in liquidity problems for some products and threatens financial loss for some trust companies where implicit guarantee was a marketing tool used by the trust companies to attract investment.

Real estate funds provide another major financing and investment vehicle. They started in the tight credit conditions from 2003, but grew after 2010 when the development sector needed large amounts of funding. Real estate funds provided large amounts of finance to both housing and commercial property development, in particular, to HOPSCA. However, these funds are still short-term finance to real estate development rather than long-term finance to hold real estate as investments.

The Chinese institutional environment could not provide sufficient long-term finance to developers and property investors, prompting them to list in Hong Kong or overseas; or to merge or enter into strategic partnership with banks; and to raise debts in foreign monetary markets. It needs institutional innovation to allow investors to hold property as an investment vehicle so that the thousands of commercial properties in Chinese cities can be held by investors using domestic long-term finances.

Note

1 China Urban Realty Association was formed in 1999 to address issues concerned by developers.

12

FOREIGN INVESTMENT IN CHINA AND CHINA'S EMERGING OVERSEAS INVESTMENT

This chapter investigates international investment in real estate to and from China. It first examines foreign real estate investment as foreign direct investment (FDI) in China, and then total foreign money invested in China's real estate. The chapter then investigates rising real estate investment from China. The informal channels through which a lot of money has been transferred in and out of China are discussed to reveal that Chinese real estate investment abroad is susceptible to crackdown on those informal channels by the Chinese government.

12.1 Foreign investment in real estate in China

Since the reforms and the opening-up campaign started in 1979, foreign investment has played an important role in China's real estate market development. It started as a major source of investment in real estate development and holding in the early stages of market development, and contributed to the establishment of current market practices in development, professional services and investment. As the Chinese economy became increasingly an integral part of the world economy, the Chinese property market has attracted more international investment.

12.1.1 The development stages

The participation of foreign investment in China's real estate market can be divided into three stages (Table 12.1). The first stage was from 1979 to 1990 when foreign investment was introduced into China under strict control. In 1979, the first batch of commodity housing was sold in Guangzhou to buyers with remittances from their Hong Kong relatives. In the early 1980s, foreign investors, who were mainly from Hong Kong, leased land, offices, shops, factories and housing, and bought housing under LUR in Shenzhen Special Economic Zone. This kick-started real

TABLE 12.1 Stages of foreign real estate investments in Mainland China

Stages	*Main developments*	*Time scale*
Stage I: initiation	Participation in the real estate market in a few coastal cities under strict control	1979–1990
Stage II: expansion under favourable policies	Participation extending to most major cities encouraged by national and local polices; rapid market expansion crucial to the establishment of the commercial real estate market in those cities; setting good examples in property development and services; an important role in fuelling housing price inflation	1991–2005
Stage III: expansion under unfavourable policies	Participation in development financing, buy-to-let investment and professional services provision; large investment in buildings; policies enacted to limit the scale of investment due to perceived inflationary impact and market risk impact	2006 onwards

estate market transactions after a long suspension from 1955 (Section 1.1). Foreign investment at this stage was crucial in financing some major projects, for example the International Trade Centre in Beijing and the World Trade Centre in Guangzhou. Techniques and practices in planning, design, construction, marketing, leasing and sales came with foreign investment. From 1985 to 1987, US$1.2 billion of foreign investments was used, accounting for 20–30 per cent of total investment in property development in China at the time (Nie, 2011). The economic slowdown in 1990 led to a fall in foreign investment in real estate (Figure 12.1).

The second stage, from 1991 to 2005, was a period with policy sanctions for foreign investment in China's real estate market. During this period, opening up was given high priority in the country's economic development strategy and foreign investment brought much-needed capital, technology, management knowhow, advanced business models and matured market practices. In May 1990, the State Council enacted the *Interim Measures for the Administration of Foreign Businesses Developing and Managing Tracts of Land*, which allowed foreign businesses to service large tracts of undeveloped land. The single plot size was up to 66.67 hectares for arable land and 133.33 hectares for other land. After the provision of required services, the foreign developer could sell or let the land, or construct buildings for sale or letting. Along with Decree 55, the Interim Measures boosted the confidence of foreign investors and resulted in a surge of foreign investments in real estate in 1991 and 1992 (Figure 12.1). In November 1992, the State Council enacted the *Circular on Issues of Developing the Real Estate Industry*, which indicated that foreign investment could make up the shortage of investment funds in real estate and bring competition to the sector. The circular called for directing foreign investment to real estate development associated with approved projects for foreign investment, and to high-end, high-specification real estate development projects. Another rapid surge in the volume of foreign investment in real estate ensued in 1993 and 1994. In BSGS, dozens of high-end and high-specification office and mixed-use projects were developed by foreign developers and investors, leading to the first office development boom in China.

During the period from 1995 to 2005, FDI in real estate stagnated at around US$5 billion a year (Figure 12.1) while domestic investment became five times bigger (Figure 4.2). In 2005, FDI in real estate was equivalent to 2.75 per cent of the total investment in property development, which was less than the same measure of 24.28 per cent in 1994. The proportion of FDI in real estate in the same year was only 8.98 per cent of total FDI into the country, a fall from 22.19 per cent in 1994 (Figure 12.1). During this period, foreign developers had successfully developed a number of large commercial and residential projects, contributing to the maturity and sophistication of the domestic development sector.

However, from 2004, large amounts of foreign money entered China but not as FDI. The 2004 *Report on Real Estate Finance*, published by the People's Bank of China, first raised the alarm. The report noted that foreign money came into China through four channels. First, foreign money was used to set up real estate investment firms or to buy shares in domestic firms. Second, foreign money was used to purchase bonds issued by domestic developers, and to purchase multiple housing units through foreign intermediary companies for resale later. Third, foreign banks issued loans to developers and individuals for real estate investment. Fourth, money

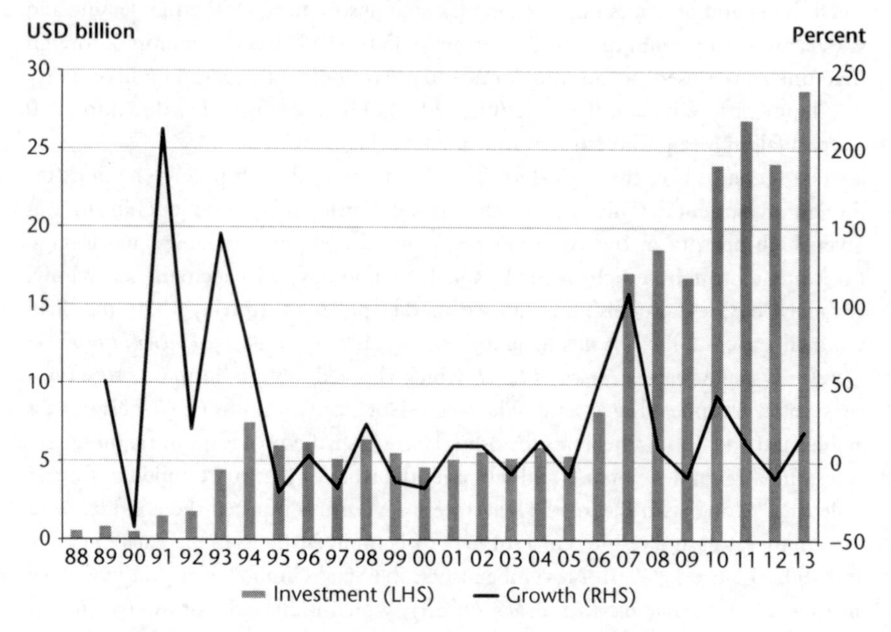

FIGURE 12.1 Total foreign direct investment in real estate in China, 1988–2013
Source: NSBC (2013).

was simply remitted to purchase housing. The report indicated that in Shanghai, foreign money accounted for 23.2 per cent of total funds spent on housing purchase in quarter 4 of 2004, rising from 8.3 per cent in quarter 1 of 2003.

Increasingly, the entry of foreign money to purchase high-end housing was viewed negatively. Foreign money was considered an important factor in fuelling housing price inflation in Shanghai, which initiated macro control in March 2005 aimed at mitigating housing price inflation (see Section 4.6). Furthermore, the preference of foreign investors to purchase high-end housing was deemed to contribute to the structural imbalance of housing development, with a bias on high-end and luxury housing (see Section 7.3). This was regarded as incongruent with the preferential treatment afforded to foreign investors.

The third stage, from 2006 to present, witnessed the rapid increase of FDI and other sources of foreign funds invested in real estate along with the shift of attitudes to and treatment of foreign investment in real estate by the government. In July 2005, Renminbi, the Chinese currency, ended its decade-long dollar peg and started its unilateral appreciation against the US dollar, which attracted more foreign money into China through FDI (Figure 12.1) and other channels. For example, in the first quarter of 2006, foreign purchases of real estate in China were US$4.5 billion, more than that for the whole of 2005 (Wang, 2006). In May 2006, Deutsche Bank spent approximately CN¥400 million on purchasing 175 villas in a Beijing project as a buy-to-let investment, which was the first residential purchase by a foreign institution in Beijing (*People*, 2006b). Meanwhile, Morgan Stanley spent over CN¥5 billion to purchase ten residential towers and a failed commercial project in Shanghai (Wang, 2006). These purchases were sensational at the time, reinforcing the negative perception of foreign investment taking advantage of housing price inflation and Renminbi appreciation.

The rapid increase in foreign money became disruptive to macro control. Such money reinforced the purchasing power that resulted from exchange control (see Section 2.3) and fuelled continued housing price inflation. Furthermore, much of the money was for short-term investment, imposing risks on the country's real estate market and financial system: if housing prices fell, the exit of hot money from the property market would deepen the fall; if such money departed the country in large quantities over a short time span, the country's financial stability would be undermined.

The initial response from the government was the enactment of the *Opinions on Normalisation of Entry and Management of Foreign Investment in the Real Estate Market* in July, 2006, by several ministries led by MOHURD. This regulation imposes a number of conditions that proved cumbersome to foreign investment in real estate. First, a one-year residency requirement is required for foreigners to buy residential property. Second, foreign real estate companies have to register an independent company for each project with the local governments and obtain approval for the company by the Ministry of Commerce. Third, the minimum registered capital for the company is 50 per cent of the total investment.

To contain FDI in real estate, in October 2007, the central government amended the *Catalogue of Industries for Guiding Foreign Investment*, which divides industries for foreign investments into encouraged, restricted and prohibited categories. Prior to the 2007 amendment, property development was in the encouraged category and received governmental incentives. However, the 2007 catalogue put FDI in property investment (including land preparation, real estate development, trading, and service provision) into the restricted category. The same category was updated in 2011 and is still effective at the time of writing, with foreign real estate investments still in the restricted category.

Since 2006, FDI in real estate has maintained its growth, albeit slowing down in 2009. In addition, other forms of foreign money entered the country in significant amounts in 2010. When purchase restrictions limited the potential for housing price inflation in BSGS, foreign investment went into commercial property and played a role in financing developers. Many foreign real estate firms choose to collaborate with local companies, providing capital rather than participating directly in property development.

12.1.2 Origins and destinations

Different from the overall FDI picture, FDI specifically in real estate in China is mainly from a selected number of countries and regions. The main source is Hong Kong, which has contributed to over 60 per cent of FDI in real estate from 2002 to 2012 (Figure 12.2). It is worth noting that a part of the funds from

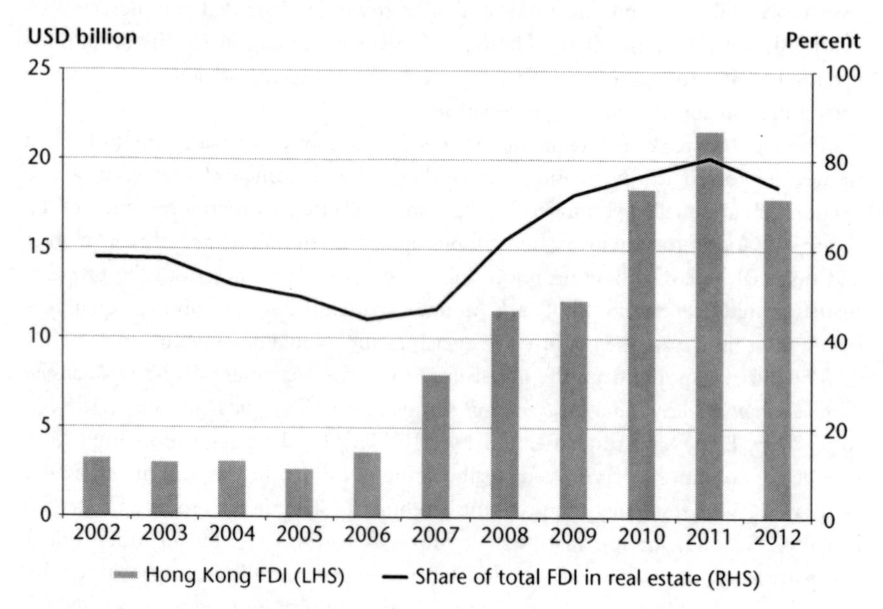

FIGURE 12.2 FDI in real estate from Hong Kong
Source: Ministry of Commerce (2013).

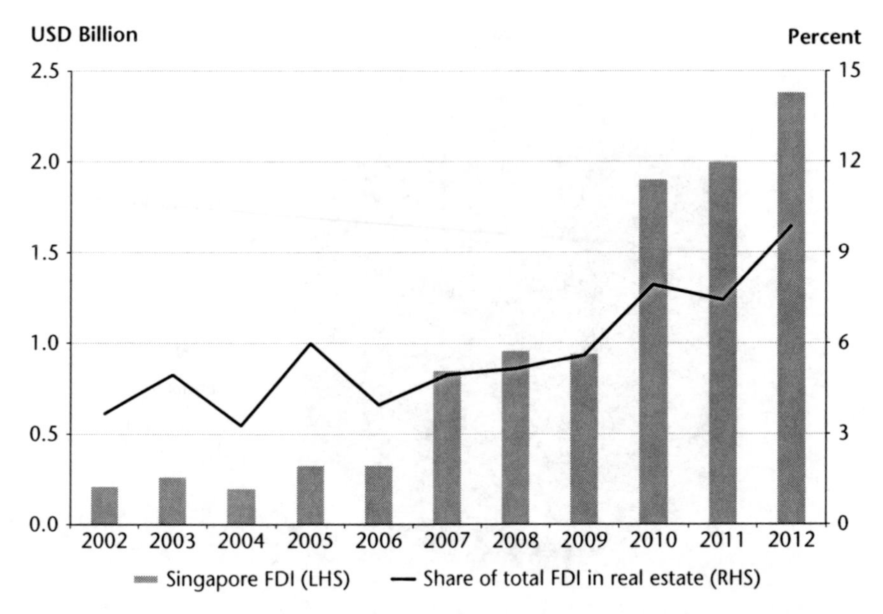

FIGURE 12.3 FDI in Chinese real estate from Singapore
Source: Ministry of Commerce (2013).

Hong Kong were from various sources in the USA, South-East Asia, Europe, and Mainland China. The reason why funds from Mainland China are invested back into Mainland China is in order to obtain the preferential treatment offered by the government. Another major source is Singapore, whose contribution has increased significantly in recent years (Figure 12.3). It is worth noting that Chinese real estate is the largest investment sector for Singaporean money. Significant amounts of FDI were also from the British Virgin Islands and the Cayman Islands, both tax havens and just collecting points for money that could be coming from anywhere. Figure 12.4 illustrates the geographical distribution of FDI into Chinese real estate in 2012.

FDI in real estate mainly went to East China,[1] which includes most of the coastal provinces. Figure 12.5 shows the amount of FDI in real estate invested in the three regions in China from 2002 to 2012. Central China[2] received the least amount of FDI in real estate, which corresponds to the status of those provinces having the least mature real estate markets in the country. West China,[3] on the other hand, received almost twice the amount of FDI in real estate as Central China. It is worth noting that FDI in real estate to West China was the largest compared to all other sectors, including manufacturing. This is because there has been strong interest among foreign developers in Chengdu, the capital city of Sichuan Province, and Chongqing, the largest provincial-level municipality in China.

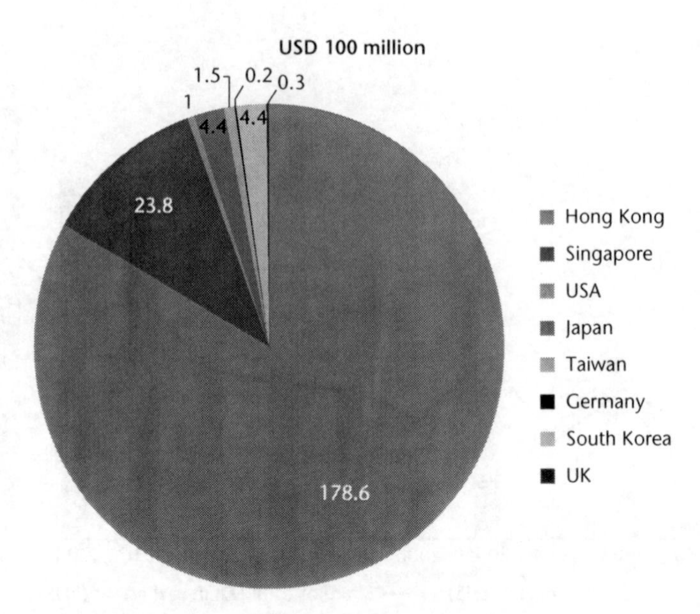

FIGURE 12.4 Sources of FDI into China's real estate in 2012
Source: Ministry of Commerce (2013).

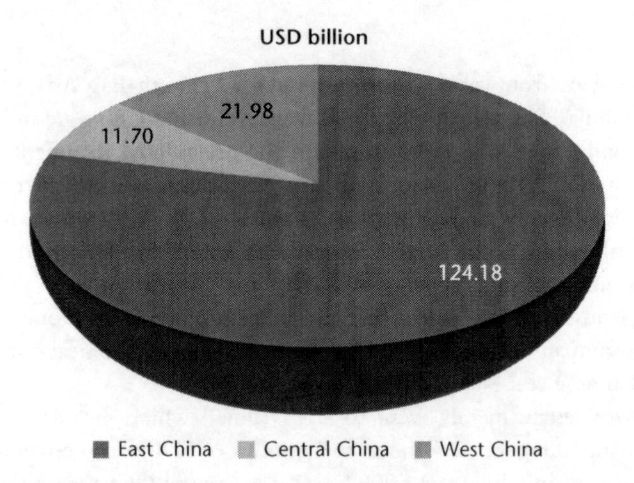

FIGURE 12.5 Geographical distribution of FDI in real estate in China from 2002 to 2012
Source: Ministry of Commerce (2013).

12.1.3 Can foreign real estate firms compete with the domestic counterparts?

Competition in the real estate market in China is intense. In the 1990s foreign developers had the competitive advantages in product design and financing, but in the 2000s domestic developers had largely caught up. Domestic developers, on the other hand, have the competitive advantages of economy of scale, knowing

TABLE 12.2 Top ten foreign-funded development firms, 2012–14

Rank	*2014*	*2013*	*2012*
1	RK Properties	RK Properties	Kaisa Group
2	The Wharf (Holdings)	Coastal	RK Properties
3	Huchison Whampoa	The Wharf (Holdings)	The Wharf (Holdings)
4	New World Properties	Yanlord Land	Sun Hung Kai Properties
5	Shui On Land	Capitaland	Huchison Whampoa
6	Yanlord Land	Shui On Land	Shui On Land
7	Sun Hung Kai Properties	Sun Hung Kai Properties	Yanlord Land
8	Henderson Land	Huchison Whampoa	Capitaland
9	Capitaland	Tomson	Tomson
10	Keppel Land China	CITIC Pacific	CITIC Pacific

Note: Yanlord Land, Capitaland, and Keppel Land China are Singaporean firms; all the rest are Hong Kong firms.

Source: China Real Estate Association (2014).

their clients, and having better links with local governments and local business communities. Employing local staff can only partly make up the gap in knowledge and links for foreign developers.

The position of foreign developers in China can be explained in part by the annual league table produced by China Real Estate Association, the largest semi-governmental industry body. Real estate development firms are assessed on operational scale, risk management, profitability, growth potential, performance, innovation and corporate social responsibility. Table 12.2 lists the top ten foreign development firms according to the 2012 to 2014 league tables of best foreign real estate firms. From 2011 to 2013, a total of 11 developers from Hong Kong and three from Singapore were included in the league tables. Two Hong Kong developers, RK Properties and Kaisa Group, made it to the 2014 top 100 league table for developers in the country, and were ranked 17th and 30th, respectively. Kaisa Group was also ranked fifth in the top ten firms in terms of operational efficiency, and second in the top ten firms in growth potential. RK Properties was ranked fifth in the top ten firms on stability of operation. However, none of the non-Hong Kong foreign firms made it into the various top ten league tables. For example, Capitaland from Singapore owns ten shopping malls in China, but could not make the top ten of the commercial real estate composite league table. The number one in that table, Dalian Wanda, the second largest owner of commercial real estate in the world, owns 94 Wanda Plazas (HOPSCAs), with 84 Wanda Department Stores and 1,247 cinema screens.

In a different league table, there were five Hong Kong developers among the top 50 on sales volume and on floor area sold (Table 12.3). This relative success of Hong Kong developers explains why there are more firms from Hong Kong than from the

TABLE 12.3 The biggest foreign developers in China in terms of sales of new property in 2013

Sales volume (CN¥ billion)			*Floor area sold (million m²)*		
Rank	*Firms*	*Volume*	*Rank*	*Firms*	*Area*
25	Kaisa Group	23.90	26	Kaisa Group	2.46
32	Huchison Whampoa	21.59	43	The Wharf (Holdings)	1.59
33	The Wharf (Holdings)	21.40	46	RK Properties	1.27
43	New World China Land	15.63	50	Huchison Whampoa	1.18

Source: CRIC and China Real Estate Evaluation Centre (2014).

rest of the world. In 2012, there were 3,451 development firms from Hong Kong, Macau and Taiwan, with the great majority from Hong Kong, and 1,713 from the rest of the world.

Nevertheless, foreign developers have made a significant contribution to real estate development in China in the past 30 years. The most notable example, Xintiandi, a 52-hectare mixed use urban redevelopment project in the commercial heart of Shanghai, was an exemplar in conservation, retail, tourism and environmental quality. It preserves the old stone gate houses built around 1860s instead of razing them for high-rise development, and creates a green space, centred with an artificial lake. With skilful refurbishment, the old houses combine history and modern luxury, and are distinguished from the surrounding new and high-rise development. Xintiandi made a strong statement to Chinese cities obsessed with total rebuilding: conservation and environmental improvement can bring commercial success.

Through more straightforward equity deals, many foreign investors had success in the acquisition of Chinese real estate firms and buildings. For example, foreign investors accounted for 86 per cent of the CN¥26.4 billion prime commercial property transactions in Shanghai in 2013. American and European investors have competitive advantages in financing and asset management. Recently they are active not just in tier 1 and tier 1.5 cities, where risks are lower but yields are lower, but also in tier 2 or even tier 3 cities, where risks are higher but the upside is that potential is stronger.

12.2 China's emerging overseas real estate investment

Since 1979, China has received large amounts of inward FDI in manufacturing, real estate and many other sectors (Figure 2.6), but made only small outward investments. Since 2003, Chinese companies, led by the mining sector, have started to invest abroad and FDI from China rapidly increased (Figure 2.6). By 2013, China's non-financial outward FDI grew from 5.4 per cent of non-financial inward FDI in

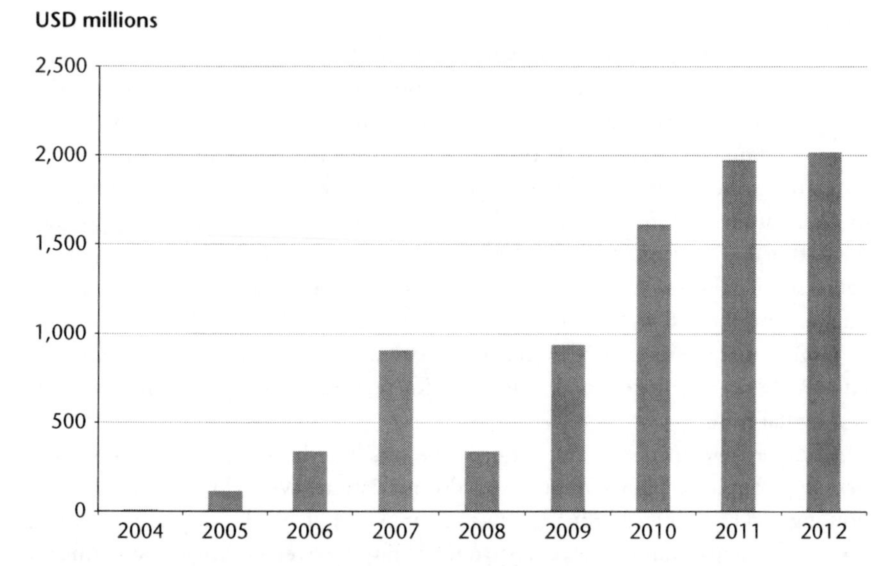

FIGURE 12.6 China's outward FDI in real estate since 2004
Source: NBSC (2013).

2003, to 76.7 per cent in 2013. China's real estate sector joined the outward FDI recently, but still played a small part (Figure 12.6).

Yet outward FDI in real estate, as shown by the official data in Figure 12.6, is only a small part of China's real estate investment abroad. After 2008, Chinese investment in foreign real estate rose rapidly, particularly after 2010. One estimate puts total outbound investment from China into foreign real estate at roughly US$3 billion in 2010, but over US$16 billion in 2013 (Colliers International, 2014d). The National Association of Realtors in the USA indicates that Chinese buyers spent US$22 billion buying housing in USA from April 2013 to March 2014 (CNN, 2014). Although some buyers were from Hong Kong and other places, Mainland Chinese buyers accounted for most of the purchases, often paying cash to buy homes with an average price of $591,000.

Real estate investment by Mainland Chinese outside China has developed in the last few years. Housing investment by individuals used to dominate most of the investment. However, purchase of commercial property for owner occupation and investment purposes by corporate investors and institutions rose from 2012. A landmark case was the purchase by Ping An Insurance of the Lloyd's Building in London in 2012. China Investment Corporation, China's sovereign wealth fund, purchased Chiswick Park, the London headquarters of Deutsche Bank, and Sydney Centennial Plaza in 2013. Chinese developers, from 2012, started to acquire sites for development in overseas markets. The first case of such investments was the US$54.2 million deal by Xinyuan to purchase land in Brooklyn, New York, for a residential development project.

Chinese developers became confident enough to start development projects rather than relatively straightforward equity deals.

The motivations of individual investors, corporate investors, institutions, and developers are different. Individual investors are mainly high-net-worth-individuals, whose traditionally preferred destination for overseas real estate investment is often the same as the destination for overseas study for their children and immigration for their family. Yet investment purchase as a new trend is on the rise. For example, the author has seen more buy-to-let investment in the UK. In the USA, only 39 per cent of Chinese buyers indicated that they would stay in the homes they bought, and the rest would let the homes to tenants (CNN, 2014).

Corporate investors, who are not real estate companies, buy commercial property for owner occupation and sometimes for prestige, and expect savings on rent and capital gains.

Chinese developers and institutional investors have been looking for new business opportunities, higher returns and diversification. With domestic markets of housing and commercial properties in oversupply, some big developers are looking for possible expansion overseas. The trend of buying overseas homes by Mainland Chinese provides an opportunity for Chinese developers to conduct housing development overseas and sell those housing to Chinese investors. At present, these developers are targeting popular investment destinations, for example New York, Vancouver and London, for Chinese home buyers. In addition, they treat the overseas markets as new channels of financing. For example, developers conducted backdoor listing by acquiring firms in Hong Kong, and they use these firms to issue shares and bonds, raising money at lower costs than in their domestic markets. Some developers choose some projects to collaborate with local partners to learn international practices. Institutional investors, on the other hand, focus on diversification and higher returns. For example, prime commercial properties in Beijing and Shanghai fetched yields as low as under 5 per cent but those in other global gateway cities can be bought at higher yields.

This surge of Chinese investments to overseas real estate is in line with the 'go abroad' strategy, first formulated in 2000 by the central government in its Tenth Five Year Plan (2001–05). Before 1997, the Chinese government had adopted a policy to restrict domestic firms from carrying out overseas investment due to the lack of hard currency the government had, and the low capability of domestic firms. The 1997 East Asian Financial Crisis was a turning point, when the government encouraged firms to invest abroad to set up overseas production bases that required exports from the country; a measure to combat the harsh competition in exports when the currencies of a number of South-East Asian countries fell substantially against hard currencies but the Chinese currency stood unchanged against the US dollar. From 2000, the government provided favourable policies for companies to invest abroad so that the Chinese economy could integrate better into the world economy and have greater access to overseas markets. In March 2010, 'Go Abroad' became a campaign as the government announced a policy boost in the Third Plenary Session of the National People's Congress to provide support and substantially simplify the required administrative approval procedures.

International real estate investment by Chinese corporate investors and institutions is still at an early stage. China is currently in a different property cycle than the USA and Europe, with oversupply looming for virtually all property types and the amount of development work is likely to stagnate, if not to fall. This provides incentives to Chinese developers to invest overseas. Although obstacles such as familiarity with overseas market and regulatory regimes remain, Chinese developers have acquired skills and ability to develop large and sophisticated property development projects and are looking forward to globalising their operations.

BOX 12.1 DIRECT INVESTMENT IN LONDON'S PROPERTY BY CHINESE INVESTORS

Richard Zhang

In 2012 and 2013, Chinese direct investment in Central London properties was US$1.24 billion and US$2.48 billion.

In the first six months of 2014, Chinese direct investment in Central London properties has exceeded US$3.31 billion, with US$2.1 billion being brokered through CBRE.

Recently China's outbound investment has increased dramatically, making the country the world's third largest investor behind the USA and Japan. Total outbound investment reached a historic peak in 2013, totalling US$80.24 billion for the first 11 months in 2013. Two factors underpinned this development. On the one hand, the financial crisis brought opportunities to China for overseas investment. As the holder of the largest foreign exchange reserves in history (nearly US$4 trillion), China can support its companies to invest overseas. On the other hand, government policy has been favourable to outbound investment and a 'Go Abroad' campaign has been in place since 2010. The procedures for applying for overseas investment have been simplified and private enterprises have been encouraged and supported in overseas investment.

By contrast, domestic investment channels in China are quite limited. As one of the pillar industries of the Chinese economy, the real estate industry in China has been under restriction since January 2011. Taking into consideration international diversification and the asset allocation, and the shortage of land resources in China, more and more Chinese property investors have started to join the 'Go Abroad' campaign.

The UK has been ranked one of the top ten Chinese outbound investment destinations. Chinese investments in the Central London commercial property market have grown by 94 per cent year-on-year since 2007 because Chinese investors were attracted by the diverse range of properties, opportunities and dynamic activity that no other destination can offer, taking advantage of the size, liquidity and transparency of the Central London market, as well as the relatively easy legal structure, friendly policy environment, and the perceived 'safe haven' status. The Central London investment turnover is expected to reach US$25 billion by the end of 2014, and Chinese direct investment is expected to achieve US$5 billion.

The initial success of Chinese investments can be illustrated by two important deals. In 2009, the Bank of China purchased One Lothbury, EC2; a new build behind the existing Portland stone façade. The building provides 117,500 sq ft of prestigious offices in the heart of the City, directly opposite the Bank of England. The purchase occurred at the bottom of the market during the financial crisis. The current capital value of this building has now increased by 80 per cent and the rental has increased by almost 25 per cent.

As the first Chinese real estate developer to start construction in the UK, Dalian Wanda is going to build two mixed-use landmark towers in Nine Elms, currently the largest redevelopment area in London. The whole project will occupy 20,000 m^2 of a 105,000 m^2 site at One Nine Elms in Vauxhall. The two towers, measuring 200 and 160 metres in height and providing 439 private residential flats, 52 affordable housing units and a 187-room five star hotel, will undoubtedly become the iconic residential building in the future. A year after acquiring the site, the target internal rate of return for the project has already doubled because house prices have recorded a significant increase in the area.

After Premier Li Keqiang's visit in June 2014, the relations between China and the UK have further improved. The investment deals brought by Premier Li were worth up to US$30 billion, showing a great future between China and the UK in business relations. With London becoming a centre of RMB business, more and more Chinese investment will come to London and other UK regional cities.

12.3 Discussion: informal funding channels for inward and outward investments

China has an exchange rate control regime that restricts free movement of capital from and to China. There is a quota, which is US$50,000 a year, for a Chinese citizen to remit money abroad. Foreigners investing inside China and Chinese investors investing outside China, need to obtain approval from government ministries and agencies, in particular SAFE, the State Administration of Foreign Exchanges. Such approval imposes risks and increases transaction costs substantially. For example, it can take up to six months for a Chinese investor to obtain the approval to invest overseas, where deals are typically closed in a shorter time frame. As a result, both foreign investors and Chinese investors seek alternatives to bypass the restriction. For foreign investors, one way is to use money already in China. For

Chinese investors, an equivalent approach is to use money already outside China. For example, there are over 50 development companies listed on the Hong Kong Stock Exchange and more companies have investments throughout the world. These companies can reallocate their overseas money. The other alternative, however, is to make use of informal channels, as explained below.

One way a foreign investor can bypass China's foreign exchange control is to manipulate the import and export system, i.e. the commodity account. For example, an illegal private bank acts as an intermediary between a foreign buyer and a Chinese exporter, with the foreign buyer prepaying a certain amount of cash. Then the Chinese exporter asks for delayed delivery due to operational difficulties, with endorsement from the foreign buyer. Such exchanges of request to delay exports by the Chinese exporter and acceptance of such delays by the foreign buyer go on until the Chinese side asks for a cancellation of the contract several months later and then repays the foreign buyer with a penalty. In fact, the prepayment is 'hot money' that avoids capital control, and the penalty the 'profit' of the hot money. All the exchanges between the two sides are to avoid checks by the regulator.

For Chinese investors, similar informal channels abound. For example, there are numerous foreign exchange bureaus in Hong Kong, some of which could provide an exchange facility for very large sums of money. Chinese state-owned banks are also involved in providing informal channels. It is no wonder that Macao, the world's largest city of casinos, is filled with Mainland Chinese money to feed the casinos, even though the Chinese government strictly bans its citizens from gambling inside and outside China.

The informal channels, nonetheless, are subject to government crackdown from time to time. In 2014, the government tightened the exchange control. For example, the Bank of China's investment abroad scheme, You Hui Tong, to help customers to transfer their money outside China, was exposed in early 2014 by China Central Television (*The Wall Street Journal*, 2014). The crackdown on such schemes by state-owned banks and similar deals by illegal private banks has affected Chinese developers' real estate development schemes overseas. For example, Malaysia has become a hot destination for Chinese home buyers, with several developers purchasing land there for housing development. However, difficulty in moving funds abroad in the middle of 2014 has led to some cancellations by Chinese buyers for some residential projects that relied on Chinese buyers rather than local buyers (ibid.).

Crackdown on informal channels will affect real estate investment in and out of China. Before China opens up the capital account, the informal channels will play a part in international real estate investment related to China, with the risks of crackdown by the government.

12.4 Conclusion

Foreign direct investment in real estate has contributed to the development of China's real estate industry and real estate market. In particular, it was important in the provision of funding to property development and generation of demand for the property market in the 1980s and the early 1990s. FDI in real estate stagnated for a decade after 1995 and became less significant in terms of its proportion of total FDI and of total investment in real estate development. From 2005, foreign money came into China through non-FDI channels to take advantage of expected Renminbi appreciation and rising property prices, which led to a change in attitude by the government towards foreign investment in real estate. In 2006, restrictions on FDI in real estate were imposed. In 2007, FDI in real estate was given a restricted status in the *Catalogue of Industries for Guiding Foreign Investment*. Nevertheless, foreign investment has been rising rapidly through formal and informal channels since 2005 and has mainly targeted the East of China. With a similar cultural background and speaking the same language, Hong Kong developers seem to be able to compete with their domestic counterparts. Singaporean developers also fared well. Western investors, on the other hand, have been more successful in equity investment and lending.

Real estate investment from China has been rising since 2004 as Chinese businesses started to 'go abroad' following the government's policy support. It used to be individuals purchasing houses but since 2010 corporate and institutional investors have become active in this arena. Many developers and institutions such as insurance companies and the sovereign wealth fund, CIC, have made long-term plans for overseas real estate investment, targeting mainly English-speaking developed countries and South-East Asian countries. The increasing number of development projects committed by developers in the USA, the UK and other countries heralds a wave of Chinese investment in real estate in the near future. However, funding for international investments in and out of China is made difficult by the exchange control regime in China. As a result, informal channels, which are subject to government crackdown from time to time, are used extensively by individuals as well as corporate investors.

Notes

1 East China contains 11 provinces and provincial level municipalities: Beijing, Tianjin, Hebei, Liaoning, Shandong, Jiangsu, Shanghai, Zhejiang, Fujian, Guangdong and Hainan.

2 Central China contains 80 provinces: Shangxi, Jilin, Heilongjiang, Henan, Hubei, Hunan, Anhui and Jiangxi.

3 West China contains 12 provinces, autonomous regions and municipalities: Guangxi, Yunnan, Sichuan, Chongqing, Guizhou, Shaanxi, Ningxia, Inner Mongolia, Gansu, Qinghai, Xijiang and Tibet.

PART V
Future directions and conclusion

13

REBALANCING THE HOUSING PROVISION

State and market

This chapter discusses the provision of affordable and social housing (ASH) in China after the housing reforms introduced commodity housing and phased out welfare housing. It first examines the vehicles of ASH in China, which are designed to help households unable to buy or rent commodity housing, and the property rights issues involved. The chapter then reviews the evolution of Chinese housing policy up to 2006, detailing the 1998 housing policy with the focus on affordable housing, and the policy shift in 2003 to concentrate on commodity housing. Explanation is later provided on the housing policy shift from 2006 to 2010, which set the foundation of a new system of ASH in China. The 2011 programme to start 36 million ASH units from 2011 to 2015 is investigated and assessed. The chapter makes use of the recent fieldwork data in Guangzhou and Wuhan to provide case studies of China's ASH regime. Finally the chapter analyses the efficiency of China's ASH system and identifies its main problems, with recommendations to solve these problems.

13.1 Affordable and social housing vehicles and property rights issues

As the ASH system evolves, five components or vehicles have emerged in China's ASH system:

- ECH: literally *Jingji Shiyong Fang*, or Economic and Comfortable Housing
- LRH: literally *Lianzu Fang*, or Low Rent Housing
- PRH: literally *Gongzu Fang,* or Public Rental Housing
- LPH: literally *Xianjia Fang*, or Limited Price Housing
- RIH: literally *Penghuqu Gaizao Fang*, or Redeveloped and Improved Housing.

ECH is a type of affordable housing for sale to households qualified by the income and wealth requirements in China. It is mainly developed by developers of commodity housing, but can also be developed by government agencies. Under state approval and planning, ECH is heavily subsidised by: provision of allocated land at zero land costs; normally 50 per cent discounts in taxes and fees chargeable on commodity housing development; and a cap on profits. Applicants need to be registered households of the cities and towns where ECH is provided, and to satisfy an income and wealth test. Conditions for resale include a minimum duration of owner occupation, which is normally five years, and payment of a land premium and a share of value appreciation to the local government.

LRH is provided in two forms: public rental housing let at low rents; and rental subsidy paid to beneficiaries, with social security considerations to rent private housing. Different from ECH that only applies to newly built housing, LRH consists of new builds and existing housing. Applicants need to pass an income and wealth test at the application stage and, in some cities, at regular intervals during their occupation. There is no right to buy offered to tenants. However, tenants can remain in LRH indefinitely if their incomes and wealth remain below the qualification thresholds.

PRH is public rental housing with tenants paying affordable rents that are closer to market rents rather than the heavily subsidised social rents for LRH. It consists of new build and existing buildings. Applicants need to pass an income and wealth test and can only stay for a term of up to five years. Once the term expires, occupiers who want to remain need to reapply. There is no right to buy offered to tenants, who are supposed to be able to buy commodity housing during or after occupation of PRH.

LPH is a form of subsidised commodity housing. Applicants need to be registered households and pass an income and wealth test. Prices of LPH are normally 30 per cent to 50 per cent cheaper than commodity housing. Conditions for resale include a minimum duration of owner occupation, which is normally five years.

RIH is designed to benefit existing homeowners in substandard and usually dilapidated housing. Provision of RIH involves purchasing those substandard and dilapidated housing units on rural collective land that are owned by local rural households, or in some cases rented by migrants. In urban areas, such housing mostly occurred in urban villages, i.e. rural villages encircled by urban development (see Figure 6.3 for an example in Guangzhou). Because the collectively owned land is not converted to state-owned land, housing development on such land has not been subject to building regulation control, leading to the occurrence of substandard housing, which quickly becomes dilapidated. In rural areas, there are large numbers of such housing in mining areas and large state-owned farms. The solution is normally the nationalisation of the collectively owned land and demolition of the substandard and dilapidated houses for redevelopment. Owners are compensated for their demolished housing at negotiated values to enable them to buy new housing or are given new housing as replacement *in situ* or elsewhere. Tenants are provided with subsidised new housing such as LRH or PRH.

Unlike commodity housing with clearly delineated property rights, ASH, in particular LRH and PRH, have less clear property rights. There is no national law or ordinance on ASH to delineate the rights and obligations of an ASH beneficiary. Traditionally, qualification criteria for ASH have been set by local governments. This insufficiency of institutional arrangements has the following consequences. First, rights to ASH applied only to registered urban households unable to buy or rent commodity housing until very recently. Trials have started to allow migrant workers who are not urban registered households to apply for LRH and PRH, but ECH, LPH and RIH remain exclusively for qualified registered urban households. Provision of housing for them has been the local government's responsibility: before housing reform, they were covered by the welfare housing system; after the housing reform, they are covered by the ASH system if they cannot afford commodity housing. Rural residents do not qualify for government's housing assistance: they have residential land parcels allocated free from the collective land ownership system.

Current national housing policies require local governments to take responsibility for migrants who settle but are not registered households. The problem is that the central government has made only *ad hoc* transfer payments to local authorities, mainly in Central and West China, to build ASH. Local governments in the relatively well-off East China, where migrants like to settle down, often have to foot most of the bills for ASH provision. On the other hand, local governments are given the discretion to decide on local ASH policies. As a result of these factors, they are reluctant to take the level of responsibility envisioned by the national housing policy. Cao and Keivani quote an interviewee's comment on that matter:

> It is not that simple. It is not that the municipal government is not willing to reward those migrants who work for the city. What follows providing LRH or PRH to migrants are expenditures on education, healthcare, pensions, social security … the city has limited resources.
>
> *(2014b: 33)*

What the local governments are afraid of is that LRH and PRH are gateway benefits that open up for migrants the range of benefits that registered residents are entitled to: education, healthcare, pensions, social security, etc., without explicit responsibility of the state.

Second, the rights and obligations of urban households as ASH beneficiaries, and local governments as providers are not clear. Urban households passing ASH qualification thresholds can apply for ASH, and local governments have the obligation to provide ASH without subjecting applicants to unreasonably long waiting periods. However, urban households, as ASH beneficiaries, are not bound by defined obligations. For one thing, there has been no tough penalty for dishonesty exhibited by ASH applicants on incomes and assets, leading to the capture of ASH by unqualified households. For another, there have been no binding behavioural requirements on LRH and PRH tenants: some delaying rental payment and other fees. A small proportion of tenants exhibit anti-social behaviour[1] without serious sanctions. Local governments, on the other hand, have no effective authority to set rules to deal with such behaviour.

13.2 The missed opportunity

Prior to 1980, urban residents in China were allocated public housing at very low rents. Very limited resources were devoted to housing construction due to tight government budgets, and housing standards were very low (see Section 4.2). Housing reforms started in 1980 to relieve the government of the burden of housing provision and to mobilise non-state resources, including foreign and private resources, for house building. In 1988, the Central Government decided that the direction of housing reforms was the marketisation of housing provision (State Council, 1988). At the same time, the central government at the constitutional level allowed local governments at the collective choice-level to relax spending rules and so enable state-owned enterprises to allocate more funds for house building.

As state sector employees' housing conditions improved, some major cities such as Guangzhou and Wuhan started to solve the housing problem of urban households in extreme housing poverty (with less than 2 m^2 per capita living space[2]), most of whom did not work in SOEs and local government agencies and thus were not able to have access to new welfare housing. In Guangzhou, a housing conditions survey conducted in 1985 disclosed that 11,642 households had a per capita living space less than 2 m^2, a very desperate housing condition (Lin, 2013). From 1987 to 1989, the municipal government offered *Jiekun Fang* (poverty relief housing), a form of subsidised housing, for those families to purchase. Some 10,864 households bought *Jiekun Fang* by the end of 1991 (ibid.). In Wuhan, the municipal government started offering *Jiekun Fang* from 1988 to help 6,600 households with per capita living space less than 2 m^2 in five years (Cao and Keivani, 2014b). In 1990, provision of *Jiekun Fang* became a national policy (MoC, 1990), with half a million urban households provided with the option to purchase housing at heavily subsidised prices.

The 1988 national housing policy (Table 13.1) called for the establishment of housing cooperatives to speed up improvement of housing conditions (State Council, 1988). This followed local initiatives in some cities, such as Wuhan. In 1987, China's first non-governmental housing cooperative, Chang Dock Housing Cooperative, was established by a private company, Yangzi Housing Industry Corporation (Yangzi), in Wuhan. The cooperative built housing and sold housing units at RMB280 per m^2, which was less than half of market prices, to members. However, urban land is owned by the state, and housing cooperatives had to rely on administrative allocation of land by the government. In 1991, Yangzi created the Fuxingcun Housing Cooperative and built the first housing estate, Fuxingcun, for 1,700 household members (Cao and Keivani, 2014b). The success of housing cooperatives in

TABLE 13.1 Major national policies on ASH provision by the State Council

Years	*Policy documents*	*Key policies*
1988	Circular on Conducting Housing Reform in Cities and Towns in Stages and Batches	Reform of welfare housing provision; marketisation in housing provision; housing cooperatives
1994	Decisions on Deepening Urban Housing Reform	A two tier housing provision system: • Commodity housing • ECH
1998	Circular on Further Deepening Urban Housing Reform and Accelerating Housing Construction	A three tier housing provision system: • Commodity housing • ECH • LRH ECH is the core component.
2003	Circular to Promote Healthy Development of the Real Estate Market	Ordinary commodity housing is the core component of housing provision
2006	Opinions on Adjusting the Structure of Housing Supply and Stabilising Housing Prices	Establishment of LRH system by local governments
2007	Opinions on Solving the Hardships of Urban Low-Income Households	Establishment of an ASH system to parallel the commodity housing system: • LRH • ECH • LPH

Source: adapted from Cao and Keivani (2014b).

Wuhan and other cities prompted the enactment of a national regulation to provide regulatory guidance on housing cooperatives in 1992 (MoC, 1992). However, the rapid rise of ECH quickly replaced housing provided by housing cooperatives. For example, in 1994, Wuhan Municipal Government allocated 180.6 hectares of land, exempt from land prices and at reduced taxes, for developers to build housing for sale at low prices. Such housing was referred to as ECH in the 1994 policy (Table 13.1). In the same year, national regulation on ECH was enacted (MoC, 1994).

In the 1990s, China's housing policy for the socialist market economy was formed in two steps (Table 13.1). The first step was the proposal of a two-tier housing provision system (State Council, 1994a), comprising ECH for low- and middle-income households to buy and commodity housing for high-income households to purchase. Subsidised renting as a tenure choice was not mentioned. The 1997 Asian Financial Crisis, a contextual factor, prompted the Chinese government to enact the 1998 policy (State Council, 1998a; Ho and Kwong, 2002) to put forward a three-tier housing provision system, i.e. commodity housing, ECH and LRH; with ECH for low- to middle-income families and LRH for the lowest income households. The 1998 policy recognised subsidised renting (LRH), and regarded ECH as the main pillar of the housing provision system. LRH was different from welfare housing, also a form of public subsidised rental housing, with the former for very low-income households, and the latter for all households.

A major problem in the 1994 and 1998 policies was that no rules for the collective choice level were provided, leaving local governments to interpret the policy and set their own rules of implementation. This is effectively a delegation of power from the constitutional level, the central government, to the collective choice level, which makes the rules of implementation according to local priorities. As a result, local polices deviated from the 1998 national policies, leaving the housing provision system biased towards commodity housing, resulting in a rapidly declining stock of public housing for renting.

First, ECH was marginalised in practice. The 1998 policy resulted in the rapid expansion of ECH construction from 1998 to 2003, accounting for 29.35 per cent of all sales of housing in the market in 2000 (Figure 13.1). Nevertheless, ECH did not become the main housing choice of urban residents. From 1998 to 2003, sale of commodity housing was 964 million m^2, but that of ECH was only 202 million m^2 (NBSC, 2011; see Figure 13.1). One reason is that much of the ECH was captured by higher-middle- and high-income households, because many lower-middle-income households could not afford ECH. At the time national regulations (MoC, 1994) at the constitutional level or local regulations at the collective choice level provided no rules on the maximum sizes of ECH. As a result, developers built ECH to the same sizes as commodity housing, with some ECH units over 150 to 200 m^2

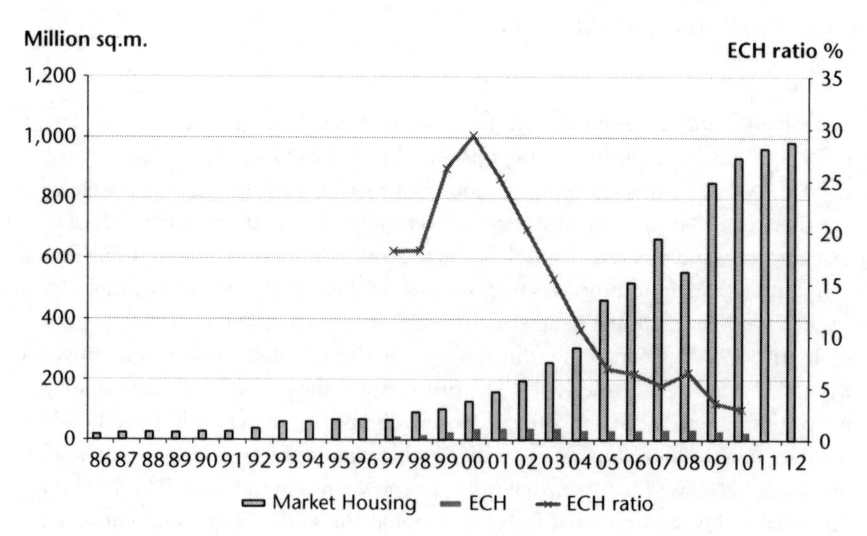

FIGURE 13.1 Sale of commodity housing, including Economic and Comfortable Housing, in China, 1986–2012

Source: NBSC (2012; 2013).

(Ren, 2013; data from interviews), far bigger than most of the commodity housing units. Interviews conducted by the author in 2000 found that some developers were delegated authority by the local government to conduct the screening of applicants. To promote sales, they simply sold ECH to anyone who could afford it. Thus a lack of proper institutional arrangements resulted in a significant number of ECH units being bought by wealthy households. Furthermore, many ECH quotas were used by various government agencies and state-owned enterprises as cheap housing for their employees to buy, despite their incomes being at higher-middle and high levels. During the period from 1998 to 2003, the total completion of ECH was 477 million m^2 (Ren *et al.*, 2012), but only 202 million m^2 was sold openly. The rest were taken up by SOEs and government agencies. This obvious low efficiency of ECH was widely criticised, partly contributing to the decline of the ECH programme from 2000 (Figure 13.1).

Second, LRH was ignored in practice. Interviews conducted by the author indicated that local response to the provision of LRH was unenthusiastic, because there was no immediate gain from implementing LRH. With discretion on LRH provision, local governments generally paid lip service to the national regulation on LRH promulgated in 1999, which required local governments to set operational rules without any targets and obligations (MoC, 1999). By 2005, only 221 out of 280 cities at or above prefecture level had set up an LRH system, with 329,000 households accommodated in LRH or given rental support for private renting (Cao and Keivani, 2014b). There was very limited investment in converting existing public housing into LRH and building new LRH.

Third, constitutional-level decisions were also responsible for the under-development of ASH. The success of the housing market from 1998 to 2002 led national policy-makers to redesign housing policy. The *de facto* main housing choice, commodity housing, received formal recognition by the central government in 2003 (Table 13.1). The 2003 policy stipulated that ordinary commodity housing should be 'for the majority of households to buy or to rent' (State Council, 2003), formally marginalising ECH. On the other hand, in 2002, the Ministry of Land and Resources and Ministry of Supervision required all land for market-oriented property development to be sold in the open market (see Section 3.4; Wu *et al.*, 2007). Subsequent market bidding significantly raised land prices, increasing land sale revenues to the local governments. With land provided free, ECH and LRH could not generate land sale incomes and were not favoured by local governments. Both regulatory changes resulted in a decline of ECH and reinforced the neglect of LRH.

13.3 Establishment of the affordable and social housing system

Housing price inflation and deterioration of support for the urban poor prompted the constitutional level to modify collective choice-level rules, to enable local governments to improve ASH provision. From 2005, the central government started to reshape the country's housing policy by intervening in the housing market to slow down housing price inflation and re-emphasise the provision of ASH (Cao and Keivani, 2014a). The 2006 Measures (Table 7.2) required cities without an LRH system to build such a system by the end of 2006, and stipulated that at least 10 per cent of the land sale revenues should be used to support the construction of LRH (State Council, 2006). It urged further ECH provision through better distribution arrangements to prevent capture by government agencies and state-owned enterprises. Furthermore, the 2006 Measures allowed local governments to create LPH to help families not qualified for ECH. LPH is less of a burden to local governments because it generates a significant part of land sale revenues, as land is sold with a price cap, not provided free as in the case of ECH. To keep LPH prices low, sizes of housing units and building specifications were controlled by an operational rule designed to reinforce the income and assets check on applicants. The small sizes and suburban locations make LPH less attractive as housing investments.

In 2007, the central government enacted the *Opinions on Solving the Hardships of Urban Low-Income Households* (Table 13.1; hereafter referred to as the 2007 Policy), which started the rebalancing of the roles of state and market in housing provision (State Council, 2007b). The 2007 policy focuses on LRH, with ECH playing a supplementary role. Such an arrangement is different from the 1998 policy that focused on ECH, with LRH playing a supplementary role; and from the 2003 policy that focused on ordinary commodity housing, with ECH and LRH playing supplementary roles. A notable change in the 2007 policy was the provision of RIH for families living in substandard and dilapidated housing, and extension of ASH to rural migrant workers who seek better paid jobs in urban areas. It sets an ambitious target for the ASH sector to meet the basic housing needs of low-income households in the urban areas and replace substandard or dilapidated housing by RIH by the end of 2010. Different from previous housing policy documents that left excessive discretion to local governments, the 2007 policy sets maximum sizes for LRH at 50 m^2 and for ECH at 60 m^2, removing much of the incentives of investors to capture LRH and ECH. However, these size standards were found to be too small and unattractive and were ignored by some cities.

The implementation of the 2006 Measures and the 2007 policy between 2006 and 2010 has resulted in a fundamental change in ASH provision. First, local governments were bound by rules set by the central government to deliver ASH. An administration network was installed at all levels of government, with a Department of Housing Security set up in the Ministry of Housing and Urban-Rural Development (MOHURD) for policy development and supervision. Offices of Housing Security were set up by local governments to address local policy development and implementation. Second, large-scale LRH construction and conversion began because of the requirement by the central government to expand the stock of LRH rather than relying on rental subsidies (State Council, 2008). In most cities, funding and land for building LRH were made available to meet completion targets set by higher levels of government. Interviews by the author indicate that new LRHs were necessary because there was a lack of small housing units in most Chinese cities, and rental subsidies could not entice more market provision of small units (Cao and Keivani, 2014b).

Third, ECH provision was significantly improved in terms of fairness with the implementation of size standards and transparency requirements. Interviews conducted by the author found that interviewees believed that ECH became truly for low- to lower-middle-income families. Fourth, new vehicles emerged to expand the coverage and choices of the ASH regime. The provision of RIH and the increase in ASH stock helped both lower-middle-income families not on the housing ladder, and families living in substandard accommodation but unable to change or improve their housing conditions. RIH formally became a vehicle of ASH in 2009 (MOHURD, 2009b). In 2010, LPH was also included in ASH provision (State Council, 2010a). PRH, originally a local initiative, was admitted into the ASH system (MOHURD, 2010) and played an important role in helping people not qualified for LRH. Let at affordable rents, it benefited new employees working in both the private and public sectors, and lower-middle-income migrant workers in some cities. With sizes ranging from 40–60 m^2, PRH is not attractive as permanent residence, but only a bridge for its tenants to commodity housing.

The 2006 Measures and 2007 policy resulted in rapid development and expansion of ASH. By using new build, acquisition, existing stock and investing in improvements, the government was able to house nearly 22 million low- and lower-middle-income families in cities and towns by the end of 2010, accounting for 9.4 per cent of all registered urban resident families. In addition, close to 4 million families with registered urban residence were given rental subsidies to rent private sector housing (Ren, 2013). However, an ambitious target of building 5.8 million ASH in 2010, put forward by the premier in his 2010 *Report of the Work of the Government*, was not completed.

13.4 The great building programme

In 2010, containment of housing price inflation became a priority of the central government. To alleviate the impact of rapid housing price inflation and worsening affordability, the central government put forward a very ambitious plan for ASH construction: starting 36 million units of ASH during the Twelfth Five Year Plan period (2011–15). The overall targets were: (1) building 36 million ASH units to cover 20 per cent of urban households; (2) extension of ASH to graduates and rural migrant workers as well as low-income and lower-middle-income households; (3) PRH to become the main form of ASH; and (4) RIH to take a larger share of the overall ASH building. Wen Jiabao, the premier, announced the 2010 plan to start 10 million new ASH in 2011 in his 2011 *Report of the Work of the Government*.

Despite widespread scepticism on the possibility of completing the 2011 target, the central government made full use of its power to set binding targets for local governments to achieve. Consequently, 10.43 million new starts were achieved in 2011 (Table 13.2). Among them, PRH accounted for 22 per cent, LRH for 16 per cent, ECH and LPH for 20 per cent, and RIH for 42 per cent. In terms of

TABLE 13.2 Progress of ASH building in the Twelfth Five Year Plan (million units)

	2011–2015 target	*2011*	*2012*	*2013*	*2011–2013*	*2014 plan*	*2015*
New starts	36.00	10.43	7.81	6.66	24.90	7.00	4.10*
Completions		4.32	6.01	5.44	15.77	4.80	

Note: *Remaining units to start to hit the target of starting 36 million units.
Source: MOHURD.

geographical distribution, 44 per cent of the new starts in 2011 were in the relatively poor Western China, 21.7 per cent in Central China, and 30.4 per cent in the relatively wealthy Eastern China. In 2012, 22 per cent of new starts were PRH (Ren, 2013). By early 2014, it was clear that the target of 36 million new starts could be achieved because 24.90 million units had been started from 2011 to 2013 (Table 13.2). In 2014, 7 million new starts were planned, leaving just 4.1 million in 2015 to hit the 36 million target from 2011 to 2015. It is worth noting that RIH takes an increasing share of ASH. In the 2014 plan, RIH overtakes the other four vehicles in the number of units (Table 13.3). This indicates that the government has to address the likely oversupply of PRH and LRH.

In the period from 2013 to 2014, a number of institutional changes in ASH occurred. The most significant development is the consultation of the *Ordinance for Security of Housing in Cities and Towns* in March 2014. If enacted, the Ordinance will change the situation that there is no national statute on ASH. The second change is the merger of LRH and PRH in 2014 (MOHURD, 2013), which will simplify the management of the two public rental housing vehicles. The rents charged for PRH will depend on the income and wealth of tenants, with different subsidies paid to different tenants. The third change is the phasing-out of ECH, a policy announced by MOHURD at the end of 2013. Many cities and provinces are exiting ECH due to the level of subsidy paid to successful applicants and the incidence of corruption. The fourth change is the trials by many local governments in supplying shared ownership housing to replace ECH. The difference between ECH and shared ownership housing is that the buyers of shared

TABLE 13.3 Breakdown of ASH new starts and completions (million units)

		ASH	*RIH*	*PRH, LRH, ECH & LPH*
2011–2013	New starts	24.90	10.84	14.06
	Completions	15.77	6.68	9.09
2014 plan	New starts	7.00	4.70	2.30

Source: MOHURD.

ownership housing have to share the value appreciation with the government, allowing funds to be recouped for additional ASH building and management; while ECH buyers obtain large amounts of subsidy in reduced prices and a far greater share of appreciation in value. The fifth change is the gradual inclusion of qualified migrant workers in local ASH systems in some cities, such as Wuhan and Beijing. Inclusion of migrant workers was first proposed in the 2007 policy, but there has been strong local resistance because extending ASH to migrant workers has implications for local financial and other resources.[3] The increase in ASH stock prompted some local governments to extend PRH to migrant workers to avoid vacancies. The sixth change is the sourcing of PRH from private housing. For example, in Wuhan and Guizhou Province, local governments have carried out experiments to provide PRH by *en bloc* renting of private housing as PRH, which opens up another channel of PRH provision in addition to new build, acquisition and conversion. This approach avoids overbuilding of ASH due to temporary population increase. The seventh change is in sophistication of management and quality control, with construction, distribution and management information systems to offer transparency in housing standards, in the applications made and in approvals by the authority. In more and more cities, approved lettings are made public online and regular reviews of the qualification of tenants are conducted by some local governments.

13.5 Case studies

Guangzhou and Wuhan are chosen as case studies for the current status of ASH. In addition, ASH projects in Shanghai and Urumqi are presented to provide a fuller picture of the state of development of ASH in China.

13.5.1 Guangzhou

Guangzhou started to solve housing poverty from 1985, focusing on families with per capita living space less than 2 m². By 1998, the entry threshold to ASH was raised to per capita living space less than 5 m². Since 2006, the entry threshold has been raised to per capita living space less than 10 m² (Table 13.4).

TABLE 13.4 Qualification thresholds for ASH in Guangzhou

*Income entry thresholds for LRH**					
Years	1998	2007	2010	2012	2013
Per capita disposable income (CNY)	4,680	7,680	9,600	15,600	20,663
Floor area entry thresholds for ASH					
Years	1985	1998	2004	2006 onwards	
Per capita living space less than (m²)	2	5	7	10	

Note: *LRH ceased to exist from 2014.
Source: adapted from Cao and Keivani (2014b).

TABLE 13.5 Types of ASH and qualification thresholds in Guangzhou in early 2014

Types	*Target population*	*Eligibility*		*Unit size standards*	*Rents & prices*
		Space	*Income*		
PRH	Low- to lower-middle-income households	<10 m² per capita living space; or <15 m² GFA	Annual disposable income lower than: RMB24,795 for one-person family; RMB45,458 for two-persons family; RMB61,989 for three-persons family; RMB74,388 for four-persons family;	40-60 m²	Reference to market rents; subsidies provided to low-income families
ECH	Low-income registered residents	<10 m² living space	<RMB18,287 per capita disposable income	<60 m²	Sold at costs; <3% profits for non-state builders
LPH	Lower-middle-income registered residents	Non-homeowners	Annual income: <RMB200,000 for family; <RMB100,000 personal	<90 m²	About 70% of market prices

Source: adapted from Cao and Keivani (2014b).

Table 13.5 illustrates entry thresholds in floor area of existing housing, income, unit sizes available, rents and prices of ASH in Guangzhou.

New ASH estates have been planned and constructed to high standards. For example, Fanghe Garden is a purpose-built ASH estate on Dongjiao Street, Liwan District; one of the inner city districts. Fully funded and built by the municipal government, it was occupied in November 2011 by approximately 18,000 people living in 3,988 units of ECH and 1,947 units of LRH. The site area is 119,600 m² and total floor space is 477,800 m², with a 40 per cent coverage of trees and lawns. The ECH at the estate was pre-sold in October 2010 at about 30–35 per cent (approximately RMB4,000 per m² of gross floor area) of market prices. ECH owners at Fanghe Garden registered for ECH in 2008 and became qualified to purchase ECH in 2009 (Cao and Keivani, 2014b).

The estate has three functional areas, i.e. residential, school and business, and culture districts. The residential district is filled with 21 high-rise apartment

FIGURE 13.2 Residential district of Fanghe Garden, with greenery and sports facilities
Source: photo by Albert Cao (2013).

buildings measuring 28–32 storeys (Figure 13.2). The school district accommodates a kindergarten (Figure 13.3), which is a branch of a reputable kindergarten chain accredited as a class 1 kindergarten by the municipal education bureau; and a primary school, which is a branch of a highly respected primary school chain accredited as a class 1 primary school by the provincial educational department. The business and cultural area has a municipal exhibition hall for ASH in Guangzhou on the ground floor and a community service centre on other floors. The estate has gates, which were not guarded by management staff at the time of the site visit.

When starting to build ASH units in large numbers, the Guangzhou Municipal Government carried out a design competition to choose the designs that provided the best layout and functions within the specified size standards stipulated by the central government. The final designs were chosen from over 80 submissions from the country's reputable architects and academics. Figure 13.4 shows a model of the two-bedroom flat design, with a gross floor area of 66.35 m^2. The flat has a living room, two bedrooms, a kitchen, a toilet/shower and a balcony. All units are provided with air-conditioner hangers, and most households use split-part air conditioners to cope with the city's hot and humid summer.

FIGURE 13.3 The kindergarten inside the Fanghe Garden
Source: photo by Albert Cao (2013).

FIGURE 13.4 A model of typical two bedroom ECH flat at Fanghe Garden
Source: photo by Albert Cao (2013).

13.5.2 Wuhan

Wuhan was the pioneer in some of China's ASH trials in the 1990s, such as housing cooperatives and ECH. Yet housing conditions for the poor were rather low by 1998, with 11,000 households having per capita living space under 4 m^2. In August 2002, Wuhan Municipal Government tabled a regulation on LRH to cover all households with per capita living space less than 6 m^2. By moving into the LRH, households achieved a minimum per capita living space of 8 m^2 (Cao and Keivani, 2014b). Since the enactment of the 2007 policy, Wuhan has built an ASH system with ECH and LRH as core vehicles, and PRH, LPH and RIH being new vehicles. According to an internal government document, by the end of 2012, the city's ASH sector had completed 20.49 million m^2 of ECH benefiting 218,100 families; it spent about RMB4.0 billion to provide LRH support to 82,000 families. Increased ASH stock enabled the city to cover all households with a per capita disposable income lower than RMB600 per month and per capita GFA lower than 12 m^2 by LRH at the end of 2012. This included 28,000 families living in purpose-built LRH units and 54,000 families who were offered rent reductions and rental subsidies. With the participation of enterprises, 17,754 PRH were completed. In addition, 17,290 units of LPH were completed for those who qualified. Furthermore, 110,620 families were relocated using the RIH vehicle, improving their housing conditions (*China Daily*, 2013).

Table 13.6 lists the types of ASH available and the qualification thresholds at early 2014. Interestingly, LRH is still listed as an ASH vehicle at the time of writing (August 2014) by the Wuhan Housing Security and Management Bureau though MOHURD has officially called for the merger of PRH and LRH.

Similar to Guangzhou, new ASH estates have been planned and constructed to high standards. For instance, Wenhuiyuan Estate is a mixed ASH estate in Baibuting Garden, a very large ECH and low-end commodity housing area, in the Jiangan District. The Estate has an area of 80,800 m^2, with a plot ratio of 2.72 and green space accounting for 38 per cent of the site (Figure 13.5). The estate offers 2,640 LRH, 770 ECH and 55 RIH units and can accommodate about 12,000 residents. Sizes of ECHs are from 50 to 88 m^2. ECHs were pre-sold at an average price of RMB2,950 per m^2 in 2010, and management staff on site indicated that commodity housing nearby was offered at above RMB6,000 per m^2 at that time. All buildings were between 16 and 28 storeys, with common areas like entrance halls and community centres found at ground and first floors. Due to hot weather, many families living in LRH use air conditioners (Figure 13.6), some of which were given by estate management staff through donation campaigns. The estate was wheelchair-friendly and run as a gated community for safety and comfort. The LRH units in the estate were occupied in October 2012.

Another example of ASH in Wuhan is the Huiminju PRH Estate, a purpose-built estate situated in Jiangan District and developed by the municipal government (Figure 13.7). The estate occupies a 36,500 m^2 site and has 2,145 units of PRH,

TABLE 13.6 Types of ASH and qualification thresholds in Wuhan at early 2014

Types	*Target population*	*Eligibility*		*Unit size standards*	*Rents & prices*
		Space	*Income*		
LRH	Low-income households	<12 m^2 GFA	<RMB600 per month	<50 m^2	RMB0.75-1.5/m^2
	Lower-middle-income households	<8 m^2 GFA	<RMB3,000 per month	<60 m^2	Reference to market rents
PRH	Migrant workers with required skills	Not owning housing	<RMB3,000 per month	<60 m^2	Reference to market rents
ECH	Low-income registered residents	<16 m^2 GFA	<RMB824 per month	60 m^2	Sold at costs; <3% profits for non-state builders
Enterprise housing cooperative	Employees of qualified enterprises	None	None	60 m^2	Sold at costs; <3% profits for non-state builders
LPH	Households affected by relocation	None	None	60 m^2	About 70% of market prices

Source: Wuhan Housing Security and Management Bureau.

FIGURE 13.5 The central open space at Wenhuiyuan Estate
Source: photo by Albert Cao (2013).

FIGURE 13.6 Entrance, common area and air conditioners at a building in Wenhuiyuan Estate

Source: photo by Albert Cao (2013).

FIGURE 13.7 Huiminju at the final phase of construction

Source: photo by Albert Cao (2013).

which were occupied in November 2013. When applications were opened in October 2013, 4,641 families applied (*Changjiang Daily*, 2013). Huiminju Estate provides three types of accommodation: units with one bedsit room, normally 40 m^2 in GFA; and units with one living room and one bedroom, normally 50m^2 in GFA; and units with one living room and one bedroom, normally 61 m^2 (GEA) for families with disabled persons. At the time of occupation, monthly rent was CN¥15 per m^2 in GFA, about half the market rent for comparable housing. Estate management charge per month was RMB1.4 m^2 in GFA.

13.5.3 Other cities

Standards of new ASH in many cities are comparable to those in Guangzhou and Wuhan. For example, Boyayuan, Puhang New City is an ECH estate in Minxing District, Shanghai (Figure 13.8). It is one of first ECH estates in Shanghai's suburbs, offered for sale in 2010. The estate mainly contains two-bedroom flats at 70 m^2, with a smaller proportion of one-bedroom flats at 45 m^2 and three-bedroom flats at 90 m^2. Like its counterparts in Guangzhou and Wuhan, it has a green coverage of 35 per cent.

Another example of ASH is a RIH estate in Tianshan District, Urumqi, Xinjiang Uighur Autonomous Region (Figure 13.9). In 2010, the dilapidated housing on the site was redeveloped to a mixed use building complex with retail space at the first two floors and 240 flats in the two tower blocks.

FIGURE 13.8 Boyayuan, Puhang New City, Minxing District, Shanghai
Source: photo by Albert Cao (2013).

FIGURE 13.9 Wanheju RIH project, Yuejin Street, Tianshan District, Urumqi, in 2014
Source: photo by Yuanze Gao (2014).

13.6 Discussion: efficiency issues in China's ASH regime

The rapid development of ASH from 2007 in China has fundamentally altered the bias towards commodity housing. The central and local governments spent large amounts of money to provide price subsidies to home buyers by offering ECH and LPH and to offer rental subsidies to renters. Furthermore, they redeveloped or improved the substandard and dilapidated housing to housing owners and renters. Due to problems in institutional arrangements, however, inefficiencies are found in the Chinese ASH regime.

First, there is insecure funding for ASH. Local governments are bound by constitutional rules to provide ASH, but the central government only provides funding on an *ad hoc* basis. For relatively well-off provinces in East China, such funding is often small in relation to the expenses incurred. Consequently, local governments have to choose the least costly options for ASH provision. For example, ASH is sited in remote areas, such as Boyayuan, Puhang New City in Shanghai, with insufficient infrastructure, leading to slow take-up of the completed ASH units and thus many vacant units. Interviews by the author reveal some of the construction quality problems are due to the use of cheap materials under budget constraints. Without an agreed funding formula, inefficiency in ASH can persist for the long term.

Second, there are insufficient institutional arrangements at the collective choice level and operational level to guide behaviour and penalise wrongdoings. Although ASH was provided from 1994 in the form of ECH, there was insufficient regulation of size standards and ways of distribution. ECH in large unit sizes were too expensive for the target population of lower–middle-income families, leading to their capture of the units by higher-income families. Government agencies and SOEs often built housing under ECH schemes to benefit from the preferential treatments. Since 2007, there have been more rules and regulations on ASH. However, some rules are not reasonable. For example, the national size standards of ECH, LRH and PRH are too small, causing vacancies of completed ASH built according to national size standards because of lack of demand. As a result, local governments, such as Shanghai, often build ASH at bigger unit sizes to reduce the chance of vacancy.

Third, there is no national ordinance or law on the status and property rights of ASH. As a result, both providers and beneficiaries do not know their rights and responsibilities clearly. There are management problems that are not easy to solve due to management staff's lack of authority. For example, in Guangzhou, the municipal government has to increase rents on units occupied by wealthy tenants because it has no power to evict them, keeping those units away from families in need. It has no legal authority to throw out those wealthy tenants, who often prefer the location or amenities in ASH estates.

Finally, there is a lack of choice for LRH and PRH tenants. The only provider of ASH is the local government. Housing cooperatives, once a third provider in addition to market and the state, could not survive due to lack of land and credits, all of which are controlled by the government. Large ASH estates are built to increase the ASH stock, but some tenants do not want to live in ASH estates in remote locations. Concentration of ASH units and thus low-income households in estates of thousands of units has the potential to generate social problems.

Cao and Keivani (2014b) offer recommendations to increase efficiency in ASH provision. First, a funding formula needs to be agreed between the central government and local governments. With certainty of funding, local governments can choose options that reduce inefficiency currently caused by insufficient funding. Second, the supply of institutional arrangements for the ASH needs to be speeded up. In particular, the rights and obligations of occupiers and management companies and their staff should be clearly defined, with sanctions and rewards made clear. Commodity housing estates obliged to include a percentage of ASH need to be given a clear definition of property rights and management rules to ensure efficient management. Third, small ASH estates and mixed communities should be the direction of future ASH development. Finally, the private and voluntary sectors should be encouraged to participate in ASH provision so that more choices are available and competition exists between providers.

BOX 13.1 HANDLING IMPORTANT RELATIONSHIPS IN ASH IN CHINA

Hongyu Liu

Housing issues and housing security are complex and long term in nature, requiring government policies on housing issues to think about the long term and for sustainability. Housing policies should not deal with current problems at the expense of the future.

China's ASH provision has a short history. It is important to recognise the long-term nature of provision of housing security. Even in developed countries, ASH shortages can occur due to changes in circumstances. Thus China should not attempt to solve the housing problems in one go, with a transition from a campaign-style provision to long-term institutionalised provision. It is also important for China to understand the focus of ASH needs to change to meet new challenges. At present, the country has built ASH on a very large scale. In the future, the focus may change to the development of mixed communities and provision of rental subsidies. Factors affecting ASH provision are multi-faceted and include population trends; the sizes of families; employment and incomes; culture and traditions; the fiscal positions of governments and taxation policy; the market capacity and financial instruments.

The following points need to be considered in building a new ASH regime. First, the relationship between commodity housing and ASH needs to be carefully studied and adjusted to achieve the best outcomes for the people. Housing is both a necessity and an investment asset, and a functioning housing provision system must combine both market and state provision. Furthermore, quasi-market provision should be formed in addition to state provision and market provision. The principle is not to exclude the market and not to leave the market unregulated. The current trend is to give the market more opportunities and responsibilities, with the government collaborating with market forces.

Second, the economic and social functions of housing should be properly balanced. Commodity housing should not be relied upon too much for economic growth. The subprime mortgage crisis in the USA is an example of the negative consequences of amplifying the economic function of housing. Provision of ASH, on the other hand, should be linked to fiscal capacity of the government. Families should be guided to prioritise the social function of housing, and leave the economic function of housing to the second place. This means policies to discourage investment and speculation are required.

Third, people's unlimited housing need and limited fiscal ability and the government's management capability should be properly balanced. In the era of rapidly rising housing prices, many people hope to obtain assistance on housing from the government, leading to pressure to expand ASH provision.

However, the government's fiscal and management capacity, the impact of ASH on the housing market and the welfare trap of ASH should impose a limit to the scope of the ASH regime. In China, the ASH system should cater for basic, rather than high-end, housing needs.

Fourth, housing security should be at proper levels without causing welfare traps. A welfare trap forms when housing security results in beneficiaries losing motivation to find jobs, which may lead to loss of housing welfare. Housing policy should be designed to incentivise people to solve their housing problems. The policy to provide housing coverage for all those considered in need of housing assistance is not as ideal as providing coverage for all those considered in need of housing at the right time. Housing policy should motivate housing beneficiaries to move from welfare to work. Housing subsidies should be combined with employment, with priority given to key occupations at low wages.

13.7 Conclusion

China used to have a system of public rental housing for all urban residents. Housing reforms were conducted to introduce market mechanisms and non-state resources to speed up housing provision to improve very poor housing conditions. *Jeikun Fang* and ECH provisions in the 1980s and 1990s reduced the state's burden and raised housing standards. A blueprint for a housing system in a market economy was developed in 1998, with affordable housing, ECH, as the core for the middle-income households and LRH for low-income households. However, ECH provision declined after 2000 due to widespread take-up of ECH by wealthy buyers and the affordability problems of lower-middle-income households. New housing policy in 2003 focusing on ordinary commodity housing led to the marginalisation of ECH. Furthermore, lack of central government funding and binding collective choice-level rules led to the neglect of LRH by many local governments. As a result, more and more urban households were unable to improve their housing conditions.

Macro control measures from 2006, in particular, the national ASH policy in 2007, have established a new ASH regime with public rental housing, namely LRH and PRH, as the core of the housing security system in urban China. LPH helped families not qualified for ECH to step onto the first rung of the housing ladder. RIH, on the other hand, has become the largest ASH vehicle, improving the housing conditions of people owning substandard and dilapidated housing. The ambitious target to speed up ASH provision by starting 36 million units of ASH from 2011 to 2015 is on target and very likely to be achieved. However, the rush to rebalance housing provision by the state and the market has led to inefficiencies in ASH provision: being built at poor locations; widespread slow take-up of completed ASH and vacant units; dissatisfaction about lack of choices; and management problems. It is recommended that the central government provide funding certainty to local governments by: developing a funding formula; speeding up the provision of statutes and regulations to define rights and responsibilities; enacting rules for ASH to be provided in smaller housing estates; and encouraging non-state provision.

Notes

1 Anti-social behaviour in ASH housing estates includes throwing objects from the home, which are often tens of metres high, and noisy behaviour at night.
2 Living space includes internal floor area excluding kitchen, toilet and shower.
3 At present, migrant workers and their families are not entitled to healthcare, education and social security provided by local governments to registered households.

14

REPOSITIONING OF THE PROPERTY SECTOR

Focusing on the discussion of market changes and future directions, this chapter explores debates on the challenges brought about by the looming market downturn and responses from the property industry and policy-makers. It first surveys the indicators that point to a downturn caused by a shift in market expectation, increasing oversupply and policy change; along with the resultant challenges to market recovery and economic growth. Then the chapter evaluates macro control as a regulatory tool and the changes to local government control. The long-term adjustment and control mechanism, an institutional change to phase out macro control, is explored in principle and possible directions of change are considered. The chapter provides details on spending supported by land sale receipts and discusses expected institutional change to phase out local fiscal dependence on the property sector. Energy consumption created by the building spree in China is then examined, and the government's attempt to save energy and to promote green buildings is investigated. This chapter concludes that the forthcoming reforms will reposition the property market to play a main role in resource allocation, and the 'go green' campaign will alleviate the environmental impact of the property sector.

14.1 The challenges ahead

The international financial crisis in 2008 resulted in a sharp slowdown of China's GDP, which prompted the Chinese government to release a stimulus package at the end of 2008 to boost the economy (Section 2.3). The housing market, suffering from sharply falling housing prices, was the main target for the stimulus package, in particular at the local level. As a result of this rescue package, economic slowdown was short-lived and the GDP growth returned to double-digit territory in 2010 (Figure 2.1). Housing prices, on the other hand, rebounded quickly in 2009 (Figure 4.3). However, the excessive liquidity created by the stimulus package

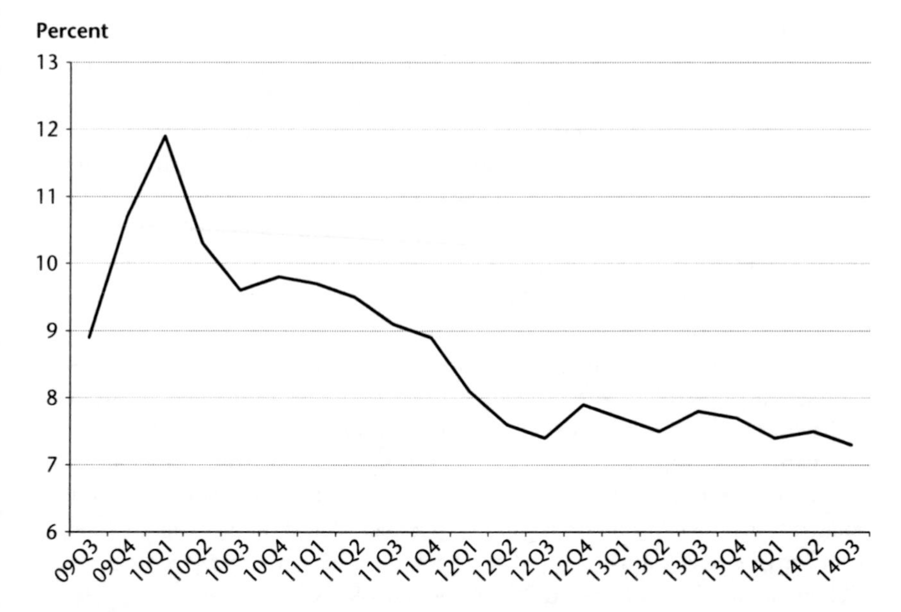

FIGURE 14.1 Quarterly GDP growth in China since quarter 3, 2009
Source: NBSC.

generated a property development boom that led to a new round of housing price inflation and the resultant toughest macro control to be installed from 2010. The surge in GDP was also short-lived, with GDP peaking in quarter 1 of 2010 and falling since (Figure 14.1). In the first half of 2014, China's GDP growth rate slipped to 7.4 per cent, which was the lowest since 1998 when the East Asian Financial Crisis struck. As discussed in Chapter 2, the slowdown of the GDP growth rate is as much structural in nature as cyclical. It is partly due to a government-engineered adjustment to control the economy and absorb excessive capacity created by the stimulus package that had led to overcapacity and asset price inflation; and partly due to the bigger size of the economy. As discussed later, the housing market started to lose momentum due to oversupply in tier 2 and tier 3 cities, and a change in market expectation on future price rises, from the first half of 2014. The question is whether the Chinese government will intervene as it did in 2008 to stimulate the economy to a higher growth rate, or to rescue the housing market to sustain the economic growth at the current level.

The message from the central government was that there will be no more stimulus package to prop up GDP growth rate. Instead, the *People's Daily*, the Chinese Communist Party's major newspaper, indicated that the current growth rate between 7 per cent and 8 per cent should be regarded as a 'new normal' growth. Rapid growth from 1999 to 2011, on the other hand, is not 'normal' due to the various stimulus packages provided by the government. Furthermore, the slowdown from 2012 is different from that during the East Asian and Global Financial Crises, which led to the loss of large numbers of jobs and to deflation. Nevertheless, the government was not idle. It conducted a targeted relaxation of reserve requirements on some medium-sized banks to benefit some sectors, and provided mini stimulus measures to promote growth in specific areas, such as in agriculture, infrastructure and the environment. Such measures are meant to alleviate the decline in growth, which slowed to 7.3 per cent in quarter 3 of 2014 (Figure 14.1), and to promote structural change in the economy for sustained growth. In the first nine months of 2014, for instance, new employment was over 10 million, hitting the target for 2014; the consumer price index rose 2.1 per cent, little changed compared from last year; and per capita disposable income for the country as a whole grew 10.8 per cent in the first six months of the year (NBSC, 2014a).

The key indicators of the property market in the first ten months of 2014 are illustrated in Table 14.1. New property starts fell by 5.5 per cent, improved from the 16.4 per cent in the first six months; similar improvement occurred in housing new starts. Land purchase by developers went up by 1.2 per cent, improved from the -5.8 per cent in the first six months. However, commodity housing sales deteriorated as the stock of commodity property for sale. The improvements in development activities are considered to be the result of a series of targeted measures, such as allowing developers to borrow from the inter-bank monetary market, relaxing purchase restrictions, and the relaxation of the Housing Provident Fund for home purchases. The housing market, in particular, has high expectations about the credit loosening package for home purchases by the central bank and the Banking Regulatory Commission tabled on 30 September, which urged banks to discount mortgage interest rates by up to 30 per cent and to treat borrowers who had paid up all previous mortgage loans as first-time buyers. The measure increased home sales in October.

A scrutiny of quarterly figures and changes for the past year before mid-2014 is provided in Table 14.2. Developers' spending on land acquisition started to slow down in the first quarter of 2014 after a big surge in the previous two quarters; property new starts declined significantly in quarter 1 of 2014, though the decline narrowed in quarter 2; and both property sales and investment in property development declined continuously from quarter 1 of 2014. Despite a glut in commodity

TABLE 14.1 Key indicators of the property market in first ten months of 2014

Indicators	*Volume*	*Change (%)*
Investment in property development (CN¥ billion)	4,201.9	14.1
Buildings under construction (million m²)	6,114.1	11.3
Housing new starts (million m²)	566.7	–19.8
Housing sales (million m²)	424.9	–9.2
Land purchase (million m²)	148.1	–5.8
Commodity property for sale (million m²)	544.3	24.5

Source: NBSC (2014c).

TABLE 14.2 Key indicators of the property market in quarterly breakdown

Indicators	*13Q3*	*13Q4*	*14Q1*	*14Q2*
Land purchase by developers (CN¥ billion)	395.70	402.40	205.20	490.40
Year-on-year change	12.2%	29.8%	28.5%	25.0%
Property new starts (million m²)	490.00	563.08	290.90	510.36
Year-on-year change	14.9%	33.1%	-25.2%	-10.5%
Property sales (million m²)	329.50	461.68	201.11	282.54
Year-on-year change	15.7%	7.7%	-3.8%	-7.5%
Investment in property development (CN¥ trillion)	2.43	2.49	1.53	2.67
Year-on-year change	18.9%	19.9%	16.8%	12.6%

Source: NBSC (2014c).

housing and office space on the market, a large number of new commodity housing and office space is being built, as indicated by the large amount of space under construction in Table 14.1. Oversupply in housing, offices and retail space has led to negative market expectations on the future of the property market.

The debate on whether the housing market has reached a turning point was concluded when the 70 cities housing price index indicated that housing prices in BSGS all fell in July 2014. In fact, the number of cities in the group of 70 cities with a month-on-month fall in housing prices rose from 4 in March to 64 in July. In August and September, it was 69. Figure 14.2 illustrates the number of cities with year-on-year rises in housing prices and the weighted average rate of housing price inflation for the 70 cities. In September 2014, only ten cities recorded rises in prices on a year-on-year basis and two recorded unchanged prices. A situation similar to that occurred in July 2012, when only 12 cities experienced price hikes (Figure 14.2). In particular, the fall in commodity housing prices in 2012 was a result of the imposition of home purchase restriction on 46 cities. The economy then still grew at a faster speed and housing demand was still strong with strong market expectations of further price hike. The current fall in housing prices, however, is mainly the result of a change in market expectation on future price change under economic slowdown, housing oversupply, and central government's determination not to offer rescue packages.

The start of the downturn in the property market has led to a slowdown in land sales by local governments. Land sales receipts in 300 economically important cities in the first and second quarters of 2014 were CN¥717.5 billion and CN¥524.3 billion, with average prices per m² of permitted floor space falling from CN¥1,245 to CN¥1,135 per m². Nationally, in the first seven months in 2014, land sale revenues and property-related taxes hit record highs, but the pace of growth slowed down significantly. According to data from the Ministry of Finance, total receipts from land sales were CN¥2.4961 trillion for the first seven months, up 23.9 per cent from

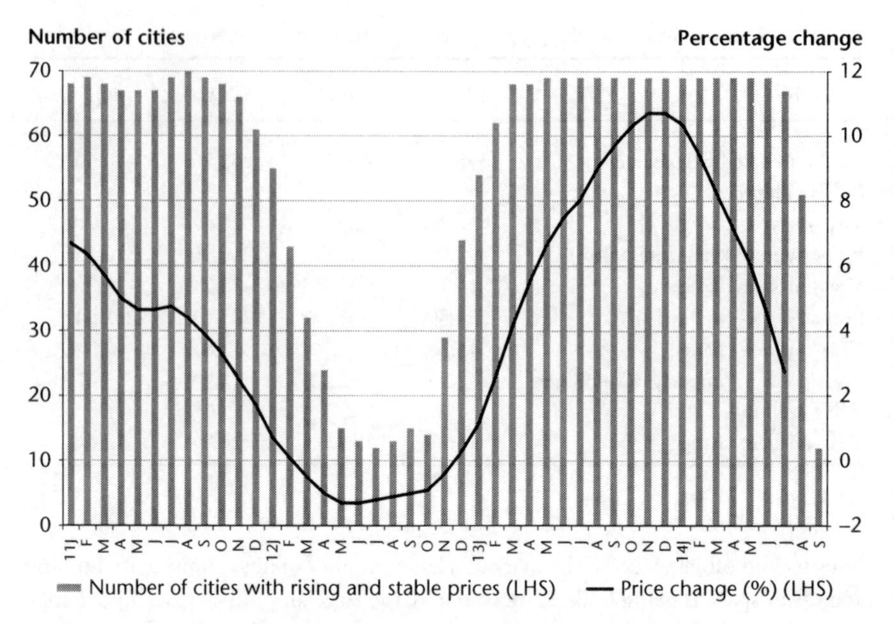

FIGURE 14.2 Number of cities with rising annual housing prices and the housing price inflation rate

Source: NBSC.

the same period in 2013. The growth rate in the same period in 2013 was 49.4 per cent. Measuring the receipt of land sale revenues rather than the contract prices, the figure from the Ministry of Finance lags behind the actual occurrence of land sales in the market. However, the receipts from land sales show a trend of rapid deceleration: growth in the first quarter was 40.3 per cent; and that for the first half of the year was 26.3 per cent. Meanwhile, Corporation Tax from real estate firms grew 5.8 per cent in the first seven months of the year, down from 27.4 per cent in 2013 for the same period; Business Tax grew 6.4 per cent, down from 43 per cent in 2013 for the same period; and Deed Tax grew 10.5 per cent, down from 40.2 per cent in 2013 (Yicai, 2014). Revenues from the property market are expected to shrink if the contraction of the property market continues, with implications for the payment of local government debts, the continuation of infrastructure projects, and local spending on affordable and social housing (ASH) and other social security spending.

With rising vacancy rates in the office market (Chapter 10) and falling housing prices, the property market is in contraction, with negative impacts on economic growth and local government fiscal health. The sagging property sector has become the biggest risk to economic growth. Slower economic growth will lead to further reductions in confidence in the housing market, and in turn to a further fall in prices. A smooth transformation of the property sector has become the biggest challenge to China's short-term economic growth, the abatement of local government debts, the health of the country's financial system, and the success of reforms promised by the CCP (Section 2.1).

BOX 14.1 FUTURE DIRECTIONS OF THE CHINESE HOUSING MARKET

Qiang Chai

To analyse and make a judgement on the future directions of the Chinese housing market, we need to take into account two groups of factors. The first group consists of generic factors affecting China's real estate industry, which include urbanisation, population growth, change of family structure and the demand generated by such change. The second group comprises the peculiar factors in China's institutional arrangements, which include the central government's adjustment and control policies, and the local governments' fiscal dependency on the real estate market. To explore the potential of the Chinese real estate industry we need to examine the problems in China's housing stock.

An analysis of the generic factors indicates huge potential for further development of China's real estate industry. First, per capita housing space is still insufficient in China, which was partly due to the very low level and low quality of housing construction between 1950 to 1978. In 2012, per capita housing space in Chinese towns and cities was 32.9 m^2, which is still far from the standards of developed countries, which ranges from 40–50 m^2. There is room for further growth of per capita housing space using European or American standards, which denotes potential for house building. Second, China is still in the process of rapid urbanisation. In 2013, China's urbanisation level was 53.7 per cent, which included all households. If only registered households were accounted for, China's urbanisation level was 35 per cent, much lower than the average of 80 per cent in the developed countries. New urbanisation focusing on towns and small cities is still underway. Migration continues from rural areas to small cities, to medium and large cities, and to super-large cities such as Beijing and Shanghai, providing immense opportunities for the real estate industry. Third, the economic life of China's buildings is short, necessitating building renewal in a shorter period than other countries. If the average economic life of buildings is 50 years, then every year, 2 per cent of the buildings need to be redeveloped. At present, the average building life is only 25 years in China. Fourth, China's population will continue to rise for some time, peaking above 1.4 billion. Fifth, the average size of Chinese families has been falling, with the number of one-person and two-person households rising rapidly.

Analysis of the peculiar factors in China shows continued policy support is available for the real estate sector. The real estate industry is one of the key industries in the country, with economic growth and local government revenues depending on the real estate industry in the foreseeable future. As a result, the central government will support the real estate industry for economic growth, while local governments buttress the industry for fiscal revenues. The rapid withdrawal of purchase restrictions indicates the support by local governments for the housing industry.

After 30 years of development, the real estate industry in China has made significant progress in the quantity of output. Yet there is a problem in the quality of output. Housing built before 2000 lacks amenities such as sports and other service facilities, and green spaces. Other facilities such as waste-water treatment and solid waste disposal are insufficient. New housing being developed includes good neighbourhood facilities, sufficient services and good transportation, and will attract demands to upgrade.

Based on the above analysis, China's real estate industry will continue to grow and the housing market will not collapse. However, such growth will differentiate according to the nature of cities and neighbourhoods. With high-quality education and medical services, tier 1 and tier 2 cities will continue to attract home buyers, resulting in continued housing price inflation. In contrast, tier 3 and tier 4 cities, if with a net outward migration, will experience falling housing prices. The other differentiation occurs among neighbourhoods in the same city. Housing in neighbourhoods with good amenities will demand higher prices, while their counterparts with a lack of amenities will experience slower or depressed housing prices.

BOX 14.2 THE FUTURE DEVELOPMENT PATHWAYS OF CHINA'S REAL ESTATE INDUSTRY

Hongming Zhang

The future of China's real estate industry is mainly determined by the country's economic growth. A discussion of the future of the industry thus starts with the economy: Economic growth in China at a medium speed in the future was heralded from 2011 when GDP growth strayed from the long period of double-digit growth. In 2012 and 2013, GDP growth fell abruptly to 7.7 per cent. In 2014, China's GDP is struggling to maintain a 7.5 per cent growth. It is estimated that GDP growth will remain between 6 per cent and 7 per cent for a long period, a view accepted by the authorities. In the past, the growth of the housing market followed the path of GDP growth. Economic slowdown inevitably affects the growth of housing market activities. However, a reduced growth rate still brings growth opportunities, which are produced by many factors including urbanisation, further industrialisation, regional differences and continued reforms.

With a medium speed of growth, what are the pathways that China's real estate industry can take in the future to remain a key industry? Three likely pathways are discussed as follows.

First, China's real estate industry should follow the principle of thrift and ecologically sound development and adopt development models that economise on land and energy. The government has been promoting such development models to the industry for years, but this fell on deaf ears during the development boom. When the market cools down, the real estate industry should start exploring ways to meet market demand and maintain investment returns. This pathway is applicable to all real estate firms and policy guidance is necessary to achieve the desirable outcome.

Second, Chinese real estate firms should maintain their traditional business model to achieve an economy of scale and fast circulation of funds. Such a model allows most firms to grow rapidly, but is based on a booming market. At present the boom period is at its end for many cities, with two exceptions. The first exception is that some cities in Central and Western China could have accelerating housing markets. The second exception is that some large cities still have large tracts of housing in poor condition suitable for redevelopment. Thus this model can still be applied for five to eight years.

The third pathway is diversification, which is suitable for some large real estate firms. As growth slows down, returns to capital for real estate firms fell from 10 per cent around 2010 to 4 per cent to 5 per cent, which is the average return on capital for the country as a whole. New businesses, such as internet, finance, environmental protection and overseas investment, should be found if firms are to maintain high returns on capital.

14.2 From macro control to a long-term control mechanism on the housing market

The key problem in China's property market is that the market mechanism plays a limited role in deciding resource allocation; a situation shaped by China's institutional arrangements on economic growth and governance. Local governments are required to promote economic growth. However, after the tax-sharing deal in 1994 (Section 2.2), local governments have to undertake more and more tasks with limited fiscal capacity. The lack of sustained funding for affordable social housing (ASH) is an example of a major task undertaken by local governments without a match between responsibility and funding (Chapter 13). To promote economic growth, local governments widely adopt a property-led growth model to generate revenues by formulating plans for phased urban development, which are driven by land sales and subsequent market-oriented property development. Thus the real estate market is effectively under the direction of the local economic development plans.

To maximise revenues from the real estate market, local governments at the collective choice level exploited the lack of binding rules made by the constitutional level. On one hand, they did not stick to the three-tier system of the housing market stipulated by the 1998 housing policy to promote affordable housing represented by Economic and Comfortable Housing (ECH) and social housing represented by Low Rent Housing (LRH). Rather, the lack of size standards was exploited to allow the housing supply to increasingly cater for the high-end market and investment demand, which generated higher land sale revenues and taxes. On the other hand, local governments pushed through large-scale commercial property development because there was no central government policy on commercial property. As a result, there was a campaign of building international metropolises and designating CBDs by over a hundred cities. As a result, prosperity of the property market is important to local governments, especially as vast numbers of huge commitments are tied up with various infrastructure and other projects.

14.2.1 An evaluation of macro control by the central government

Macro control represents direct intervention by the central government in the housing market, in the form of reactions by the constitutional level to the consequences of local practices in the property-led development model. Rather than an *ad hoc* approach to control housing prices and adjusting housing demand and supply, macro control was in fact used as a regular approach from 2005 to 2013 after the housing market became the main source of housing provision from 1999. Macro control only targeted the housing market, where price hikes and worsening affordability generated public discontent. It did not touch the commercial property market where resource allocation is also distorted by local governments' plans and preferential treatment.

On the positive side, macro control on the housing market resulted in the formation of an ASH regime in 2007, which corrected the excessive marketisation in housing provision since 2000 and achieved the rapid expansion of ASH, lifting tens of millions of low- and lower-middle-income households out of housing poverty (Chapter 13). Moreover, it led to the establishment of a set of institutional arrangements that improved the efficiency of the housing market. These institutional arrangements include: housing market regulation as a local government responsibility; a land supply system that becomes responsive to market demand; housing size standards that increase the supply of low-end housing; lending prudence that reduces the risk exposure of banks to property development and home mortgages; and rules of conduct on market behaviour for developers, intermediaries and professionals.

On the negative side, macro control reduced the role of markets in the allocation of resources and introduced inefficiencies in market operation and outcomes. First, macro controls represented a *de facto* retreat of marketisation, and have disrupted resource allocation by the market. They sent out distorted signals to the market, which shaped the market expectation of further price hikes. In the past ten years, macro measures such as mortgage deposit requirements, down payments for land purchase, Business Tax exempt periods, and interest rates on mortgages all fluctuated with housing prices; going up when prices were up and going down when prices were down. Investors perceived this pattern of adjustment as attempts by the government to maintain housing prices so that sufficient land sale revenues and property-related taxes were generated to safeguard local economic growth and infrastructure development. When rigid demand for housing existed due to population growth and urbanisation, investors invested in property in the expectation of capital appreciation. Under macro control, investors prioritised policy examination over economic research, rendering market signals less important than they should be.

Second, macro control failed to dampen housing demand and to control prices in the longer term. Sources of demand are mainly from contextual factors, i.e. rising incomes, availability of credit, ease of domestic migration, and entry of foreign investors. Availability of credit came from the monetary policy that expanded the money supply to encourage economic growth. The need to maintain a managed exchange control further increased the domestic money supply. As a result, prices were controlled only for a short period for each wave of macro control, or not at all. Once controlled, prices then jumped up high when control became weak or non-existent. Thus a pattern emerged: the more control, the higher the prices.

Third, widespread intervention in the market led to inefficiencies in the market operation and outcomes. For one thing, the tightening of the land supply in 2004 caused a sense of land shortage, leading to an expectation of further housing price hikes. For another, controls on bank financing of property development and investment forced developers to resort to more expensive sources of funding, such as lending from trusts and funds, and extensive use of presale receipts. High financing costs and a lack of long-term financing led to strata title sales of commercial properties, such as offices and shops, with negative consequences on use, management and maintenance.

Fourth, macro controls raised transaction costs in housing purchase and impinged on the property rights of developers, investors and homeowners. The imposition of Business Tax raised the monetary costs in transacting housing for both buyers and sellers. Long queues formed outside registration offices before tax hikes. Families and investors had to match the time to the frequency changes stipulated by the macro control measures for buy and sale decisions.

Fifth, perverse innovation and market behaviour developed under macro controls. For example, twin flats were developed to avoid the 70/90 rule (Section 7.3), leading to sales of single flats with title deeds for two component flats. Both developers and buyers of twin flats suffered losses when purchase restrictions were imposed in 2010 and 2011 in 46 cities in China. Other examples include advice supplied by estate agents to sign yin and yang contracts and carry out false divorces (Section 8.5). Furthermore, buyers and sellers entered into informal arrangements without the protection of contract, in order to avoid the Business Tax on housing transaction.

14.2.2 Adjustment and control according to local conditions

After the 2013 Measures were enacted in February (Table 7.2), the new central government led by Premier Li Keqiang, taking office in March that year, stopped issuing new macro control measures, even though housing prices rose sharply in the year. Instead, the task of macro control was delegated to the collective choice level, ending the era of central government-led macro control. Local governments were delegated power to conduct adjustment and control of housing prices according to local conditions. Meanwhile, the 2013 Measures delegated power to local branches of the central bank to adjust lending according to local market conditions.

Faced with year-on-year housing price inflation at 19.7 per cent in October 2013, as indicated by the 70 Cities Housing Price Index, Shenzhen acted to impose local control measures on top of the 2013 Measures to rein in housing price inflation. On 31 October 2013, the Shenzhen Branch of the People's Bank of China adjusted the down payment on mortgages for second homes from 60 per cent to 70 per cent, and continued to refuse to issue mortgage loans to families buying third homes. This is in addition to upward adjustments at 20 per cent and 30 per cent of mortgage loan interest rates from the set interest rate for first-time buyers and second home buyers, respectively. However, the down payment requirement remained at 30 per cent for mortgage loans for first-time buyers. On the same day, the Shenzhen Municipal Government announced a series of measures to control housing prices, including an additional supply of 40,000 ASH units within the 2013 plan.

With annual housing price inflation at 16 per cent in October 2013, Beijing followed Shenzhen to introduce, over a two-year period, 70,000 units of Owner Occupation Housing, which is a form of Limited Price Housing with a 30 per cent discount on market prices. Shanghai, where annual housing price inflation was 17 per cent in October of the same year, raised the down payment for second home mortgages from 60 per cent to 70 per cent, and required non-registered households to contribute to social security insurance for 2 years rather than 1 year to qualify for home purchase. With annual housing price inflation at 20 per cent in the same month, Guangzhou went further, requiring three years of social security contribution, and 70 per cent down payment for second home purchase. By early December, 17 cities had issued local measures to cope with housing price inflation. All the local measures to control housing prices followed the same approach of macro control carried out by the central government, achieving a slowdown in the housing market (Table 14.2).

In March 2014, more power was delegated to local governments to conduct adjustment and controls according to local conditions. When housing prices started the across-the-board decline in the second quarter of 2014, local governments reacted quickly to table local policies to prop up the housing market. By 25 June, 18 cities had relaxed parts of their home purchase restrictions. Holhot, the capital city of Inner Mongolia, became the first city to formally withdraw home purchase restrictions on 26 June 2014. With tacit approval from the central government, 37 out of the 46 cities that had imposed purchase restrictions withdrew such restrictions by the end of August 2014. In addition to withdrawing purchase restrictions, local governments tried many other measures to boost housing sales to support the property market. Such measures include withdrawal of strict down payment requirements on mortgage loans, adjustment of mortgage interest rates, reduction of Deed Tax, and the provision of bonuses or subsidies to home purchasers.

However, relaxation and withdrawal of purchase restrictions have had a limited impact on stopping the contraction of housing sales and the decline of housing prices. In July, housing transactions in Holhot, where some large housing projects were found with no occupiers, increased 55 per cent month-on-month in June, and 56 per cent more than in March. However, many cities that withdrew purchase restrictions experienced only a short period of increase in transactions, while others had no big changes in the volume of transactions. This indicates that without a national stimulus package, local support to the housing market is futile. Local governments could not adjust and control many of the contextual factors of the housing market: general economic growth, general availability of credit and cost of credit (interest rates).

This round of housing market contraction is determined by the monetary policy set by the central government, which seems determined to restructure the economy and phase out market intervention at the expense of an economic slowdown. The only favourable policy tabled by the central government so far was to encourage banks to lend to home buyers who are owner-occupiers. Thus, central government will allow the market to determine the movement of housing prices.

14.2.3 Institutional change: long-term adjustment and control mechanism

With macro control unable to solve the affordability problem, the constitutional level acted to rebalance market provision and state provision of housing in 2007, which resulted in the establishment and rapid expansion of the ASH regime. Reform of housing market administration has also been under consideration. The new approach, suggested by the central government, is to establish a long-term adjustment and control mechanism (LTACM), which first appeared in central government policy documents in 2013.

The rationale of establishing an LTACM is not to curb housing price inflation when the housing market enters the upturn, such as what the central government did in February 2013 (2013 Measures) and what some local government did at the end of 2013 (see Section 14.2.2). Nor is it to prop up housing prices when the market enters a downturn, such as what the local governments did from the second quarter of 2014 (see Section 14.2.2). An LTACM is to ensure the credibility of the housing policy and offer a stable expectation to investors to stabilise the impetus to invest, so that housing prices can return to a normal long-term average level. To put it simply, an LTACM is simply a set of housing policies that allows all urban residents to understand what the country's housing policy is. An LTACM is a set of institutional arrangements which give the property market the main role in resource allocation and limit direct intervention by the government. It contains several elements, namely: rule of law; unified and comprehensive property registration; levy of a housing holding tax; a market-responsive land supply system; a market-based finance and credit system; a bigger role for ASH; and indirect market regulation. Supporting the institutional arrangements includes the reform of the existing GDP-based performance assessment for officials that thus far has pressurised local governments to boost economic growth; and reform of the tax sharing system that thus far has imposed a mismatch between responsibilities and fiscal capacity on local governments.

At the time of writing, no official explanation has been provided on LTACM, which is in the formulation process. First, Chen (2013) explains that an LTACM:

- allows the market to allocate resources, and supply and demand to self-adjust;
- ensures different levels of housing demand are met so that everyone has a home to live in;
- adjusts the interests of multiple sectors of society, long-term and short-term interests, the interests of the rich and the poor, the balance between government and market, and tensions between central and local governments; and
- promotes the application of the latest innovations and technologies, efficiency in resource uses, and balancing societal needs with those of the natural environment.

He argues that an LTACM is not a mechanism to reinforce government's power, to adjust prices at will, to build a housing system on egalitarian basis, or to protect monopoly and vested interests.

There are a number of issues that the LTACM should be designed to address. First, an LTACM will scrap many permits issued by a market administration regime; for example, a permit on the reasonable prices for new housing. Prices go up and down, and the LTACM should not attempt to adjust those short-term fluctuations in the market (Liu, 2014).

Second, an LTACM should provide channels to housing for everyone and every group in the society. These channels should be accessible by personal decisions; provide choices at each of the channels; be affordable; and be sustainable. This means that an LTACM is a housing system that offers security to lower-income households; support to middle-income households; and market choices to high-income households. The current culture of home purchase is peculiar in China, with a firm belief that a late purchase is a loss and no purchase is a big loss. With such a belief, people who can afford it but do not need more housing, rush to buy housing as an investment, often leaving the housing idle. Rising prices, however, justify this behaviour. As a result, housing demand is amplified, causing shortages of housing (ibid.).

Third, land supply, a part of an LTACM, should be from multiple sources rather than the current monopoly by local governments. One option is to reuse developed land, and other is to explore the sale of Land Use Rights of collectively owned land to the market. Recycling used land offers great potential. For example, industrial land in China occupies an area of close to 50,000 km^2, which was the result of low industrial land price policy. In contrast, industrial land in Japan was 1,600 km^2. The ratio of industrial land to residential land in China was 1 to 1.4, much higher than that of 1 to 6 in Japan (Yu, 2013). Thus the recycling of used land offers great potential to increase the land supply.

Fourth, financial system reforms are to be conducted to contribute to the establishment of an LTACM and the management of financial risks derived from the property market. An important change to be made is to deregulate interest rates, i.e. allowing the market to set interest rates according to supply and demand at the monetary market. At present, a base mortgage interest rate is set by the People's Bank of China, and commercial banks are given discretion to go 20 per cent to 30 per cent above or below the base rate. As a result, the mortgage interest rate was not sensitive to market demand. For example, the Consumer Price Index in 2011 was 5.4 per cent, but the mortgage lending rate was 6.4 per cent until 9 February 2011, then 6.6 per cent until 6 April, 6.8 per cent until 7 July, and 7.05 per cent for the rest of the year. Developers, on the other hand, had to borrow from trusts, funds and other non-bank institutions at above 15 per cent to 20 per cent for many of their projects. Should the commercial banks have required higher interest rates on mortgage loans, housing prices would have fallen, and the macro control would have been unnecessary.

Fifth, the levy of a property holding tax should not increase the tax burden of the property sector, which is already heavily taxed. The aims of the taxation reform should serve the LTACM, rather than intensify macro control. Calls for the property holding tax to be established to aid short-term control of housing prices are unnecessary because the existing taxes, if strictly implemented, are sufficient. Rather, taxation reform should focus on adjustment of taxation without increasing the overall tax burden (Liu, 2014).

Sixth, an LTACM should alleviate local governments' fiscal dependency. Local governments rely on local financing platforms to raise money to pay for infrastructure projects and ASH. Such projects have low returns, but local governments pay higher interest for funds raised from local financing platforms, with some loans demanding interest rates over 10 per cent. Land sale revenues and property-related taxes serve as collateral for such loans and as a source of money to repay such loans. New financing methods, such as loans from policy-based financial institutions, should be used to support infrastructure and ASH, and to relieve local governments from fiscal dependency on the property sector (ibid.).

As part of the attempt to build an LTACM, the draft *Interim Real Property Registration Ordinance* was released for public consultation on 15 August 2014. The change to a unified registration will reduce the transaction costs and the confusion in property registration, by offering one-stop registration rather than the current two stops for urban real estate (one for land registration at the Bureau of Land and Resources and the other for building registration at the Building Administration). The ordinance also includes registration of rural real estate, whose registration is currently rudimentary, and the registration of cultivated land, forests, grassland, etc. High hopes have been pinned on unified registration to reduce multiple ownership of housing, in particular by investors and corruption officials. This would occur by making information available from the register, to the authorities and qualified members of the public, on the housing wealth of officials; and by increasing the holding costs for empty housing through the more comprehensive levy of property holding tax following unified registration (Zhu, 2014).

The reform to allow owners of collectively owned land to enter the market, that was set by the Third Plenary Session of the Eighteenth Congress of the CCP, has not been started. There are many issues to be resolved before the reform can be carried out. One of them is the lack of property rights certification in the countryside. Li (2013) emphasises the importance of property rights certification in the countryside through his fieldwork in Hangzhou, Jiaxing and Huzhou in Zhejiang Province. The three cities carried out property rights certification for several years. Farmers were very satisfied with certification because they were assured that the land leased to them belong to them for the duration of the contract. They can re-let their land to others, enter their land into a company for shares, or rebuild their home to let part of it for rental income. With certification of property rights, all the above uses were legal and farmers substantially increased their investment of time, money, and commitment in the land, increasing agricultural productivity. In Jiaxing, the per capita incomes ratio of urban residents to rural residents was 3.1 to 1 before the certification of property rights; after the certification, the ratio became 1.9 to 1. Thus a series of such reforms should be conducted before the reform for collectively owned land to enter market can be carried out.

BOX 14.3 THE FUTURE TREND OF ADJUSTMENT AND CONTROL ON CHINA'S HOUSING MARKET

Hongming Zhang

China's housing market is a topical issue around the world. The adjustment and control by the Chinese government on the housing market are closely linked to this issue.

The housing market originated in 1986 and has been subject to adjustment and control by the government since then. In 2002, adjustment and control became a key national policy for the first time. This dominated housing policies in the next ten years. The ten years, dubbed the 'decade of adjustment and control', have been controversial in their effectiveness and achievement. My opinion is that the adjustment and control, based on good intentions and sound objectives, are necessary. However, its low effectiveness and unsatisfactory achievement are rooted in a myriad of factors typical for China in its fast transition from a command to a market economy and rapid transformations in many aspects.

What is the future trend of adjustment and control for China's housing market? An analysis of official documents on the housing market offers insights into this issue. First, the reforms package put forward by the Third Plenary Session of the Eighteenth Congress of the CCP proposed an organic combination of adjustment by the market mechanism and administrative control. The package does not mention adjustment and control directly, but indicates that the market should play a decisive role in resource allocation and the government's role should be better performed. This statement indicates that regular adjustment should be made by the market and government intervention only carried out when market failure is expected or has occurred.

Second, the *Report on the Work of the Government*, released in March 2013 at the annual conference of Parliament, provides details on how to coordinate the market mechanism and administrative control on the housing market. The Report stated that adjustment and control on different cities should be differentiated, with increased supply of small to medium-sized commodity housing and restriction on housing investment demand. This indicates administrative intervention is still carried out because the market fails to provide sufficient small to medium-sized commodity housing and investment demand is too strong. However, such an intervention is subject to local conditions.

Third, administration in 2014 has indicated a new approach to adjustment and control on the housing market. From April 2014, the housing market started its fall in prices and transaction volumes. However, the central government did not table policy or directives, allowing local governments to make adjustments, such as relaxation of credit supply, ending purchase restrictions and the provision of incentives. While many local governments tried to rescue the housing market, major cities such as Beijing and Shanghai did not intervene in the housing market. This divergence of actions by local governments illustrates the future trend of adjustment and control, and is in line with the requirements made in the Report.

To achieve effective adjustment and control, reforms need to be conducted in housing supply, land supply, fiscal and taxation system, financing and capital markets. It has a long way to go.

14.3 Fiscal system reforms and fiscal dependency on the property market

Fiscal dependency on the property sector has been criticised widely in China as an important reason for housing price inflation, because local governments boost land prices. It was also regarded as a major reason for the occurrence of forced demolition and relocation. A scrutiny of local government expenditures supported by land sale revenues provides a clearer picture of fiscal dependency on the property market.

Table 14.3 provides a breakdown of local government spending supported by land sale receipts in 2011. Local governments spent 48.5 per cent of the receipts

TABLE 14.3 Spending supported by land sale revenues

	CN¥ billion	*Share of expenditures (%)*
Total land sale receipts	3,150.0	
Expenditures funded by land sale revenues		
Compensation for land acquisition and building demolition	1,435.9	46.2
Land preparation and servicing	532.5	17.1
Urban infrastructure construction	556.5	17.9
Rural infrastructure construction	76.0	2.4
Subsidies to farmers whose land is acquisitioned	69.0	2.2
Expenses on land sales operation	21.7	0.7
Expenses on building affordable and social housing	52.0	1.7
Other expenditures	361.6	11.6
Total spending funded by land sale revenues	3,105.2	100.0

Source: China Finance (2012).

to compensate land and building owners and provide subsidies to those with difficulties. They spent 17.1 per cent on land preparation and services before land sales, and 0.7 per cent on organising land sales. Thus the total costs for land sales amounted to 66.3 per cent, or two-thirds of total land sale receipts. Local governments spent 20.4 per cent of the receipts to invest in urban and rural infrastructure, but only 1.7 per cent, which is much lower than the required 10 per cent, on ASH. The remaining 11.6 per cent includes a myriad of items, but the details are not available. The actual use of this 11.6 per cent of land sale receipts is likely to be disclosed after the national audit on land sale receipts, starting in August in 2014 and lasting for two months, is completed. This audit is the third of its kind, with the first two covering social security and local government debts, with important findings and reforms to follow (Economic Information, 2014).

It is obvious that land sale receipts provide local governments with a major income stream to finance infrastructure, ASH and other expenditures, and collaterals for local governments to borrow money from banks, or other sources. To relieve local governments of this fiscal dependency, major reforms are needed to match local governments' responsibilities and funding. The blueprint of reforms for the country's finance and taxation system, a reform task set by the Third Plenary Session of the Eighteenth Congress of the CCP, was approved by the CCP on 30 June 2014. The reforms proposed by the blueprint include improvement of budgeting management systems, restructuring of taxation that includes the creation of a property holding tax, and the establishment of a system that matches responsibilities and expenditures (Lou, 2014). According to the plan, major reforms will be completed in 2016 and a modern fiscal system will be established by 2020.

BOX 14.4 REPOSITIONING THE PROPERTY SECTOR IN CHINA

Stanley Chin

The property industry in the past two decades had been characterised by a period of explosive growth across all sectors, particularly the residential sector. Due to a combination of slower GDP growth, restrictive government policies (to cool down real estate demand and prices), tighter credit conditions, the historically successful residential-centric model of high velocity and volume, pursued by the majority of industry participants, is increasingly under strain. This is coupled with the advance of e-commerce which is giving physical commercial space 'a run for its money'.

In 2014, a commonly asked question is: Is the golden age of the Chinese property sector over? A natural extension of that question is: what will be the future of the property sector?

Due to continuing urbanisation, land will remain an increasingly scarce resource particularly in tier 1 cities like Shanghai and Beijing, which are already categorised as two of the world's 'megacities'. The real estate landscape will also be shaped by nontraditional industry drivers such as e-commerce, social media, technological advancement and lifestyle changes. End users' expectations, recognition and acceptance of such drivers are fundamental to the success or failure of any real estate development.

It is anticipated that there will be an inevitable shift from 'pure play' developments to mixed-use developments particularly in urban locations of tier 1 and tier 2 cities as land becomes increasingly scarce and expensive.

Mixed-use development entails a multi-disciplinary model, which is radically different from that of a residential development model of 'build and sell'. In addition to a design emphasis with strong aesthetics and functional features, the developer is also required to have strong development management, leasing, sales, marketing and asset management capabilities. Given the multitude of competing schemes in China, a successful mixed-use development needs to possess the necessary content and 'software' to provide an attractive hub for all forms of functions (life, work, play, etc.).

Despite the rapid advance of e-commerce in all aspects of the socio-economic landscape, the basic human needs for social interaction in physical spaces have not changed since the beginning of civilisation. However, developers have to be cognisant of the need to develop physical spaces with coherent forms, flow and content to cater for and attract increasingly sophisticated and discerning end-users.

The impending shift towards a more holistic and higher value creation development model as described above, inevitably has its challenges, not least that it is relatively new to most local developers. The fundamental criteria of 'location, location, location', human capital, product quality, financial capital and market positioning will become increasingly critical factors that must be skilfully orchestrated for success.

14.4 Flight to commercial real estate

The downturn in the housing market has led many developers to adjust their development activities in housing and look for new business opportunities. Commercial real estate in many tier 2 and tier 3 cities still presents opportunities, and attracts a number of traditional housing builders to enter development projects. The scarcity of suitable sites in city centres results in many of the retail projects being located in the suburbs. Many of the out-of-centre retail projects are large outlets that offer a range of brands with large discounts from recommended retail prices (RRP). The first outlet was opened in Beijing in 2002, and by 2013 there were over 400 outlets in the country. Hundreds more are being planned and developed, in particular in many tier 3 cities. Logistics property, traditionally not covered by housing developers, has also received attention (see Section 10.5).

The flight to commercial property is partly a response from the operational level, i.e. firms, facing the contraction of the housing market and partly a strategy of the collective choice level, i.e. the local governments need to sustain receipts from land sales and taxes from the property market, and to maintain local economic growth. The economics of oversupply, discussed in Section 5.7, apply in many cities. This will result in upgrading of the stock of offices and retail properties in a number of tier 2 and tier 3 cities, which enhances the competitiveness of these cities. As in the past, there is a lack of attention from the constitutional level on commercial property development. Oversupply and business risks are significant in many cities due to the number and scale of commercial development projects. Greater perception of such risks has resulted in greater inputs in planning, design and operational arrangements. For example, more leisure and cultural facilities are being planned in new shopping malls and HOPSCAs to strengthen the attractiveness of new projects. However, the success of new projects may be at the expense of older projects.

14.5 Towards sustainability

Continued growth in living standards in China has resulted in rising per capita energy consumption from 1983 to 1988. In the 1990s, the improvement in energy efficiency in electricity generation and industrial processes led to a fall in per capita energy consumption until 1999. Since 2003, energy consumption grew rapidly due to increased economic activities, expansion of private car ownership, widespread use of room air conditioners and increase in per capita living space (Figure 14.3). The increase in annual completion of buildings (Figure 14.4), in particular higher-quality buildings excluding farmers' buildings, poses a huge challenge to energy supply in China and has implications also for global warming. Room air-conditioning is becoming a standard building facility in southern China with the long summers (Figure 14.5).

Building energy consumption as a percentage of total energy consumption has been high in China. Dong (2013) indicates that buildings account for 39.88 per cent of total energy consumption in China, with production of building materials

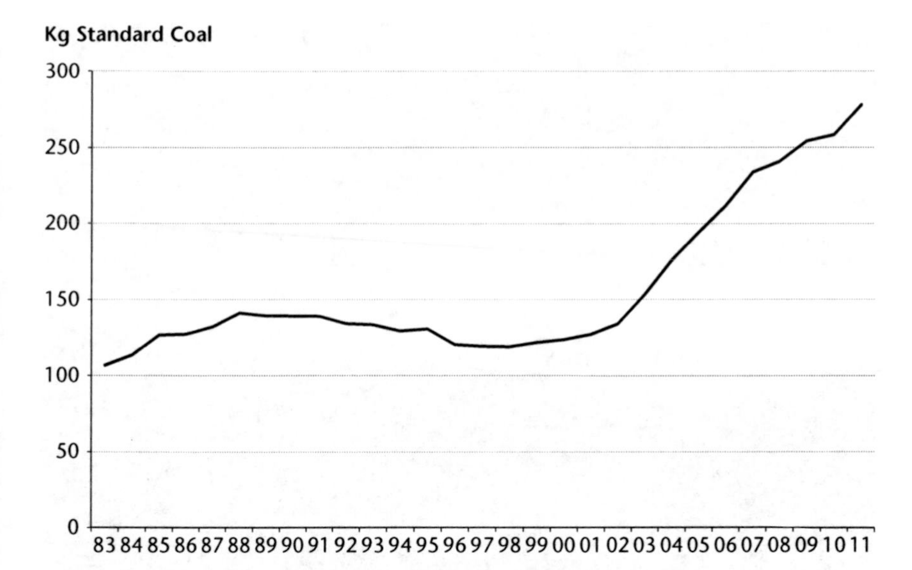

FIGURE 14.3 Rising per capita energy consumption in China, 1983–2011
Source: NBSC.

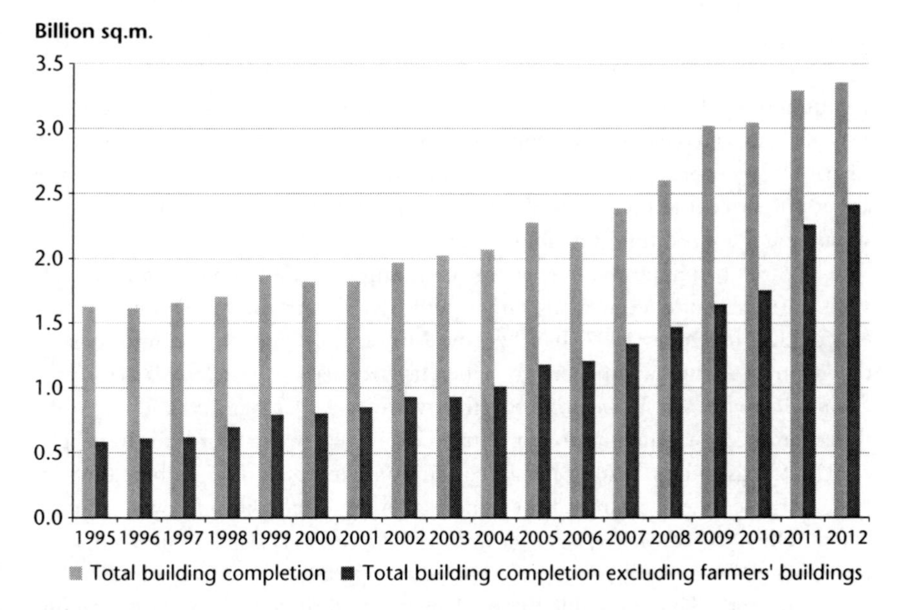

FIGURE 14.4 Total building completions in terms of gross floor area in China
Source: NBSC.

FIGURE 14.5 Zhuhai in Guangdong Province – indoor air conditioning is required for the long summer

Source: NBSC.

accounting for 15 per cent; energy consumption in construction accounting for 1.49 per cent; and building operations and maintenance accounting for 23.39 per cent. In particular, energy consumption in building operations and maintenance was about 10 per cent at the end of the 1970s when there was no air conditioning in summer and less room heating in winter.

The earliest attempts to lower energy consumption by buildings started in 1986, with a 30 per cent energy saving target put forward.[1] An Energy Saving Law was enacted in November 1997. In 1999, the Ministry of Construction promulgated the country's first building energy efficiency regulation, *Civil Buildings Energy Saving Administration Measures*, to be effective from 1 October 2000. This regulation required new buildings to conform to energy-saving standards for heating in civil buildings and for heating and air conditioning in hotels. In October 2000, the first energy-efficiency standard for buildings using air conditioning was promulgated. The *Design Standard for Energy Efficiency of Residential Buildings in Hot Summer and Warm Winter Zones*, promulgated in 2003, upgraded the energy saving target to 50 per cent. The Renewable Energy Law, promulgated in 2005, prompted further energy efficiency regulations on buildings. Since 2005, the Ministry of Construction has tabled a set of policies and regulations to promote energy efficiency. The first *Design Standards for Energy Efficient Public Buildings* and the *Notice on Strict Compliance on Energy Efficient Design Standards for New Residential Buildings* were both enacted in April 2005. Tabled in 2006, the *Evaluation Standards for Green Building* (GB/T 50378–2006) provides the standards for green buildings, which cover land saving and outdoor environment; energy saving and usage; water saving and usage; building materials saving and usage; indoor environmental quality; and operation management (for residential buildings) and full life-cycle performance (for public buildings). By 2010, the *Design Standard for Energy Efficiency of Residential Buildings in Severe Cold and Cold Zones* was released, requiring a 65 per cent energy saving target.

The attempt to promote energy-efficient buildings gradually paid off. In 2004, only 50 per cent of new buildings adopted energy-saving design standards and 23 per cent adopted energy-saving standards in construction. In 2011, almost all new buildings adopted the 50 per cent energy-saving design standards and 95.5 per cent adopted the 50 per cent energy-saving standards in construction (Li, 2012). However, an upgrade to 65 per cent energy-saving standards in design and construction for hot summer and warm winter zones was slow, with only Shanghai and Chongqing enacting local standards. The reasons for this slow progress are differentiated climate conditions, efficiency of energy-saving techniques, difference in measurements and local economic conditions.

The campaign to promote energy-saving buildings has made significant progress. MOHURD (2014) indicates that by 2013 the country had built 8.8 billion m^2 energy-saving buildings, equivalent to approximately 30 per cent of the total building stock and saving 80 million tons of standard coal every year. In 2013 alone, 1.44 billion m^2 energy saving buildings were constructed or converted. In the urban areas, 2.7 billion m^2 solar panels had been installed.

The promulgation of *Implementation Opinions on Speeding up the Development of Green Buildings* in 2012 represented a new round of efforts by the constitutional level to set clear targets for the progress of green buildings (MOHURD, 2012). The document requires: the proportion of green buildings to reach 30 per cent of all new buildings by 2020; all public buildings invested by the government to be green buildings; and all new ASH in provincial-level municipalities, cities designated in the national plan, and provincial capital cities to be green buildings by 2014. It also requires the total floor area of green buildings to reach 1 billion m^2 by 2015. Furthermore, the document promotes the formation of green ecological urban districts and provides financial incentives for many initiatives.

A 'go green' campaign was started in 2013. On 1 January 2013, the State Council distributed the *Green Building Action Plan*, which was jointly enacted by the Development and Reform Commission and MOHURD. The Plan puts forward an ambitious plan to build 100 new urban districts, including new residential districts, economic and technology development zones, high-tech industry development zones and ecological demonstration zones. In December of the same year, MOHURD and Ministry of Industry and Information Technology jointly issued the *Circular on Promoting Green Agricultural Buildings*, extending the green campaign to the large amount of building completed by the farmers in the country (Figure 14.4). By the end of 2013, China had approved 1,446 green building demonstration projects with a total gross floor area of 162.7 million m^2. In particular, the number of demonstration projects and gross floor area approved in 2013 were 94.9 per cent of the total number of projects, and 114.6 per cent of the total gross floor area, approved from 2008 to 2012, respectively (Li, 2014). The construction industry and the property sector has been mobilised to go green.

14.6 Conclusion

In 2014, China's property market reached a turning point, with falling housing prices, worsening oversupply in housing and commercial properties, and a commitment by the policy-makers not to directly intervene in the housing market to boost demand. Falling prices and demand have led to early indications of a contraction in land sales, which is likely to result in falling land sale receipts and property-related tax revenues received by local governments. This in turn poses a challenge to the property-led growth model and raises concerns about whether China can maintain a medium fast growth rate, which has slowed down to around 7.5 per cent in the past two years from double digits in the 2000s.

A more fundamental issue is whether macro control, as practised from 2005 to 2013, should be replaced by a new set of institutional arrangements that allow the market to make decisions on resource allocation, rather than follow decisions made by local governments. In particular, macro control interferes with decision-making based on market factors, but promotes decision-making based on government policies. Macro control, in its current form, represents a retreat in marketisation and remnants of the command economy. It creates market expectations that the government will not let housing prices fall because economic growth depends on rising housing prices. Such an expectation generated excessive demand for housing for investment, leading to a high proportion of multiple home ownership and empty housing. Excessive multiple home ownership reduces effective housing supply and aggravates housing price inflation.

The solution to the vicious circle of excessive housing price inflation and intensified intervention by the government is to build a long-term adjustment and control mechanism (LTACM). This mechanism does not involve fighting short-term fluctuation of housing prices, but aims to build a long-term institutional arrangement to guide behaviour in the housing market. The LTACM focuses on allowing the use of market mechanisms for resource allocation and the establishment of institutional arrangements to limit direct intervention by the government. It includes rule of law, unified property registration, a real estate tax on housing, a market-responsive land supply system, a market-based finance and credit system, a bigger role for affordable and social housing (ASH), and indirect market regulation.

Local governments have been using land sale receipts to finance infrastructure projects, ASH and other expenditures. With the introduction of alternative local financing platforms and reforms on matching responsibilities to expenditures, the reliance by local governments on land sale revenues and property-related taxes to fund their expenditures will be alleviated. Once their responsibilities are matched by appropriate funding, local governments will not need to push forward development projects which are not needed at current level of economic development, and can stay away from intervention in the market to force out high levels of property development activities. This change in institutional arrangements will alleviate the oversupply of commercial properties, a consequence of local governments' ambitious urban development plans.

Over the last three decades, the construction and property industries built tens of billions m^2 of buildings with low energy efficiency. The campaign to build energy-saving buildings started in the late 1990s and eventually covered over half of new buildings excluding farmers' buildings. However, the standards for energy-saving buildings are low in terms of energy conservation and fighting global warming. The 2013 green buildings standards provide a new platform with higher standards for sustainability. Currently the government has mobilised the construction and property industry to go green. Once it is successful, the 'go green' campaign will greatly alleviate the environmental impact of the property sector.

Note

1 This target was put forward by Energy Saving Design Standards for Civil Buildings 1986.

15

CONCLUSION

In a period of about 60 years, the world's most populous country, China, has traversed three worlds of real estate: two worlds of poverty and a world of plenty. A dynamic real estate market, underpinned by a productive real estate industry that emerged in the 1980s, is both the cause and the outcome of this breathtaking transformation. This real estate market has made important contributions to and benefited from China's miraculous economic growth and urbanisation.

This concluding chapter evaluates the Chinese real estate market in terms of the developments achieved and problems remaining on efficiency and equity grounds. It first evaluates the achievements made from the operation of the real estate market since the reforms and opening-up campaign to raise living standards and promote economic growth. Then it examines the remaining problems, in particular the institutional arrangements, in the real estate market that retard efficiency and equity, impose economic risks, and negatively affect the environment. Based on current developments, the chapter predicts the possible future changes following the new reform initiatives, and outlines areas that reforms need to address to improve efficiency, equity and performance of the real estate market in China.

15.1 China's real estate market: the huge strides

Since December 1978, the ongoing reforms and opening-up campaign have transformed China, making the Chinese economy the second largest in the world, and substantially lifting the living standards of the Chinese people. The average annual growth rate of GDP for the past 35 years was 9.76 per cent, and that of per capita GDP was 8.67 per cent. The development of the land and real estate markets has been an important cause, as well as a consequence, of this economic miracle. China has made a number of huge strides to raise the efficiency of land use and property provision through the newly created land and real estate markets. These are described below.

First, China has developed a dynamic and functioning urban land market to allocate urban land resources. This has been achieved by the creation of a private property rights system that is compatible with a market economy, and the development of a land sale mechanism that is based on open market disposal. A system of land use rights (LURs) was established in 1990 on state-owned land, allowing land to be exchanged for more efficient and better uses. LURs are tradable and renewable, offering incentives to invest in land and buildings. From 1991 to 2013, the LURs of 32,664 km^2 of land were sold to land users. The open market land sale mechanism, fully established by 2004, has resulted in open market bidding on land for market-oriented uses and higher efficiency in land uses. Foreign investments played an indispensable role in initiating and in expanding the urban land market.

Second, from 2000 onwards, the urban land market has generated large amounts of funds for most Chinese cities to improve infrastructure to attract inward investments and to expand social welfare coverage. It allows local governments to use land to raise more capital to push forward local economic development. From 1993 to 2013, a total of CN¥19.842 trillion has been received by local governments as land sale receipts. State ownership of urban land and the land market are unique advantages of Chinese cities in global competition; a major reason why many Chinese cities can achieve decade-long double-digit growth and build the infrastructure to rise in the world urban hierarchy.

Third, a series of reforms on urban housing provision in the 1990s successfully ended the inefficient and inequitable welfare housing allocation system, paving the way for the commodity housing market to grow to become the main provider of urban housing. The final phase of the housing reforms, imposed in 1998, ushered in the largest privatisation of public housing in the world and promoted the housing industry as a pillar industry for economic growth.

Fourth, a dynamic and innovative urban commodity housing market was established in the 1990s and expanded in the 2000s to provide new lifestyles as well as more housing space for Chinese home buyers, rapidly upgrading housing conditions in Chinese cities. The standards, choices and comfort of new housing delivered by domestic commodity housing developers have caught up with the rest of the world. The housing market both follows and leads China's large-scale urban expansion, providing homes to enable greater mobility among the Chinese people. From 1986 to 2013, a total of 8.85 billion m^2, or approximately 100 million units, of new commodity housing were sold to home buyers. Foreign housing developers, in particular those from Hong Kong, played a crucial part in setting standards and market practices in the early stage of housing market development.

Fifth, the commercial property market, which includes offices, retail, and industrial properties, prospered in China from the 1980s. It has played important roles to transform Chinese cities in respect of their functions and mode of economic growth. The office market was initiated by foreign occupier demand. Hong Kong, and to a lesser extent, Singapore developers played key roles in the early stages of the office market. The success of office and retail developments in Beijing, Shanghai, Guangzhou and Shenzhen led dozens of cities to focus on building central business districts (CBDs), greatly enhancing these cities' real estate infrastructure. Mixed-use work, living and leisure complexes (HOPSCAs) are transforming Chinese cities. Although the industrial market in general has been dictated by local governments, the logistics market, neglected by the government until late 2000s, has become a new growth area in the commercial property market, with international developers taking a major role.

Sixth, a financing market for real estate development has formed, with banks, trusts, funds and other informal financing channels providing trillions of yuan to the development sector. Banks were the dominant sources of financing for property development in the 1990s, and housing purchase from the late 1990s. Real estate trusts and funds, taking advantage of government control on bank lending, became important sources of development funding in the mid-2000s and late 2000s. Many informal channels joined in to meet the strong demand for funds in the early 2010s. Fundraising in international stock markets to finance development projects has also grown significantly.

Seventh, a professional services market, including valuation, agency, consultancy, property management, and asset management, has developed to service the fast expanding real estate market. International property consultancy firms assume a crucial role in setting standards and market practices in China. With national certification schemes, valuation, agency and property management sectors have embraced professionalism in services provision.

Eighth, an affordable and social housing (ASH) regime has been established since 2007 after an earlier failure by the government to provide affordable housing and social housing to help the less well-off when provision of welfare housing was ended by the housing reform. The ASH regime comprises five vehicles, namely: Economic and Comfortable Housing, and Limited Price Housing for purchase; Low Rent Housing and Public Rental Housing for rent; and Redeveloped and Improved Housing for existing owners and renters to part with substandard and dilapidated housing. Significant progress had been made by 2010, and an ambitious programme to build 36 million units of ASH was executed from 2011. The coverage of ASH was extended to 12.5 per cent of urban registered households in 2013. From 2011 to 2014, the programme had led to approximately 30 million new starts of ASH. As a result, the ASH sector has addressed some of the housing market failures in China by improving the housing conditions of tens of millions of households priced out of the commodity housing market. Local innovation, such as shared ownership and private sector participation, continues to widen the coverage of ASH.

Ninth, foreign direct investment (FDI) in Chinese real estate, ranked second place by sector in FDI in China, has made important contributions to property development and management in China. Foreign investments in Chinese property, surging in the 2000s and early 2010s, have increased significantly despite unfavourable policy treatment. International financial institutions and private equity funds are playing an increasingly important role in financing Chinese property development. On the other hand, Chinese investments in foreign real estate by families, institutions and developers have started to expand from 2010. With overcapacity in the domestic market, Chinese developers have started operations in overseas markets.

Tenth, a market administration system has developed in China to administer the land market, the commodity housing market and the commercial property market. Despite having many imperfections, this system has accommodated the rapid expansion of the property market, stopping the trend of diminishing affordable housing provision from 2006, and preventing a major property market fluctuation in 2008 from harming economic growth. In particular, a multi-ministerial approach has developed to administer the property market to rein in housing price inflation and ensure provision of ASH.

Eleventh, macro controls imposed by the Chinese government have prevented major corrections in the housing market, maintaining stability and avoiding huge fluctuations in housing prices and housing development activities. In fact, all the downturns in the housing market were at least partly engineered by the government, with reserve policies and resources to intervene in the market to limit the range of fluctuations. This is the case for the market downturn started in some cities in late 2013 and currently unfolding at a national scale. As a result, the housing market has maintained rapid growth.

Twelfth, China has protected its food production capacity by adopting a unique system of land administration to protect arable land and started a campaign for sustainability in property development and use. The top-down system of land use planning and control has arrested the encroachment of property development on arable land, enabling the country to be self-sustaining in food production. Several attempts to lower energy consumption in construction and operation of buildings have reduced energy consumption by the construction industry and building occupation. The 'go green' campaign from 2013 is gathering momentum in the construction and property industries.

These huge strides have helped China, in particular, the Chinese cities, to develop quality accommodation in large numbers for production, living and leisure. A comparison with the state of property provision in 1978 indicates that the progress made in the last 35 years has thrust the country into a new world of property in terms of quantity, quality, variety and comfort. China has ended its long history of housing poverty and possesses plenty of modern business accommodation, with positive implications for living standards and further economic growth.

15.2 The problems

The reforms and the opening-up campaign were initiated to improve economic efficiency in the late 1970s when the efficiency of China's economy was very low. At the onset, there were no clear goals, thus reforms were piecemeal and exploratory. By 1992, establishment of a market economy, albeit a socialist one with Chinese characteristics, became the goal of further reforms. A general blueprint for the reforms was produced in 1993 and the development of the Chinese property market has been under both top-down and bottom-up reforms. Nevertheless, many problems arise due to incomplete reforms, lack of experience, and a path-dependent nature of institutional change. These problems are interrelated and are described below.

First, the market mechanism for the property market has not been fully developed, with many resources not fully allocated by the market. This is in line with the current situation in China that marketisation in some sectors is still incomplete. In the property market, local governments have control of land resources, and monopolise the sale of new LURs. Both local governments and state-owned developers have better access to the financial resources of state-owned banks. As a result, local governments have been able to affect land prices and plan property developments according to their economic and urban development plans by mobilising land and financial resources. The institutional change here exhibits a path-dependent nature, with remnants of the command economy still found in the property market.

Macro control in the housing market typifies the use of local governments' administrative power; a feature of the command economy in the operation of the market. Its frequent use has formed market expectations that prices would go up because governments' measures were for the short term and higher land prices were necessary for local governments to obtain land sale receipts to build infrastructure and meet other expenditures. Developers and investors thus study government policy more carefully rather than focusing on market fundamentals.

Second, institutional arrangements for the property market to function have not been well established. A main problem is that the central government at the constitutional level has not set sufficient rules to bind local governments at the collective choice level to implement national policies. Without binding rules to follow, local governments prioritise local and short-term interests over national and long-term interests. The failure to implement the 1998 housing policy to build a housing provision system focusing on affordable housing is an example of a lack of constitutional rules to implement the set policies and oversee local governments. The prolonged macro control from 2005 to 2007 is another example of local governments' half-hearted implementation of national housing policies. On the other hand, macro control in 2010 was imposed with effective constitutional rules on local governments, resulting in local innovation, for example home purchase restriction, to implement macro control policies.

Another problem is that local governments at the collective choice level set rules for the market to operate in ways that local government revenues from the property market can be maximised. A property-led growth strategy has been widely adopted by local governments to prioritise property development that produces higher land sale receipts and taxes, for example, high-end housing and high-specification offices and retail property. A lack of affordable commodity housing aggravates affordability and the glut of offices has regularly led to high vacancy rates and depressed rental growth.

The above problems have resulted in a number of defects in the property market. Housing size standards were not set until 2006 when the central government found there was a lack of smaller, thus more affordable, commodity housing in the market. Yet local governments were not cooperative, weakening the implementation of the 70/90 rule to impose a size standard and tolerating violations in the market. The lack of affordable commodity housing eventually necessitated a massive ASH building campaign to address urban housing affordability. On the other hand, no policy or rules are provided by the constitutional level to guide the development of commercial properties, in particular, offices. As a result, local governments at the collective choice level were able to enact rules to subsidise office development and occupation to support prioritised projects, with severe oversupply being a regular outcome.

Third, local governments, under current institutional arrangements, have to seek revenues from the real estate market to carry out projects that lead to short-term economic growth. The constitutional level, the central government, enacts rules to require local governments to meet economic growth targets, without providing sufficient tax revenues for them to do so. Local governments have no access to long-term, low-cost borrowing channels, such as bonds. This creates a mismatch of responsibilities to financial capacity. As a result, local governments rely on land sale receipts and taxes from the property market to finance many important projects and to provide the security for borrowing from banks and non-bank institutions. Such reliance causes local governments to carry out large numbers of residential and commercial property development projects without considering the market's ability to absorb the supply brought by these projects. Reading from such behaviours, investors believe local governments will support housing prices and have made excessive investments in housing, bidding up prices. Due to the importance of the housing market in generating land sale and tax revenues, local governments prop up the housing market to prevent price falls, interfering in the market mechanism.

Fourth, a lack of central government policy guidance results in inadequate checks and balances on local master planning power and development control against over-ambitious urban development being planned and implemented. Intense inter-city competition, the need by top local officials to demonstrate achievements in office, and reliance on land sale receipts, are all factors contributing to very ambitious urban development plans, for example, CBDs, which often lead to large amounts of empty commercial property and depressed returns. Oversupply means that these buildings were built ahead of their time, representing a waste of resources and lower standards in buildings in terms of user comfort, technology and sustainability. In addition to causing a waste of resources, oversupply means too many buildings are built under local government policies to attract inward investment, particularly in manufacturing, often leading to oversupply and very low efficiency of industrial land in many cities.

Fifth, property financing in China features over-regulated bank lending and under-regulated non-bank lending. State-owned banks have been restricted in their exposure in the real estate market to reduce risks for them. Controlled interest rates on bank lending resulted in underpricing of development loans and mortgage loans at the time of macro control from 2005 to 2013, encouraging development and investment. It also led to underlending in 2013 to 2014 because the banks were reluctant to lend under low interest rates. On the other hand, lending from trusts, funds and other channels was not properly regulated, with implicit guarantees being offered to investors, heightening financial risks to real estate trusts and funds. Unregulated non-bank lending is responsible for some high-profile city-wide housing development failures, with Beihai, Wenzhou and Ordos being examples. Thus the over-regulation of the state-owned banks incurs a risk transfer from state-owned banks to the private sector although it is not engineered by the government, which only tried to protect state-owned banks from overexposure and to rein in the overheated property market. However, higher risks resulted from such a measure indirectly, as developers raise funds outside the state-owned banks and overseas, which are often more expensive, increasing the financial risks of developers.

Sixth, ASH provision has been troubled by lack of funding, incomplete institutional support and lack of choice. The central government only provides *ad hoc* financial support to less wealthy cities and provinces, forcing local governments in many cases to use sites at less favourable locations and low budgets to build ASH. No authoritative law or ordinance has been promulgated to guide and limit ASH provision, admission and management. As a result, ASH management organisations at the operational level find it hard to effectively deal with cheating by applicants, non-compliance by tenants, and lack of funds to carry out management. Capture of the subsidised ECH by unqualified applicants was very serious before the enactment of a size standard in 2006. However, the size standards for the ASH designed to prevent capture are so small that such ASH risks becoming obsolete soon. The coverage of ASH is still mostly on registered urban households, with over a hundred million migrant workers on modest pay not covered.

Seventh, macro control and other government interventions impose transaction costs on property development and transaction, and compromise the property rights of developers and owners. Imposition and frequent adjustments of taxes on housing transaction increased transaction costs, leading to perverse market behaviour such as cheating on prices reporting and delayed registration of transactions. Strict rules of purchase restriction forced some families to resort to divorce to avoid taxes and bans on lending. Developers, on the other hand, have to deal with many inspections and other requirements.

Eighth, there are political, social and environmental costs associated with the defects in institutional arrangements on the property market. Housing price inflation has become a political issue and corruption in the real estate sector is rife. Housing conditions and investment have affected the behaviour of households. To speed up property development and keep costs under control, compensation for forced demolition and relocation sometimes fails to be considerate. Urban expansion often causes the loss of very fine farmland and some valuable habitats.

Ninth, there are insufficient market-supporting institutions to prepare the Chinese society for market housing provision. Data availability and accuracy on housing have become sources of confusion and misjudgement, affecting market expectation and behaviour. The real estate tax has not been introduced for over a decade, rewarding multiple home owners who leave homes idle and developers who build too many high-end housing units.

The root of the above problems is in the institutional arrangements that do not allow the market to allocate most of the resources; and in the reliance on, and reluctance to depart from, administrative power. Further institutional change is needed to break such path dependency. A new and informed institutional change is needed to build a set of market-supporting institutions to allow the Chinese real estate market to allocate more resources and local governments to regulate rather than participate in market operations.

15.3 The future

The property market entered a downturn in 2014. Although local governments reused some of the market-supporting measures in 2009 to prop up the housing market, the central government opted to stay away from tabling a rescue package. Instead, the central government is adopting more indirect measures to induce rather than to enforce changes in the market, with mini-stimulus measures targeting certain parts of the economy to carry out structural adjustment. The time for macro control has passed.

The Third Plenary Session of the Eighteenth Congress of the CCP in November 2013 set out a new blueprint for comprehensive reforms on many fronts, with the key reform in the economic arena being further marketisation to allow the market to allocate resources. This is in line with the reforms being considered in the housing market, i.e. the establishment of a long-term adjustment and control mechanism (LTACM). Such a mechanism intends to stop intervention on short-term price fluctuations and provide the necessary supporting institutional arrangements to allow the property market to make decisions on resource allocation. The LTACM aims: to build a system of housing provision giving all members of the society a place in housing; to increase the sources of land supply, including recycling of industrial land and the marketisation of collectively owned land; to establish a real estate tax to discourage multiple home ownership and empty housing; to further enhance ASH, including non-governmental provisions to increase choice; and to allow property financing to operate on market principles. The central message from the government is that to promote further urbanisation, urban housing costs need to come down. This should not be achieved by direct market intervention. Rather, sufficient provision of ASH and an emphasis on supply are the focus to satisfy the different needs and demand for housing.

The Third Plenary Session plans other major reforms that have an impact on the property market. A key reform in finance and taxation is to end many of local governments' insufficiently funded responsibilities, removing the need to rely on the property market for additional revenues. A reform of household registration aims to remove restrictions for rural residents seeking registered household status in small to medium-sized cities: a move helpful to solve housing oversupply in those cities.

Reforms have also started to reduce the link of performance assessment of top local officials to economic growth, reducing the incentive to push ahead with unnecessary development projects.

Nevertheless, a number of issues have not received attention from the policymakers. First, the lack of policy on commercial property needs to be addressed to overcome oversupply in many cities that result in waste and risks, and intense competition among commercial property providers. Second, there is still insufficient intention and effort to increase the supply of rules on market operations, including those on size standards for commodity housing, landlord and tenant relationship, and housing management. Third, private and voluntary sector provision of ASH is not available in most parts of the country, limiting choice to residents who qualify for ASH.

There is no doubt that the Chinese real estate market has undergone profound development and massive changes in a remarkably short period and on such gigantic scales hitherto unprecedented anywhere in world. The ongoing story to unfold in the next decade will no doubt be every bit as remarkable and engaging to policymakers, investors, developers, property professionals, academics and onlookers alike. Most of all, however, it will impact on the 1.37 billion citizens of the land we call China, and indeed the rest of the world because the property sector in China has grown to such a scale that it affects the world commodity market and even property provision in some countries.

The transition of the Chinese property sector from command to market economy and the building of institutional capacity have been a valuable lesson to other countries that seek to improve the efficiency and equity of their property markets. Some of the undesirable developments in the Chinese real estate market and some of the counter-productive macro control measures may seem readily avoidable to readers in well-established and stable market economies. Those outcomes are temporary, however, and the results of the path-dependent nature of institutional change. As further reforms are carried out, a new set of institutional arrangements is likely to result in more efficient and equitable outcomes for housing in China. If there is one lesson to be learnt, it is that the key issue is not static efficiency: adaptive efficiency through institutional change is more important. There can be no greater example of this than the remarkable history of the 'three worlds of real estate' in China in the past 100 years.

BIBLIOGRAPHY

Baidu (2014) Macro control of real estate. Available at: http://baike.baidu.com/view/2169072.htm (accessed 5 April 2014).

BBC (2014) The real costs of China's anti-corruption crackdown, *BBC*, 3 April 2014. Available at: www.bbc.co.uk/news/blogs-china-blog-26864134 (accessed 5 April, 2014).

Bloomberg (2013) Beijing orders official cars off roads to curb pollution, *Bloomberg*, 13 January 2013. Available at: www.businessweek.com/news/2013-01-13/beijing-orders-official-cars-off-roads-to-curb-pollution (accessed 15 January 2013).

BMG (2009) *Implementation Opinions to Promotion the Healthy Development of Real Estate Market in Our City*, Beijing: Beijing Municipal Government.

BMG (2010) *Circular on Implementing State Council's Circular on Steadfastly Preventing Rapid Housing Price Inflation in Some Cities*, Beijing: Beijing Municipal Government.

BMG (2011) *Circular on Implementing State Council's Document and Further Strengthening Adjustment and Control of Local Real Estate Market*, Beijing: Beijing Municipal Government.

Bu, Y. and Li, Y. (1998) *Explaining the Land Administration Law*, Beijing: The Law Press.

Cai, R. (2013) *My Views On Real Estate*, Guangzhou: Nanfang Daily Press.

Caixin (2013) Xu Shaoshi (Head of State Development and Reform Commission) answers questions on the decisions of communiqué, *Caixin.Com*, 22 November 2013. Available at: http://china.caixin.com/2013-11-22/100608100.html (accessed 5 February 2014).

Caixin (2014) Why per capita construction land needs to be controlled within 100 m^2, *Caixin.Com*, Available at: www.yicai.com/news/2014/04/3662920.html (accessed 25 May 2014).

Cao, J.A. (2009) Developmental state, property-led growth and property investment risks in China, *Journal of Property Investment and Finance*, 27(2), 162–79.

Cao, J.A. and Keivani, R. (2008) Risks in the commercial real estate markets in China, *Journal of Real Estate Portfolio Investment*, 16(3): 363–83.

Cao, J.A. and Keivani, R. (2014a) The limits and potentials of the housing market enabling paradigm: an evaluation of china's housing policies from 1998 to 2011, *Housing Studies*, 29(1), 44–68.

Cao, J.A. and Keivani, R. (2014b) *More Doesn't Mean Better: Inefficiencies in China's Affordable and Social Housing Sector*, London: Royal Institution of Chartered Surveyors.

CASS (2013) *An Investigation on China's National Balance Sheet*, Beijing: China Academy of Social Sciences.

Castells, M. (2000) *The End of Millennium*, 2nd edn, Oxford: Blackwell.

CBRE (2005) *Market View of People's Republic of China, Q3, 2005*, Shanghai: CBRE Research.

CBRE (2013) *China Retail: A Changing Landscape; Part I: Fast Fashion Brand, Q1, 2014*, Shanghai: CBRE Research.

CBRE (2014a) *Market View of People's Republic of China, Q2, 2014*, Shanghai: CBRE Research.

CBRE (2014b) *China Retail: A Changing Landscape; Part II: Department Stores, Q1, 2014*, Shanghai: CBRE Research.

CCP (2013) *Decision of the Central Committee of The Communist Party of China on Some Major Issues Concerning Comprehensively Deepening the Reform*, Beijing: Communist Party of China. Available at: www.china.org.cn/china/third_plenary_session/2014-01/16/content_31212602.htm (accessed 28 May 2014).

CE (2014) Raising upper limits accelerates inflowing of insurance money into real estate, *Chinese Economy*. Available at: http://finance.ce.cn/rolling/201401/03/t20140103_2042356.shtml (accessed 12 March 2014).

CEDSMC (Commission of Economic Development of Small and Medium-sized Cities) (2010) *Annual Report on Development of Small and Medium-sized Cities in China*, Beijing: Social Science Academy Press.

Chai, Q. (2008) 30 years of real estate system reforms and market development, in D. Zhou and R. Ouyang (eds) *Report on China's Economic Development and Institutional Reform: China: 30 Years of Reform and Opening Up*, Beijing: Social Sciences Academic Press.

Changjiang Daily (2013) 4,641 families chasing 2,145 PRH units, *Changjiang Daily*, 4 October 2013. Available at: http://news4.cnhan.com/content/2013-10/04/content_2305457.htm (accessed 12 February 2014).

Chen, H. (2013) A long-term adjustment and control mechanism is not just for adjusting housing prices, *The Time-Weekly*. Available at: www.ce.cn/cysc/fdc/fc/201310/17/t20131017_1629577.shtml (accessed 12 August 2014).

Cheng, X. (2014) Saving and intensifying land use require participation of full society, China Land and Resources Net. Available at: http://news.mlr.gov.cn/xwdt/jrxw/201406/t20140627_1321935.htm (accessed 12 July 2014).

CHFS (2014) *2014 Report on Housing Vacancy Rate and Housing Market Development Trend*, Changdu: China Family Finance Survey of Southwest University of Finance and Economy.

China Daily (2013) Wuhan to build 170,000 ASH to achieve a 20% coverage in 3 years, *China Daily*, 3 July 2013. Available at: www.chinadaily.com.cn/hqgj/jryw/2013-07-03/content_9485212.html (accessed 25 August 2013).

China Finance (2012) *Finance Yearbook of China 2012*, Beijing: China Finance Press.

Chinanews (2014a) China's equipment manufacture output exceeded CN¥20 trillion: the number one in the world, Chinanews. Available at: www.chinanews.com/gn/2014/04-02/6022293.shtml (accessed 5 July 2014).

Chinanews (2014b) Smog covers 1.43 million km^2: 20 cities including Beijing and Tianjin are trapped by heavy pollution, Chinanews. Available at: www.chinanews.com/gn/2014/02-23/5871271.shtml (accessed 15 May 2014).

China Plan Net (2011) *China's Urban Planning Loses Its Direction*, China Plan Net. Available at: www.zgghw.org/html/tebiezhuanti/chengshijiazhi/20110710/12567.html (accessed 17 July 2014).

China Real Estate Association (2014) *Top 500 Development Firms 2014*, China Real Estate Association. Available at: www.fangchan.com/zt/top500/index.html#3 (accessed 13 August, 2014).

ChinaVenture (2014) *Research Report on CN¥Real Estate Funds, 2013*, Shanghai: ChinaVenture.

CLSPI (2014) Land price monitoring report for main cities, Q2 2014, *China Land Surveying and Planning Institute*. Available at: http://landchina.mlr.gov.cn/tdsc/djxx/djjc/201407/t20140717_1323928.htm (accessed 15 July 2014).

CMG (2005) *Chongqing Urban Master Plan 2005–2020*, Chongqing: Chongqing Municipal Government. Available at: http://politics.people.com.cn/GB/8198/49262/49263/3489364.html (accessed 25 May 2014).

CNN (2014) Land grab: rich Chinese are snapping up America's real estate, CNN. Available at: http://money.cnn.com/2014/07/08/real_estate/home-sales-to-chinese/index.html?iid=EL (accessed 15 July 2014).

Colliers International (2014a) *Chengdu Grade A Office Market, Quarter 2, 2014*, Chengdu: Colliers International.

Colliers International (2014b) *Changsha Grade A Office Market, Quarter 2, 2014*, Guangzhou: Colliers International.

Colliers International (2014c) *Chengdu Retail Market, First Half, 2014*, Chengdu: Colliers International.

Colliers International (2014d) *A Surging Wave: Chinese Real Estate Investment Goes Overseas*, Shanghai: Colliers International.

Conference Board (2014) *2014 Productivity Brief: Key Findings*, Conference Board. Available at: https://www.conference-board.org/pdf_free/economics/TED3.pdf (accessed 1 July 2014).

CPMI (2004) *Standards on Property Management Services Classes for Ordinary Housing*, Beijing: China Property Management Institute.

CREA (2014) *China Real Estate Finance 2013 Report*, Beijing: Finance Sub-Committee of China Real Estate Association.

CRIC and China Real Estate Evaluation Centre (2014) *Top 50 in China's League Table of Sales by Real Estate Firms in 2013*, Beijing: CRIC and China Real Estate Evaluation Centre.

Dalian Evening Post (2003) 182 cities to build international metropolises: Ministry of Construction urges caution, *Dalian Evening Post*. Available at: www.southcn.com/news/china/zgkx/200311130170.htm (accessed 1 July 2014).

Dong, Y. (2013) *An Investigation into Energy Consumption by Buildings in China*, Beijing: China Centre for Energy and Development, Peking University.

DTZ (2014) *China Insight: China Office Pipeline and Dynamics*, Beijing: DTZ Research.

DYXTW (2014) *Asset Scale Data of the Trust Industry from 2011 to 2013*, DYXTW. Available at: www.dyxtw.com/news/20140709135669.html (accessed 8 August 2014).

Economic Information (2014) China to carry out audit on land sale receipts, *Economic Information*. Available at: http://finance.people.com.cn/n/2014/0825/c1004-25528361.html (accessed 26 August 2014).

E-House China (2012) *2012 League Table of Housing Price to Incomes Ratio for 35 Cities in China*, Shanghai: E-House China Research Institute.

E-House China (2014a) *2014 League Table of Housing Price to Incomes Ratio for 35 Cities in China*, Shanghai: E-House China Research Institute.

E-House China (2014b) *Report on Cash Positions of House Builders in China, Quarter 1, 2014*, Shanghai: E-House China Research Institute.

EOP (2006) Guangzhou confiscate sites on a large scale, *Economic Observation Post*, 16 July 2006. Available at: http://finance.sina.com.cn/g/20060716/01472735463.shtml (accessed 18 May 2014).

Fan, Z. (2006) Average plot ratio of Chinese cities should be increased to 0.5, *House Focus Net*. Available at: http://house.focus.cn/news/2006-06-13/213768.html (accessed 35 May 2014).

Fenby, J. (2012) *Tiger Head, Snake Tails: China Today, How It Got There and Where It Is Heading*, London: Simon & Schuster.

Forbes (2013) World's 500 largest corporations in 2013: the Chinese are rising. Available at: www.forbes.com/sites/panosmourdoukoutas/2013/07/17/worlds-500-largest-corporations-in-2013-the-chinese-are-rising/ (accessed 30 March 2014).

Forbes (2014) China power shown with third place ranking on global innovation scale. Available at: www.forbes.com/sites/rebeccafannin/2014/04/07/china-power-shown-with-third-place-ranking-on-global-innovation-scale/ (accessed 8 April 2014).

Gang, G. (2012) *Contemporary Chinese Economy*, Oxford: Blackwell.

GBS (2013) *Guangzhou Statistical Information Handbook 2013*, Guangzhou: Guangzhou Bureau of Statistics.

GRES (Guangzhou Real Estate Society) (1994) A survey report on the charges and taxes in property development in Guangzhou, *Southern Real Estate*, 7: 30–2.

Gu, Y. and Xing, B. (2005) Ministry of Commerce proposes to the State Council to open up REIT channel for financing, China –CRB. Available at: www.cdbidding.com/policy/policy_detail.asp?ID=2961 (accessed 8 August 2014).

He, B. (1996) Legitimation and democratization in a transitional China, *Journal of Communist Studies and Transitional Politics*, 13(3): 315–42.

He, D. (2014) Legal barriers remain for levying real estate tax, *Economic Information*. Available at: http://xining.house.sina.com.cn/news/2014-05-28/09324172997.shtml (accessed 30 June 2013).

Herschler, S.B. (1995) The 1994 tax reforms: the centre strikes back, *China Economic Review*, 6(2): 239–45.

Ho, M.H.C. and Kwong, T. (2002) Housing reform and home ownership behaviour in China: a case study in Guangzhou, *Housing Studies*, 17(2): 229–44.

Hong, J. and Pan, W. (2002) Guangzhou imposes tender for all development land, *China Construction Post*, 15 April.

Hubei Daily and JLL (2013) *The White Paper of the Development of Hubei Real Estate*, Wuhan: *Hubei Daily*.

Huochepiao (2013) Beijing to Shanghai high-speed railway passenger number exceeded 100 million, *Huochepiao.com*. Available at: http://news.huochepiao.com/2013-3/201335919202.htm (accessed 18 April 2013).

Ifeng (2013) Beijing's National Agricultural Exhibition site becomes the new land king, *ifeng.com*. Available at: http://house.ifeng.com/special/expensiveland/zuixin/detail_2013_09/04/29313315_0.shtml (accessed 20 January 2014).

IMF (2014) *Global Financial Stability Report: Moving from Liquidity- to Growth-Driven Markets*, Washington, DC: International Monetary Fund.

International Comparison Program (2014) *Purchasing Power Parity and Real Expenditures of World Economies: Summary of Results and Findings of the 2011 International Comparison Program*, Washington, DC: International Bank for Reconstruction and Development/The World Bank.

Jiang, X. (2012) An investigation on the positioning of Housing Provident Fund development under China's current affordable and social housing system, *Journal of Dongbei University of Finance and Economics*, 2012(6): 12–17.

JLL (2004) *Global Real Estate Transparency Index 2004*, London: Jongs Lang LaSalle.

JLL (2006) *Global Real Estate Transparency Index 2006*, London: Jongs Lang LaSalle.

JLL (2008) *Global Real Estate Transparency Index 2008: From Opacity to Transparency*, London: Jongs Lang LaSalle.

JLL (2010) *Global Real Estate Transparency Index 2010: Mapping the World of Transparency*, London: Jongs Lang LaSalle.

JLL (2012a) *China 50: Fifty Real Estate Markets That Matter*, London: Jong Lang LaSalle.

JLL (2012b) *Global Real Estate Transparency Index 2014: Mapping the World of Transparency*, London: Jongs Lang LaSalle.

JLL (2014) *Global Real Estate Transparency Index 2008: Data Disclosure And Technology Raise Transparency Levels*, London: Jongs Lang LaSalle.

Kiser, L.L. and Ostrom, E. (1982) The three worlds of action: a metatheoretical synthesis of institutional approach, in E. Ostrom (ed.) *Strategies of Political Inquiry*, Los Angeles: Sage, pp. 179–222.

Knight Frank (2014) *Spotlight on China's Retail Market 2014*, Shanghai: Knight Frank.

Landvalue (2014) Land Price Index, *China Urban Land Price Dynamic Monitor*. Available at: www.landvalue.com.cn/L_LandPriceMonitor.aspx?Menu_ID=5andPID=1 (accessed 20 April 2014).

Li, S. (2012) How to shut the big mouth of building energy consumption, *China Youth Daily*, 24 November 2012. Available at: http://zqb.cyol.com/html/2012-11/24/nw.D110000zgqnb_20121124_1-03.htm (accessed 15 March 2013).

Li, X. (2014) From green buildings to green urban districts, China Green Building. Available at: www.archcy.com/point/yjmj/b9d7c74504edcd58 (accessed 25 June 2014).

Li, Y.N. (2013) China's macro economic situation and a new round of economic reforms, *Cnstock.com*. Available at: http://news.cnstock.com/news/sns_jd/201309/2753315.htm (accessed 1 April 2014).

Liao, M. (2013) The risk analysis of reserve land financing and value assessment of collateral, *Land Appraisal and Agent*, 2013(8): 3–9.

Lin, J.Y. (2012) *Demystifying the Chinese Economy*, Cambridge: Cambridge University Press.

Lin, S.S. (2013) *A Tale of Guangzhou City*, Guangzhou: Guangdong People's Press.

Liu, C., Shang, C. and Wang, W. (2010) An analysis of current status of land reserve and strategies in Wuhan, *Pioneering with Science and Technology Monthly*, 2010 (6): 13–14.

Liu, H.Y. (2014) The long-term adjustment and control system starts from housing, land, taxation and finance, *House.163.com*. Available at: http://gz.house.163.com/14/0110/18/9I8F6K0700873C6D_all.html (accessed 15 January 2014).

Liu, Z. (1998) The speech on the national workshop speeding up the housing reform, *China Real Estate*, 1999(12): 4–9.

Liu, Z. (1999) The speech on the national workshop on the management of the housing provident fund. *China Real Estate*, 1999(6): 11–16.

Lou (2014) A thorough transformation on modernisation of governance in China, Gov.cn. Available at: www.gov.cn/xinwen/2014-07/03/content_2711811.htm (accessed 5 July 2014).

Lu, J. (2009) Analysis on hot issues of the real estate market from the perspectives of land supply, *China Policy and Law for Land and Resources*. Available at: www.gtzyzcfl.com.cn/zjk_Article.asp?Id=127 (accessed 12 July 2014).

Lu, M. (2013) There is still room for growth in big cities, *China Reforms*, March.

LZMP (2003) A summary of the GITIC bankruptcy case, *Lanzhou Morning Post*. Available at: www.gansudaily.com.cn/20030301/504/2003301A00192005.htm (accessed 9 August 2014).

Mellows-Facer, A. and Maer, L. (2012) *International Comparisons of Manufacturing Output*, London: House of Commons Library.

Ministry of Commerce (2013) *Report on Foreign Investment in China 2013*, Beijing: Ministry of Commerce.

MLR (1999) *Idle Land Administration Measures*, Beijing: Ministry of Land and Resources.

MLR (2002) *Regulations on Tender, Auction and Listing of Land Use Rights in State Land*, Beijing: Ministry of Land and Resources.

MLR (2004) *Circular on Continuing Enforcement and Supervision on the Tender, Auction and Listing of Land Use Rights for Market-Oriented Uses*, Beijing: Ministry of Land and Resources.

MLR (2007) *Land Reserves Management Measures*, Beijing: Ministry of Land and Resources.

MLR (2010) *2010 China Land and Resources Almanac*, Beijing: Ministry of Land and Resources.

MLR (2014a) *Regulation on Saving and Intensifying Land Use*, Beijing: Ministry of Land and Resources.

MLR (2014b) *2013 Communiqué of Land and Resources in China*, Beijing: Ministry of Land and Resources.

MoC (1990) *Opinions on Solving Housing Problems of Households in Extreme Housing Poverty*, Beijing: Ministry of Construction (now Ministry of Housing and Urban-Rural Development).

MoC (1992) *Temporary Administrative Measures on Housing Cooperatives in Cities and Towns*, Beijing: Ministry of Construction.

MoC (1994) *Administrative Measures on Economic and Comfortable Housing in Cities and Towns*, Beijing: Ministry of Construction.

MoC (1999) *Administrative Measures on Urban Low Rent Housing*, Beijing: Ministry of Construction. Available at: www.jincao.com/fa/law19.23.htm (accessed 19 February 2014).

MoC (2001) *Administrative Measures on Sale of Commodity Housing*, Beijing: Ministry of Construction.

MoC (2004) *2003 Statistical Communique On Housing Conditions in Cities and Towns*, Beijing: Ministry of Construction.

MoC (2006a) *Opinions on Normalising Entry and Management of Foreign Capital in Real Estate Market*, Beijing: Ministry of Construction.

MoC (2006b) *2005 Statistical Communique on Housing Conditions in Cities and Towns*, Beijing: Ministry of Construction.

MoC (2006c) *Opinions on Implementing the Requirements on Structural Ratio of New Housing Construction*, Beijing: Ministry of Construction.

MOHURD (2009a) *Report on Housing Provident Fund Management in 2008*, Beijing: Ministry of Construction.

MOHURD (2009b) *Guidance on Improvement of Penghuqu in Cities and State-owned Industrial and Mining Areas*, Beijing: Ministry of Construction. Available at: www.mlr.gov.cn/zwgk/zytz/201001/t20100114_132728.htm (accessed 6 May 2014).

MOHURD (2010) *Guidance on Speeding Up the Development of Public Rental Housing*, Beijing: Ministry of Construction. Available at: www.MOHURD.gov.cn/zcfg/jsbwj_0/jsbwjzfbzs/201006/t20100612_201308.html (accessed 24 April 2014).

MOHURD (2011) *Our Country's Economic Growth Since Reform and Opening-Up*. Available at: www.mohurd.gov.cn/zxft/201108/t20110818_205900.html (accessed 21 June 2013).

MOHURD (2012) *Implementation Opinions on Speeding up the Development of Green Buildings*, Beijing: Ministry of Housing and Urban-Rural Development.

MOHURD (2013) *Circular on Intensified Special Administration of Real Estate Intermediary*, Beijing: Ministry of Housing and Urban-Rural Development.

MOHURD (2014) *The Country Sees Increase of 1.44 Billion M² New Buildings*, Beijing: Ministry of Housing and Urban-Rural Development. Available at: www.mohurd.gov.cn/zxydt/201404/t20140425_217769.html (accessed 24 June 2014).

MOHURD and MLR (2011) *Circular on Further Implementation of State Council [2010] Document 10*, Beijing: Ministry of Housing and Urban-Rural Development.

Nanfang (2003a) Shanghai plans to digest RMB 100 billion assets locked in fail real estate projects, *21 Century Business Herald*. Available at: www.nanfangdaily.com.cn/jj/20031120/dc/200311170917.asp (accessed 25 March 2014).

Nanfang (2003b) Sale of villas in Pearl River New City attracts applause, *Nanfang*. Available at: www.southcn.com/estate/zhuanti/zhujiangxinchengchenfu/toushizhujiangxincheng/200312021062.htm (accessed 25 May 2014).

Nanfang (2006) A secret person bought 110 units with RMB126 million in fraud, *Nangfang Daily*, 26 May 2006. Available at: http://bj.house.sina.com.cn/news/2006-05-28/1046129864.html (accessed 1 June 2014).

Nanfang (2014) *Developers with Sales over CN¥100 Billion Have Land Reserve Over 5 Million m² in Qingyuan*. Available at: http://house.qq.com/a/20140221/015438.htm (accessed 7 June 2014).

NBSC (1996) *China Statistical Yearbook 1996*, Beijing: China Statistics Press.

NBSC (1999) *Comprehensive Statistical Data and Materials on 50 Years of New China*, Beijing: China Statistics Press.

NBSC (2003) *China Statistical Yearbook 2003*, Beijing: China Statistics Press.

NBSC (2006) *China Statistical Yearbook 2006*, Beijing: China Statistics Press.

NBSC (2009) *China Statistical Yearbook 2009*, Beijing: China Statistics Press.

NBSC (2011) *China Statistical Yearbook 2011*, Beijing: China Statistics Press.

NBSC (2012) *Urban Statistical Yearbook of China 2012*, Beijing: China Statistics Press.

NBSC (2013) *China Statistical Yearbook 2013*, Beijing: China Statistics Press.

NBSC (2014a) *The Economy Stabilises in the First Half Year of 2014*, Beijing: National Bureau of Statistics of China. Available at: www.stats.gov.cn/tjsj/zxfb/201407/t20140716_581947.html (accessed 17 July 2014).

NBSC (2014b) *Statistics Communiqué on Economic and Social Developments in China in 2013*, Beijing: National Bureau of Statistics of China. Available at: www.stats.gov.cn/tjsj/zxfb/201402/t20140224_514970.htm. (accessed 24 February 2014).

NBSC (2014c) *National Real Estate Development and Sales from January to October, 2014*, Beijing: National Bureau of Statistics of China. Available at: www.stats.gov.cn/tjsj/zxfb/201411/t20141113_637139.html (accessed 15 November 2014).

NDRC (2007) Housing prices inflation speeding up showing supply not matching demand. Available at: http://finance.people.com.cn/GB/6599801.html (accessed 24 May 2014).

NEWSCHINA (2007) Government's Land Reserve gains strength, NEWSCHINA. Available at: http://news.sina.com.cn/c/2007-12-28/100014621884.shtml (accessed 24 May 2014).

Nie, P. (2011) Analysis on the changes and impact of different phases of foreign investment in China's real estate, *Practice in Foreign Economic Relations and Trade*, 2011(11): 19–25.

Niu, F., Li., J. and Shang, J. (2006) *2006 Blue Book of Real Estate*, Beijing: Social Sciences Academy Press.

Nolan, P. and Ash, R.F. (1996) China's economy on the eve of reform, in A.G. Walder (ed.) *China's Transitional Economy*, Oxford: Oxford University Press, pp. 18–36.

North, D.C. (1990) *Institutions, Institutional Change and Economic Performance*, Cambridge: Cambridge University Press.

Oi, J.C. (1996) The role of the local state in China's transitional economy, in A.G. Walder, (ed.) *China's Transitional Economy*, Oxford: Oxford University Press, pp. 170–87.

PBoC (1997) *Interim Measures on Management of Personal Housing Mortgage Loans*, Beijing: People's Bank of China.

PBoC (1998) *Circular on Expanding Housing Credit Supply to Support Housing Construction and Consumption*, Beijing: People's Bank of China.

PBoC (2001) *Circular on Standardising the Housing Finance Sector*, Beijing: People's Bank of China.

PBoC (2002) *Provisional Measures on Cash Trust Schemes of Trust and Investment Companies*, Beijing: People's Bank of China.

PBoC (2003) *Circular on Further Strengthening the Management of Real Estate Credit Business*, Beijing: People's Bank of China.

PBoC (2005) *2004 Real Estate Finance Report*, Beijing: People's Bank of China.

PBoC (2007) *Circular on Intensifying Management of Real Estate Lending*, Beijing: People's Bank of China.

PBoC (2008) *China Monetary Policy Implementation Report, Q4, 2008*, Beijing: People's Bank of China.

PBoC (2014) *China Monetary Policy Implementation Report, Q4, 2013*, Beijing: People's Bank of China.

Peerenboom, R. (2007) *China Modernizes: Threat to the West or Model to the East?* Oxford: Oxford University Press.

Peng, D. (1999) A review and the prospect of land development, reclamation and consolidation in China, *China Land*, 1999(10): 14–16.

People (2005) The Ministry of Construction and the Central Bank are at odds on presale, *People*. Available at: http://finance.people.com.cn/GB/1045/3675895.html (accessed 28 May 2014).

People (2006a) Land shortage argument is absurd by a foreign financial institution in Beijing, *People.cn*. Available at: www.people.com.cn/GB/news/37454/37462/4091814.html (accessed 8 May 2014).

People (2006b) The first big residential purchase, *People.cn*. Available at: http://mnc.people.com.cn/GB/4384784.html (accessed 19 May 2014).

People (2011a) Guangdong's Foshan announced policy to relax purchase restrictions but withdraws such policy in half a day, *People.cn*. Available at: http://politics.people.com.cn/GB/14562/15865933.html (accessed 15 May 2014).

People (2011b) Commercial home mortgage loan borrowed by individuals can be converted to Housing Provident Fund loan in Ningbo, *People.cn*. Available at: http://nb.people.com.cn/GB/200892/15991212.html (accessed 15 May 2014).

People (2013) Real estate tax different to housing tax: legislation goes first, *People.cn*. Available at: http://politics.people.com.cn/n/2013/1125/c1001-23647144.html (accessed 15 May 2014).

Qi, J. (2007) *Pushing Through Special Administration on Real Estate Market Order*, MOHURD. Available at: www.mohurd.gov.cn/jsbfld/200712/t20071206_165563.html (accessed 11 August 2014).

Qi, L. (2013) Beijing's land reserve scale returns to CN¥100 billion, *Beijing Business Post*. Available at: http://bj.house.sina.com.cn/news/2013-06-19/08022255366.shtml (accessed 15 March 2014).

Qianlong (2005) Guangzhou will make all failed projects alive in 3 years, *QianLong.com*. Available at: http://house.qianlong.com/33/2005/12/09/1980@2918799.htm (accessed 4 May 2014).

Qu, B. (2006) Analysis of Beijing's real estate market in 2005 and prospects in 2006, in J. Li, *China Real Estate Development Report No 3*, Beijing: Social Science Academy Press.

Ramo, J. (2004) *The Beijing Consensus*, London: Foreign Policy Centre. Available at: http://fpc.org.uk/fsblob/244.pdf (accessed 7 August 2013).

Rawski, T.G. (2001) What is happening to China's GDP statistics? *China Economic Review*, 12: 347–54.

Ren, X. (2013) The establishment and basic evaluation of China's security housing system, *China Economic Times*, 19 June 2013. Available at: www.cet.com.cn/ycpd/sdyd/881462.shtml (accessed 1 February 2014).

Ren, X., Liao, Y. and Liu, W. (2012) Further perfection of the security housing system in China, in State Council Development Research Centre (ed.) *The People's Livelihood Is The Foundation: The Road to Perfect China's Essential Public Service*, Beijing: China Development Press.

SAT (1999a) *Circular on Adjusting Some Of the Taxation Policies Real Estate Market*, Beijing: State Administration of Taxation.

SAT (1999b) *Circular on Issues of Income Tax on Housing Sales Incomes*, Beijing: State Administration of Taxation.

SAT (2006) *Circular on Income Tax Levy on Sale of Second-Hand Housing Held by Individuals*, Beijing: State General Bureau of Taxation.

Savills (2002a) *Beijing Office, Warming Up, April 2002*, Beijing: Savills.

Savills (2002b) *Shanghai Office Briefing, April 2002*, Beijing: Savills.

Savills (2008) *Beijing Office Sector Briefing, November 2008*, Beijing: Savills.

Savills (2013) *Guangzhou Office Sector Briefing, April 2014*, Beijing: Savills.

Savills (2014a) *Beijing Office Sector Briefing, April 2014*, Beijing: Savills.

Savills (2014b) *Shanghai Office Sector Briefing, April 2014*, Shanghai: Savills.

Savills (2014c) *Shenyang Office Sector Briefing, April 2014*, Shanghai: Savills.

Savills (2014d) *Dalian Office and Sector Briefing, April 2014*, Beijing: Savills.

Savills (2014e) *Hangzhou Office and Retail Sector Briefing, February 2014*, Shanghai: Savills.

Savills (2014f) *Shenzen Office and Sector Briefing, April 2014*, Beijing: Savills.

Savills (2014g) *Xi'an Office and Sector Briefing, April 2014*, Beijing: Savills.

Sciencenet (2014) State Forestry Bureau: China's forest cover reaches 21.63%, *Sciencenet*. Available at: news.sciencenet.cn/htmlnews/2014/2/289007.shtm (accessed 13 March 2014).

Shanghai Chorogophy (2014) *Old Housing, Shanghai*: Shanghai Chorogophy Office. Available at: http://shtong.gov.cn/node2/node2245/node75091/node75095/index.html (accessed 10 June 2014).

Shanghai Security News (2006) Beijing and Shanghai in the wave of confiscating idle sites, *Shanghai Security News*, 18 January 2006. Available at: http://news.xinhuanet.com/house/2006-01/18/content_4065472.htm (accessed 6 June 2014).

Sharma, R. (2014) China's debt-fuelled boom is in danger of turning to bust, *Financial Times*. Available at: www.ft.com/cms/s/0/cfcb0568-8749-11e3-9c5c-00144feab7de.html#axzz37R2gnZtA (accessed 14 July 2014).

Shenzhen Municipal Government (2004) *Stipulations on Renewal of Expired Real Estate*, Shenzhen Municipal Government. Available at: www.sz.gov.cn/ghj/zcfggfxwj/ghl_5/201204/t20120428_1846778.htm (accessed 11 May 2014).

Sina (2005) Anniversary for the century's first auction. Available at: http://bj.house.sina.com.cn/rule/summarize/2005-01-01/114858554.html (accessed 6 May 2014).

Sina (2011) Nanjing's Housing Provident Fund adjusts its lending limit. Available at: http://finance.sina.com.cn/money/bank/guangjiao/20111026/110010693435.shtml (accessed 7 May 2014).

SLAB (1992) *Provisional Rules on Administration of Allocated Land Use Right*, Beijing: State Land Administration Bureau.

SLAB (1995) *The Yearbook of the Land of China, 1994–1995*, Beijing: People's Press.

SMG (2005) *Opinions on Strengthening the Adjustment and Control of the Property Market to Promote the Healthy and Sustainable Development of the Property Market*, Shanghai: Shanghai Municipal Government.

Social Science Survey Centre of Peking University (2014) *China Family Panel Studies 2014*, Beijing: Beijing University Press.

Song, Y. (2014) *The Heavy Tax Burden*, China-crb. Available at: www.china-crb.cn/resource.jsp?id=20405 (accessed 6 August 2014).

Soufang (2014) National residential land supply plan not completed for 4 years and land prices in 105 cities continue to soar, *Soufang*. Available at: www.soufun.com/news/2014-02-12/12073773.htm (accessed 6 August 2014).

South China Morning Post (2013) China net national wealth totals CN¥300 trillion, with debts CN¥28 trillion, *South China Morning Post*, 24 December 2013. Available at www.nanzao.com/sc/china/18506/zhong-guo-guo-jia-jing-zi-chan-chao-300mo-yi-zhai-wu-28mo-yi (accessed 20 April, 2014).

SSB (2010) *Suzhou Statistical Yearbook, 2010*, Beijing: China Statistics Press.
SSB (2013) *Suzhou Statistical Yearbook, 2013*, Beijing: China Statistics Press.
State Council (1988) *Implementation Plan on Pushing Forward Housing System Reforms by Phases and Batches*, Beijing: State Council.
State Council (1990a) *Provisional Ordinance for Granting and Transferring Land Use Rights over State-Owned Land in Cities and Towns (Decree 55)*, Beijing: State Council.
State Council (1990b) *Provisional Measures for the Administration of Foreign Investors to Develop and Operate Plots of Land (Decree 56)*, Beijing: State Council.
State Council (1991) *Circular on Proactively and Steadily Continuing Housing Reform in Cities and Towns*, Beijing: State Council.
State Council (1992) *Decisions on Speeding Up the Development of Tertiary Industry*, Beijing: State Council.
State Council (1993) *Opinions on Current Economic Circumstances and Strengthening of Macro Control*, Beijing: State Council.
State Council (1994a) *Decisions on Deepening Urban Housing Reform*, Beijing: State Council.
State Council (1994b) *Circular on Continued Strengthening Macro Control of Fixed Asset Investment*, Beijing: State Council.
State Council (1995) *Circular on Strictly Banning Development of High-End Real Estate Projects*, Beijing: State Council.
State Council (1998a) *Circular on Further Deepening of Urban Housing Reform*, Beijing: State Council.
State Council (1998b) *Ordinance of Urban Real Estate Development and Management*, Beijing: State Council.
State Council (1999) *Housing Provident Fund Administration Ordinance*, Beijing: State Council.
State Council (2001) *Circular on Strengthening Management of State-Owned Land Assets*, Beijing: State Council.
State Council (2003) *Circular to Promote the Sustainable and Healthy Development of the Real Estate Market*, Beijing: State Council.
State Council (2004) *Decision to Deepen Reforms and Strengthen Land Administration*, Beijing: State Council.
State Council (2005a) *Circular to Stabilise Prices of Housing in Earnest*, Beijing: State Council.
State Council (2005b) *Circular on Effectively Working to Stabilise Housing Prices*, Beijing: State Council.
State Council (2006) *Opinions on Adjusting the Structure of Housing Supply and Stabilising Housing Prices*, State Council, May.
State Council (2007a) *Regulation on Realty Management*, Beijing: State Council.
State Council (2007b) *Opinions on Solving the Hardships of Urban Low-Income Households*, Beijing: State Council.
State Council (2008) *Opinions on Promoting the Healthy Development of the Real Estate Market*, Beijing: State Council.
State Council (2009a) *Circular on Adjusting Capital Ratio of Fixed-Asset Investments*, Beijing: State Council.
State Council (2009b) *Circular on Adjustment And Invigoration Plan for Logistics Industry*, Beijing: State Council.
State Council (2010a) *Circular on Promoting the Healthy Development of the Real Estate Market*, Beijing: State Council.
State Council (2010b) *Circular on Steadfastly Preventing Rapid Housing Price Inflation in Some Cities*, Beijing: State Council.
State Council (2011a) *Circular on Relevant Issues to Further the Control of the Real Estate Market*, Beijing: State Council.
State Council (2011b) *Opinions on Speeding Up the Healthy Development of the Logistics Industry*, Beijing: State Council.

State Council (2013) *Circular on Continuously Implementing Macro Control of the Real Estate Market*, Beijing: State Council.
State Council (2014) *National Plan on New Urbanisation*, Beijing: State Council.
The Economist (2014) Double bubble trouble: China's property prices appear to be falling again, *The Economist*. Available at: www.economist.com/news/china/21599395-chinas-property-prices-appear-be-falling-again-double-bubble-trouble (accessed 16 July 2014).
The Wall Street Journal (2014) China acts as cash departs, *The Wall Street Journal*. Available at: http://online.wsj.com/news/articles/SB30001424052702304099004580091051488831726?mg=reno64-wsjandurl=http%3A%2F%2Fonline.wsj.com%2Farticle%2FSB30001424052702304099004580091051488831726.html (accessed 15 August 2014).
Tianfu Morning Post (2004) Public prosecution of the country's first criminal case in urban planning in Chengdu, *Tianfu Morning Post*. Available at: www.china.com.cn/chinese/law/494458.htm (accessed 6 July 2014).
Topspur (2014) *Analysis Report on Fiscal Dependency on Land Sales Revenues in 45 Cities with Purchase Restrictions*, Shanghai: Topspur.
Urban and Social Survey (1991) *China City Statistics Yearbook 1991*, Beijing: China Statistics Press.
Urban and Social Survey (2013) *China City Statistics Yearbook 2013*, Beijing: China Statistics Press.
Walder, A.G. (1996) China's transitional economy: interpreting its significance, in A.G. Walder (ed.) *China's Transitional Economy*, Oxford: Oxford University Press, pp. 1-16.
Wang, Q. (2006) Foreign funds change the genre of China real estate, *China Entrepreneur*, 2006(19). Available at: www.iceo.com.cn/zazhi/2006/1019/189337.shtml (accessed 15 March 2014).
Wang, W. (2013) China urban households owns 1.02 units of housing: farewell to housing shortage, *People.cn*. Available at: http://politics.people.com.cn/n/2013/0412/c1001-21108345.html (accessed 12 July 2014).
Wang, Wei (1999) *Housing Reform*, Beijing: Renmin University Press.
Wang, Y. (2012) Revisiting HOPSCA, *CRIC Real Estate Observation*, 2012(4): 14–17.
Wikipedia (2014a) *List of Countries by Past and Future GDP (Nominal)*. Available at: http://en.wikipedia.org/wiki/List_of_countries_by_past_and_future_GDP_%28nominal%29. (accessed 24 January 2014).
Wikipedia (2014b) *List of Tallest Buildings in the World*. Available at: http://en.wikipedia.org/wiki/List_of_tallest_buildings_in_the_world (accessed 16 May 2014).
Williamson, J. (2004) *The Washington Consensus as Policy Prescription for Development*, Washington, DC: Institute for International Economics. Available at: www.iie.com/publications/papers/williamson0204.pdf (accessed 8 August 2013).
WIPO (2012) Global IP filings continue to grow, China tops global patent filings, World Intellectual Property Organisation, 11 December, 2012. Available at: www.wipo.int/pressroom/en/articles/2012/article_0025.html (accessed 18 August 2013).
WIPO (2014) US and China drive international patent filing growth in record-setting year, World Intellectual Property Organisation, 13 March, 2014. Available at: www.wipo.int/pressroom/en/articles/2014/article_0002.html (accessed 8 April 2014).
World Bank (2002) *World Development Report 2002*, New York: Oxford University Press.
Wu, F., Xu, J. and Yeh, A.G.O. (2007) *Urban Development in Post-reform China: State, Market, and Space*, London: Routledge.
Wu, J.L. (2011) The most important issues in current reforms, *China Reforms*, 2011(12). Available at: http://magazine.caixin.com/2011-11-30/100332903.html (accessed 13 July 2014).

Wu, X. and Sun, J. (2000) An investigation on land use in development zones, *China Land Science*, 14(2): 17–21.

Xie, C., of Colliers International Shanghai Office, personal communication, 15 July 2014.

Xinhua (2003a) Are there bubbles again in Beihai's property market, Xinhuanet. Available at: http://news.xinhuanet.com/focus/2003-03/12/content_759705.htm (accessed 5 May 2014).

Xinhua (2003b) Failed real estate projects in Shanghai in perspective, Xinhuanet. Available at: http://news.xinhuanet.com/newscenter/2003-11/03/content_1156941.htm (accessed 5 May 2014).

Xinhua (2004a) Traps everywhere: easy to buy but not easy to stay with peace of mind, Xinhuanet. Available at: http://news.xinhuanet.com/focus/2004-07/16/content_1596607.htm (accessed 12 May 2014).

Xinhua (2004b) China has revoked 4735 development zones, Xinhuanet. Available at: http://news.xinhuanet.com/fortune/2004-07/26/content_1652373.htm (accessed 9 July 2014).

Xinhua (2010) Relevant authorities cannot be silent on the message that 65.4 million housing units were empty, Xinhuanet. Available at: http://news.xinhuanet.com/fortune/2010-07/30/c_12389386.htm (accessed 2 April 2014).

Xinhua (2013) Sun Hung Kai Properties grabs Xujiahui Centre project site for RMB 21.77 billion, Xinhuanet. Available at: http://news.xinhuanet.com/house/sh/2013-09-05/c_117238776.htm (accessed 2 April 2014).

Xinhua (2014a) China to help 100 million to settle in cities. Available at: http://europe.chinadaily.com.cn/china/2014-07/30/content_18216332.htm (accessed 2 August 2014).

Xinhua (2014b) Nanjing forecast new real estate policy: 50% units are small to medium sizes. Available at: http://news.xinhuanet.com/house/sh/2013-09-05/c_117238776.htm (accessed 2 April 2014).

Xinhuanet (2013) China builds 'Green Great Wall' to achieve retreat of the desert, Xinhuanet, 18 August 2013. Available at: http://news.xinhuanet.com/politics/2013-08/18/c_116986875.htm (accessed 19 August 2013).

XKB (2012) Tax rates adjustment for property letting in Guangzhou, XKB. Available at: http://news.xkb.com.cn/guangzhou/2012/0214/184319.html (accessed 15 May 2014).

Xu, G. (2014) Local debts: not too much, but not right, *China Finance and Economic News*, 8 February, 2014. Available at: www.cfen.com/cn/web/meyw/2014-02/08/Content_1052316.htm (accessed 15 April 2014).

Xu, J. and Yeh, A.G.O. (2009) Decoding urban land governance: state reconstruction in contemporary Chinese cities, *Urban Studies*, 46(3): 559–81.

Yang, H. (2014) 22% housing vacancy rate is a bit too absurd, *Caijing*. Available at: http://estate.caijing.com.cn.2014-o6-12/114256164.html (accessed 1 July 2014).

Yang, L. (2005) MLR has two approaches to vitalise idle land in 5 years, *21 Century Business Herald*, 6 July 2005. Available at: http://bj.house.sina.com.cn/news/hgdc/2005-07-06/155782762.html (accessed 5 June 2014).

Yang, L. (2006) Urban planning fails: 183 cities to build international metropolises, *21 Century Business Herald*. Available at: http://view.news.qq.com/a/20060601/000069.htm (accessed 5 June 2014).

YCWB (2012) False divorce to evade purchase restriction, *YCWB*, 9 June 2012. Available at: www.ycwb.com/epaper/ycwb/html/2012-06/08/content_1410235.htm (accessed 14 July 2014).

Yicai (2014) Land sale receipts reaches CN¥ 2.5 trillion in the first seven months of 2014: why there is still growth? *Yicai.com*. Available at: www.yicai.com/news/2014/08/4012980.html (accessed 28 August 2014).

Yin, G. (1999) New circumstances in real estate administration and thoughts on improvement, *Shanghai Real Estate*, 1999(1): 10–14.

Yu, C. (2013) The long-term adjustment and control mechanism is to operate, China-crb. Available at: www.china-crb.cn/resource.jsp?id=16673 (accessed 1 July 2014).

Zhang, H. (1998) *Housing Economics*, Shanghai: Shanghai Finance University Press.

Zhang, H. (2012) *Agitating Times: A Critique on China's Real Estate*, Shanghai: Shanghai People's Press.

Zhang, Q. (2009) Occupier the owner? An investigation of housing policy in the 1950s, *Modern China Studies*, 2009(2). Available at: www.modernchinastudies.org/cn/issues/past-issues/104-mcs-2009-issue-2/1096–1950.html (accessed 10 December 2013).

Zhang, Y. (1991) Reform and development of China's real estate industry, *Southern Real Estate* 1991(9): 2.

Zhang, Y. and Liu, Z. (2014) Alarm bell for trusts: over 70 billion yuan assets at risk, *Caixing. Com*. Available at: http://finance.caixin.com/2014-08-05/100712936.html?cx_from=news.qq.com (accessed 11 August 2014).

Zhao, Q. (2014) The real estate sector becomes a tax evasion disaster area, *Economic Information*, 17 April 2014. Available at: http://finance.eastday.com/m/20140417/u1a8036067.html (accessed 15 May 2014).

Zhao, X., Zhang, Y. and Yang, X. (2013) How many international metropolises does China need? *People Digest*. Available at: http://paper.people.com.cn/rmwz/html/2013-06/01/content_1264453.htm (accessed 19 May 2014).

Zhao, Y. (2013) Beijing's purchase restriction on housing could not restrict false divorce, *Beijing Evening News*, 28 July 2013. Available at: www.soufun.com/news/2013-07-28/10632037.htm (accessed 15 May 2014).

Zhou, L. (2006) Chongqing took back 1,579 Mu idle land, *Real Estate Times*. Available at: http://old.jfdaily.com/gb/jfxww/xlbk/fangdc/node11057/node11069/userobject1ai1515138.html (accessed 5 June 2006)

Zhou, Q. (2014) 1.65 million judicial opinions from local courts were published through internet, *Chinanews*, 10 March 2014. Available at: www.chinanews.com/gn/2014/03-10/5932454.shtml (accessed 10 March 2014).

Zhu, J. (2014) Registration of real property will not bring housing price falls, *People's Daily*. Available at: http://news.southcn.com/china/content/2014-09/01/content_107701236.htm (accessed 18 August 2014).

ZOL (2014) Chinese brand occupies half of the world's top eight mobile phone brands, Zhongguancun Online. Available at: http://mobile.zol.com.cn/454/4546730.html (accessed 10 June 2014).

INDEX

Page numbers in **bold** indicate tables and in *italics* indicate figures.

Lightning Source UK Ltd.
Milton Keynes UK
UKOW01f2155071015

260050UK00003B/172/P

9 780415 723152